Simultaneous Engineering

Integrating Manufacturing And Design

C. Wesley Allen
Editor

Robert E. King
Manager

Doris A. Skiver
Production Assistant

Published by

Society of Manufacturing Engineers
Publications Development Department
Reference Publications Division
One SME Drive, P.O. Box 930
Dearborn, Michigan 48121

SIMULTANEOUS ENGINEERING

INTEGRATING MANUFACTURING AND DESIGN

First Edition

Second Printing

Library of Congress Catalog Card Number: 90-061097
International Standard Book Number: 0-87263-382-9
Manufactured in the United States of America

SME wishes to express its acknowledgement and appreciation to the following contributors for supplying the various articles reprinted within the contents of this book. Appreciation is also extended to the authors of papers presented at SME conferences or programs as well as to the authors who generously allowed publication of their private work.

Appliance Manufacturer
Corcoran Communications
29100 Aurora Road, #310
Solon, Ohio 44139

Assembly Automation
IFS Publications
Kempston, Bedford MK42 7PW England

Assembly Engineering
Hitchcock Publishing Company
Hitchcock Building
Wheaton, Illinois 60188

Automation
Penton Publishing
1100 Superior Avenue
Cleveland, Ohio 44114

Automotive Industries
1 Chilton Way
Radnor, Pennsylvania 19089-0030

Cahners Exposition Group
999 Summer Street, P.O. 3833
Stamford, Connecticut 06905-0833

Electronic Engineering Times
CMP Publications, Inc.
600 Community Drive
Manhasset, New York 11030

The Institute of Electrical and Electronics Engineers, Incorporated
445 Hoes Lane
P.O. Box 1331
Piscataway, New Jersey 08855-1331

Machine Design
Penton Publishing
1100 Superior Avenue
Cleveland, Ohio 44114

Manufacturing Engineering
Society of Manufacturing Engineers
One SME Drive
P.O. Box 930
Dearborn, Michigan 48121

Manufacturing Review
The American Society of Mechanical Engineers
22 Law Drive
P.O. Box 2300
Fairfield, New Jersey 07007-2300

Manufacturing Systems
Hitchcock Publishing Company
Hitchcock Building
Wheaton, Illinois 60188

Mechanical Engineering
The American Society of Mechanical Engineers
22 Law Drive
P.O. Box 2300
Fairfield, New Jersey 07007-2300

Rick Norman
International Technegroup Incorporated
5303 DuPont Circle
Milford, Ohio 45150

Plastics World
Cahners Publishing Company
275 Washington Street
Newton, Massachusetts 02158-1630

Production
Gardner Publications Incorporated
6600 Clough Pike
Cincinatti, Ohio 45244

Production Engineering
Penton Publishing
1100 Superior Avenue
Cleveland, Ohio 44114

The Wall Street Journal
Dow Jones & Company, Incorporated
P.O. Box 300
Princeton, New Jersey 08543-0300

Lada Zajicek
Apple Computer, Incorporated
Fremont, California 94537

PREFACE

Simultaneous Engineering is the simultaneous design of the product and the manufacturing process. Experience has shown that a close coordination between engineering organizations is necessary to achieve the fullest function, highest quality, and lowest cost products.

Many have recognized, from different viewpoints, the need for closer working relationships between engineering functions. Several have given names to the process according to the path or view they pursued. Thus we have: Simultaneous Engineering, Concurrent Engineering, Design for Manufacturability, Design for Excellence, or Design for "X". All have the common element of the design engineering function working more closely with some or all of the other engineering functions. For this book, the term Simultaneous Engineering represents the closer working relationships between design and manufacturing engineering functions in general.

Whatever we call the process, Simultaneous Engineering continues to develop as we continue to better understand the relationships, communications, and organizations necessary to produce the best product at a competitive cost.

What is "manufacturing?" Is it the manufacturing processes of "machining, casting, forging, molding, extruding, ..." that are the basic elements of producing a part from a drawing; or is it the process of producing a "product from drawings," which would include, assembly, and test; or is it the process from "concept to product," which would include design?

The papers in this book represent a wide range of the successful application of Simultaneous Engineering principles in many different situations and industries. Wide as these applications are, the fundamental is common–a better consideration of the total product during product design, through a closer working relationship between engineering functions.

I wish to thank all of the authors of the SME Technical Papers contained in this volume for their efforts in furthering the knowledge in this field. Thanks also to the Publication Development staff at SME for their assistance in the research and development required to make this book possible.

C. Wesley Allen, Ph.D., P.E.
Consultant
Martinsville, Indiana

ABOUT THE EDITOR

C. Wesley Allen is a private consultant in the areas of simultaneous engineering, design for manufacturability, and engineering education. He has a special interest in moving such new design approaches from industry, where many such techniques originated, into the college engineering curriculum.

Dr. Allen has 30 years of industrial experience, including 26 years with IBM in product development assignments. Following his industrial work, he spent two years as a Visiting Professor in the Mechanical Engineering Department at the University of Cincinnati, where he is currently an Adjunct Professor.

Through his interest in engineering education, he has served as an ASME/ABET evaluator for mechanical engineering programs and currently he is a member of the Engineering Accreditation Commission of the Accreditation Board for Engineering and Technology.

Dr. Allen received his Ph.D in mechanical engineering from Purdue University and is a registered professional engineer. He is active in professional society work and is a member of the American Society of Mechanical Engineers, Design for Manufacturability Committee.

SME

The informative volumes of the Manufacturing Update Series are part of the Society of Manufacturing Engineers' many faceted efforts to provide the latest information and developments in engineering.

Technology is constantly evolving. To be successful, today's engineers must keep pace with the torrent of information that appears each day. To meet this need, SME provides, in addition to the Manufacturing Update Series, many opportunities in continuing education for its members.

These opportunities include:

- Monthly meetings through five associations and their more than 300 chapters and 165 student chapters worldwide to provide a forum for membership participation and involvement.
- Educational programs including seminars, clinics, programmed learning courses, as well as videotapes and films.
- Conferences and expositions which enable engineers and managers to examine the latest manufacturing concepts and technology.
- Information on Technology in Manufacturing Engineering database containing technical papers and publication articles in abstracted form. Other databases are also accessible through SME.

The SME Manufacturing Engineering Certification Institute formally recognizes manufacturing engineers and technologists for their technical expertise and knowledge acquired through experience and education.

The Manufacturing Engineering Education Foundation was created by SME to improve productivity through education. The foundation provides financial support for equipment development, laboratory instruction, fellowships, library expansion, and research.

SME is an international technical society dedicated to advancing scientific knowledge in the field of manufacturing. SME has more than 80,000 members in 70 countries and serves as a forum for engineers and managers to share ideas, information, and accomplishments.

The society works continuously with organizations such as the American National Standards Institute, the International Organization for Standardization, and others, to establish and maintain the highest professional standards.

As a leader among professional societies, SME assesses industry trends, then interprets and disseminates the information. SME members have discovered that their membership broadens their knowledge and experience throughout their careers. The Society of Manufacturing Engineers is truly industry's partner in productivity.

MANUFACTURING UPDATE SERIES

Published by the Society of Manufacturing Engineers and its affiliated societies, the Manufacturing Update Series provides significant up-to-date information on a variety of topics relating to manufacturing. This series is intended for engineers working in the field, technical and research libraries, and as reference material for educational institutions.

The information contained in this volume doesn't stop at merely providing the basic data to solve practical shop problems. It also can provide the fundamental concepts for engineers who are reviewing a subject for the first time to discover the state of the art before undertaking new research or applications. Each volume of this series is a gathering of journal articles, technical papers, and reports that have been reprinted with expressed permission from the various authors, publishers, or companies identified within the book. Educators, engineers, and managers working within industry are responsible for the selection of material in this series.

We sincerely hope that the information collected in this publication will be of value to you and your company. If you feel there is a shortage of technical information on a specific manufacturing area, please let us know. Send your thoughts to the Manager, Publications Development, Reference Publications Division at SME. Your request will be considered for possible publication by SME or its affiliated societies.

TABLE OF CONTENTS

CHAPTERS

1 THE ELEMENTS OF SIMULTANEOUS ENGINEERING

Simultaneous Engineering
by *Bill Evans*
Reprinted from *Mechanical Engineering*, February 1988 3

Designing For the Life Cycle
by *Wolter J. Fabrycky*
Reprinted from *Mechanical Engineering*, January 1987 5

Toward A Science Of Manufacturing
by *Philip H. Francis*
Reprinted from *Mechanical Engineering*, May 1986 8

Simultaneous Engineering
by *Fred Gordon* and *Robert Isenhour*
Reprinted from *Electronic Engineering Times*, January 1989 14

The Final Piece to the Puzzle
by *John M. Martin*
Reprinted from *Manufacturing Engineering*, September 1988 17

Design for Manufacture
by *Henry W. Stoll*
Reprinted from *Manufacturing Engineering*, January 1988 23

Product Manufacturability
by *John P. Tanner*, P.E.
Reprinted from *Automation*, May 1989 30

Assessing the Development/Production Transition
by *R.C. Thurmond* and *D.V. Kunak*
©1988 IEEE Reprinted, with permission, from *Transactions on Engineering Management*,
Vol.35, No.4, Pages 232-237, November 1988 34

Simultaneous Engineering
by *Gary S. Vasilash*
Reprinted by *Production*, July 1987 50

Manufacturers Strive to Slice Time Needed to Develop Products
by *John Bussey* and *Douglas R. Sease*
Reprinted by permission of *The Wall Street Journal*,
©Dow Jones & Company, Inc. 1988. All Rights Reserved Worldwide 56

2 IMPLEMENTATION OF SIMULTANEOUS ENGINEERING

Simultaneous Engineering: What? Why? How?
by *C. Wesley Allen*
Presented at the SME Simultaneous Engineering Conference, June 1989 63

Design Integrated Manufacturing
Reprinted from *Plastics World*, Newton, Mass., December 1988.
Copyright Cahners Publishing Co. 69

How Process Logistics Planning Can Enhance The Effectiveness Of Simultaneous Engineering
by *Edward J. Budill*
Presented at the SME Simultaneous Engineering Conference, June 1989 73

Make Me A Match
by *Cathy Coffman*
Reprinted from *Automotive Industries*, December 1987 82

Design for Assembly
Users Speak Out
by *John R. Coleman*
Reprinted from *Assembly Engineering*, July 1988. By permission of the Publisher © 1988. Hitchcock Publishing Co.
All rights reserved 85

Gaining Competitive Advantage By Using Simultaneous Engineering To Integrate Your Engineering, Design, and Manufacturing Resources
by *John W. Foreman*
Presented at the CASA/SME AUTOFACT '89 Conference, October 1989 92

Partners for Productivity
by *J. David Griffin* and *Charles F. Myers*
Reprinted from *Production Engineering*, June 1987 106

Early Designs for Manufacturing Quality
by *John P. Hinckley*, Jr.
Presented at the SME Simultaneous Engineering Conference, June 1987 109

Meet Two Architects of Design-integrated Manufacturing
by *Carl Kirkland*
Reprinted from *Plastics World*, Newton, Mass., December 1988, Copyright Cahners Publishing Co 127

Concurrent Product/Process Development (CP/PD)
A Concurrent Design Methodology: Making It Happen
by *Rick Norman*
Copyright 1988, 1990 International TechneGroup Incorporated, Revised 1990 132

Bridging the Gap Between Design and Assembly
by *Brian Rooks*
Reprinted from *Assembly Automation*, May 1987 141

Designers Gain Insight into the Factory
by *Nancy E. Rouse*
Reprinted from *Machine Design*, June 8, 1989 145

A Closer Coupling
by *Michael Smith*
Reprinted from *Mechanical Engineering*, September 1988 151

DFA as a Primary Process Decreases Design Deficiencies
by *P.J. Sackett* and *A.E.K. Holbrook*
Reprinted from *Assembly Automation*, August 1988 152

DFA Promises and Delivers
by *Linda K. Schuch*
Reprinted from *Assembly Engineering*, May 1989. By permission of the Publisher©1989. Hitchcock Publishing Co. All rights reserved 156

Simultaneous Engineering
by *David P. St. Charles*
Presented at the CASA/SME AUTOFACT '89 Conference, October 1989 160

Simultaneous Engineering in the Conceptual Design Phase
by *Henry W. Stoll*
Presented at SME Simultaneous Engineering Conference, November 1988 165

Worlds Apart-Bridging R&D and Manufacturing
by *Robert Szakonyl*
Reprinted from *Manufacturing Engineering*, December 1987 172

Synchronous Engineering: Implementation of Modern Techniques
by *Quentin C. Turtle*
Presented at SME Simultaneous Engineering Conference: Making it Work, June 1989 176

Applying Design for Assembly Principles
by *Robert Waterbury*
Reprinted from *Assembly Engineering*, March 1986. By permission of the Publisher.©1986. Hitchcock Publishing Co. All rights reserved 184

Analysis of Production Line Efficiency from the Viewpoint of a Japanese Production Engineer
by *Takuro Yamada*
Presented at SME Machining Systems Conference, May 1989. 187

Analyzing Product Assemblability Merit
by *Carl F. Zorowski*
Presented at the Spring National Design Engineering Show & Conference, March 7-10, 1988, McCormick Place North, Chicago, IL. 205

3 SIMULTANEOUS ENGINEERING ACCOMPLISHMENTS

Product Design for Manufacture and Assembly
by *G. Boothroyd* and *P. Dewhurst*
Reprinted from *Manufacturing Engineering*, April 1988 217

NCR Cashes In on Design for Assembly
by *John R. Coleman*
Reprinted from *Assembly Engineering*, September 1988. By permission of the Publisher.©1988. Hitchcock Publishing Co. All rights reserved 222

Design For Assembly In Action
by *P. Dewhurst* and *G. Boothroyd*
Reprinted from *Assembly Engineering*, January 1987. By permission of the Publisher.©1987. Hitchcock Publishing Co. All rights reserved 225

Design for Assembly of Electrical Products
by *Jeffrey L. Funk*
Reprinted from *ASME Manufacturing Review*, March 1989 230

Designing for Productivity Saves Millions
by *Russ Gager*
Reprinted by permission of *Appliance Manufacturer*. Copyright 1986 by Corcoran Communications, Inc. 237

Integrating Product And Process Design
by *Gina Goldstein*
Reprinted from *Mechanical Engineering*, April 1989 244

How They Brought Home the Prize
by *Tom Inglesby*
Reprinted from the April 1989 issue of *Manufacturing Systems*.
Copyright 1989, by Hitchcock Publishing Company 247

A Case Study Of Simultaneous Engineering
by *Claus Madsen*
Presented at CASA/SME AUTOFACT '89, October 1989 254

A Team Approach to Success
by *John M. Martin*
Reprinted from *Manufacturing Engineering*, August 1989 260

Design-for-assembly Slashes Cost of Hoover's All-plastic Vacuum Cleaner
by *Bernie Miller*
Reprinted from *Plastics World*, Newton, Mass., April 1989.
Copyright Cahners Publishing Co. 264

Converting Customers to Partners at Ingersoll
by *Robert N. Stauffer*
Reprinted from *Manufacturing Engineering*, September 1988 268

DFM Support for Simultaneous Engineering© Product Design For Manufacturability And Assembly©
by *Lada Zajicek*
Presented at The Third International Conference on Design For Manufacturability and Concurrent Engineering, December 1989. Reprinted courtesy of the Author.
Copyright 1986/1989. Lada Zajicek. 272

INDEX 279

CHAPTER 1

THE ELEMENTS OF SIMULTANEOUS ENGINEERING

Reprinted from *Mechanical Engineering*, February 1988

Simultaneous Engineering

Engineers are rediscovering the importance of designing the process as well as the product.

BILL EVANS
ASSISTANT EDITOR

Pioneers of the automobile industry practiced what is now called simultaneous engineering. Men like Henry Ford, Ransom Olds, Karl Benz, and Adam Opel did not limit themselves simply to designing products; they were product and process engineers who designed both cars and the factories that built them. In fact, many of the early inventors in other American industries were also craftsmen, who knew how to build the products they invented. But as these industries developed and became larger, work became specialized to the point where a designer would come up with a product but expect a manufacturing engineer to figure out how to make it.

Now, however, as U.S. companies struggle to compete in the global marketplace, the old methods are being rediscovered. Although it is now augmented by the use of computers in the design process, the practice of simultaneous engineering, or designing for manufacturability, has once again become fashionable.

FORM, FIT, AND FUNCTION

Product design is generally concerned with form, fit, and function. How to manufacture a product and how much it will cost are questions not usually asked until the later stages of product development, at which point the design may have to be reworked in order to solve problems of quality or production. In a company with an effective simultaneous engineering program, however, many of the problems associated with manufacturability and quality would never arise. Indeed, when there is effective communication between design and process engineers, manufacturing becomes important at the very moment the product designer starts doing what he is there to do—create or improve the product.

The process by which a product passes from concept to the marketplace and eventually to the grave is often referred to as a chain of "handshaking." Communication occurs in this chain, but only when a particular stage in the process is completed. First, a sales engineer becomes familiar with other products in the marketplace, and after talking with customers about their needs, he gets an idea for a product. Next, the engineer may discuss the idea with a designer or give him a description of the product. When the product is defined, the design engineer works it into a design, possibly without ever having any contact with manufacturing.

The size and complexity of the developing product influence the distribution of design responsibilities. If the product is extremely complex, like an airplane or an automobile, a design engineer may be assigned to work on only a minuscule part of it. As a result, he may not be directly concerned with how the part is manufactured and used in assembly.

Once the design is completed, however, it must go to the manufacturing people, who define the process plan, determine the costs, and work out a production schedule. Materials, machinery, and tooling are then ordered by the purchasing department, and the quality control staff develops a quality assurance program. Finally, after manufacturing, the service department takes over—assuming that there is still a market for the product at the completion of this long process.

Charles Lamb, director of advanced manufacturing and engineering technology at Emhart Corporation in Farmington, Connecticut, reports that using simultaneous engineering techniques on an introductory level has reduced by one third the time it takes for a line of Emhart door hardware to reach the market. "The whole concept of simultaneous engineering is that at the same time the sales engineer starts talking to customers and defining a product, he also starts talking with engineering and manufacturing," Lamb says. "By working together, we start learning earlier."

Lamb maintains that communication already goes on in American industry, but that most of it occurs accidentally, during chats at the coffee machine. But he cautions that competition and shorter product life mean that companies are rushing to get to the marketplace faster. "The accidental communication route is no longer adequate."

Irvin Krause, a partner at Coopers & Lybrand Management Consulting Services in Boston, has followed manufacturing issues closely and is establishing a simultaneous-design consulting service within the firm. Krause worked as head of several product design groups before joining Coopers & Lybrand. He says problems with the design process persist despite the trend toward simultaneous engineering. "The attitude on the part of product designers has always been, 'I can't really be concerned with manufacturability now; I have to design it for form, fit, and function and the manufacturing people will have to worry about how to make it.' Unfortunately, this attitude is not a thing of the past, but it has to become one."

MCAD

The evolution of CAD/CAM has given companies the tools they need to encourage simultaneous work, Krause says. "Now, we can look over the designer's shoulder without actually being there. Until now, a designer would go through a complete design of the product and transfer the design, traditionally on paper, to the process planners. And if there were mistakes in the design, changes would have to be made." Ultimately, this lengthens the period from product conception to manufacture, delays the delivery of the product, and hurts a company's competitiveness.

The degree to which data have become more centralized is a barometer of change in the design process. "In the classic handshaking method," Lamb says, "data originate in sales, and then are given to engineering, manufacturing, quality, and finally service. The computer has given us a common, central data base that we can all work with, and it doesn't belong to just this guy passing it to that guy, but to everyone involved in the process."

The computer tools used by product designers are themselves changing. Many design automation functions that were once limited to mainframe computers or powerful workstations can now be performed on a PC. Several companies have introduced MCAD software packages that solve design problems while overcoming some of the difficulties that traditional MCAD systems have had with accessibility and versatility.

These conceptual design systems, which assist in the initial, creative stages of product development, replace the engineer's traditional green-paper pad. Consisting of either a two-dimensional sketching function or a three-dimensional solid modeler, coupled with analysis programs, the systems enable engineers to quickly evaluate and modify ideas before making project commitments. Iconnex (in Pittsburgh) and Cognition (in Billerica, Massachusetts) both offer a 2-D sketching and analysis package, and Aries Technology (in Lowell, Massachusetts) offers a 3-D modeling and analysis package.

Some advanced systems allow the engineer to take a 2-D sketch and instruct the modeler to create a 3-D form of the object by selecting from commands such as extrude, bore, stamp, turn, and mill, which are derived from conventional operations in design and manufacturing. In addition to allowing the engineer to use familiar terms, some systems employ what is often called feature-based design. This gives the designer access to a library of definitions and permits him to design a part in terms of parameters, attributes, and design constraints. If a single part requires a groove, for example, once such parameters as depth, width, and length have been defined, the function of the groove and other non-geometric data would be classified separately as attributes, while certain constraints might limit the size of the groove or the methods used to manufacture it. This type of standardization and organization of data may lead to the integration necessary for simultaneous engineering to work.

The emphasis in MCAD research and development is on making the systems more intelligent. Some companies are developing their own expert programs with the aid of software that facilitates the representation of knowledge in the form of rules, facts, and deductions. The development of these programs requires a good deal of mental discipline on the part of the designers who supply the knowledge. Companies have found that the development process itself leads to a better understanding of design.

INTEGRATION

Computers are not absolutely necessary to achieving integration, particularly within a small company or project team. Lamb gives computers only part of the credit for reducing Emhart's design-cycle time. The other part is "better management of programs, getting everyone together, and seeing that everyone participates in the process. The computer mechanizes a project and makes it easier to manage."

Integration is very common in Japan, even though the use of computers is less prevalent than in the United States, notes Krause. "If you can exchange information on a timely and accurate basis—as the Japanese do with the *kanban* system—you've solved a problem.

"Japan doesn't depend on computers, and I think that when they finally decide to use computers on a much larger scale, they'll leapfrog again and go far ahead of us. Although we've been talking about using computers and integration, we haven't been doing it, while they've at least been doing the integration part. Currently, we have enough technology and effective tools, but we don't have the integration to use them effectively."

RESPECT FOR MANUFACTURING

To achieve simultaneous engineering, companies may have to make a special effort to knock down departmental barriers, real or imagined, that have developed over the years. Boundaries between design and manufacturing departments have become institutionalized over the years, according to Krause. "There has been a great emphasis in the United States on research and development, and that R&D has always been associated with new products, rather than with manufacturing. The best and brightest people went into design, where they were respected and rewarded the most." Krause also notes the lack of professional status given to manufacturing in the United States and the shortage of manufacturing graduate programs and academic journals.

FATE OF THE GURUS

In the past, a company design group often included one member who was knowledgeable in many fields. Often called a "guru," this person resisted the trend toward specialization and brought his expertise from a variety of areas—marketing, manufacturing, materials—to the design team. The new computer tools, for all their promise of improved communication and expert systems, will have to perform as well as, if not better than, those gurus.

"Data and information are eventually going to be handled better by tools that help people," Krause said. "There were once gurus who knew how to perform certain types of calculations very quickly and adroitly, but that sort of ability is no longer as important with the advent of computerized computational techniques. The same thing is going to happen to these knowledge-based or information-based gurus. But we will always have to allow for human creativity, because we'd be missing something if we didn't." ■

Reprinted from *Mechanical Engineering*, January 1987

Designing For the Life Cycle

The entire life of a product, as well as that of the manufacturing process and of the service system, should be considered in the original design.

WOLTER J. FABRYCKY
PROFESSOR OF INDUSTRIAL ENGINEERING AND OPERATIONS RESEARCH
VIRGINIA POLYTECHNIC INSTITUTE AND STATE UNIVERSITY
BLACKSBURG, VIRGINIA

Using the materials and forces of nature to satisfy human needs is the concern of engineering. Because the natural resources of the world are limited, engineering is closely associated with economics. Engineers must determine how the physical environment can be altered to create the most useful products at the least cost.

PARALLEL CYCLES

The life cycle of a product or system begins with the identification of a need and extends through conceptual/preliminary design, detailed design, production and/or construction, installation, customer use, support, and finally, decline and disposal.

The principle behind life-cycle engineering is that the entire life of the product should be considered in its original design. An engineering design should not only transform a need into a description of a product, but should ensure the design's compatibility with related physical and functional requirements, and it should take into account the life of the product as measured by its performance, reliability, and maintainability.

Life-cycle engineering goes beyond the life of the product itself, however. It is concerned simultaneously with the life of the manufacturing process and of the product service system. There are actually three coordinated life-cycles going on at the same time.

These parallel life cycles are initiated when the need for the product is first recognized. During the preparation of the conceptual design, which follows, consideration should simultaneously be given to the product's manufacture. This begins the second life cycle—the creation of a manufacturing process—which will require production planning, plant layout, equipment selection, process planning, and other similar activities. The third life cycle should also be initiated at the preliminary design phase. It involves the development of a service system for the product and a maintenance system for manufacturing.

CURRENT PRACTICES

Traditionally, engineers have focused mainly on the acquisition phase of the product's life cycle. Experience shows, however, that a successfully competitive product cannot be achieved if performance and maintenance are considered only after the design is completed.

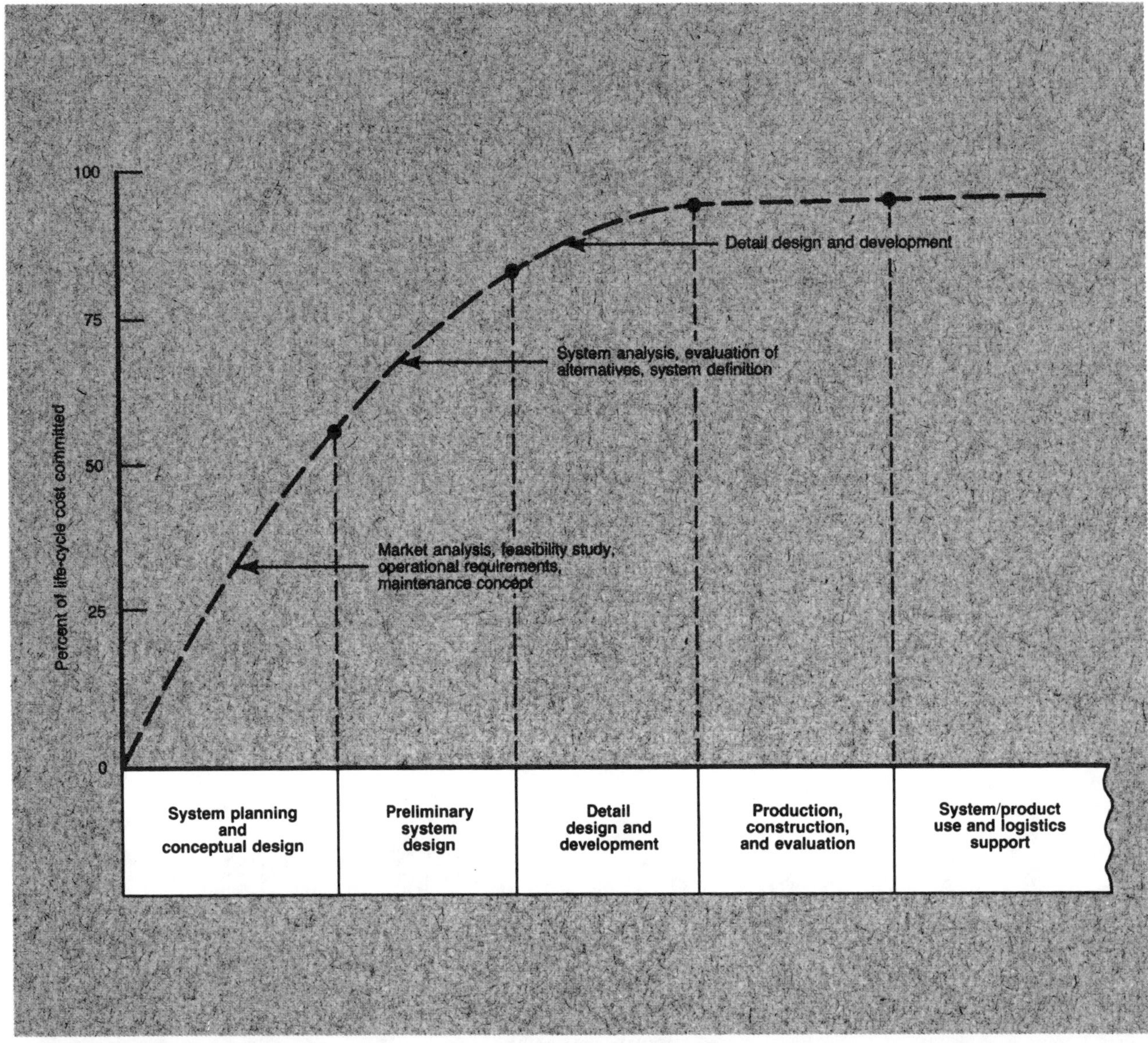

Actions affecting life-cycle cost.

Priorities for Research

1. Economic evaluation of design alternatives over the life cycle
2. Economic modeling of manufacturing processes
3. Computer-aided estimating
4. Integration of engineering with finance and accounting
5. Economic modeling of production systems
6. Timing and locating introduction of new technology
7. State-of-the-art survey and search
8. Costs/benefits of manufacturing flexibility
9. Economic evaluation of anticipated technology
10. Parametric and shortcut estimating techniques
11. Appropriate measures of effectiveness (taxonomy)
12. Economic evaluation of software development for manufacturing
13. Methods for treating risk
14. Economic design of data collection systems in manufacturing
15. Economic evaluation of alternative quality strategies
16. Comparison of cost allocation in high-tech and other industries
17. Criteria for project evaluation
18. Methods of figuring and comparing investment worth
19. Effects of design on cost in system performance
20. Effects of technical innovation on capital requirements
21. Economic consequences of alternative investments
22. Survey of function applications and needs
23. Artificial intelligence
24. Improved methods for estimating project revenues and expenses
25. Hardware and software alternatives
26. Alternatives between risk and growth
27. Economic relationship with other systems
28. Interdependence of cost elements
29. Methods of determining needs
30. Cost control (systems engineering management)
31. Data base networking
32. Noisy, fuzzy data clusters

When too great an emphasis is placed on the engineering of a product's primary function, side effects will often occur that manifest themselves in problems with operation. Although sufficient specialized knowledge exists to solve most of these problems, this knowledge is more useful if it has been integrated in a systematic way into the original design.

The Department of Defense is very concerned with the life cycles of its systems. Private firms serving as defense contractors are obliged to design and develop products in accordance with Defense Department directives, and the Defense Department is closely involved with the acquisition phase of a product's design.

There are private firms as well that are incorporating concern for the overall life of a product into the design process. Energy efficiency, for example, is now part of the design of water heaters and air conditioners, and fuel efficiency is required in automobile design. Some truck manufacturers promise that maintenance requirements over the life of their vehicles will be within stated limits. When the producer is not the consumer, however, it is less likely that potential operational problems will be addressed during development.

Well managed, high-quality life-cycle design offers many benefits. It can create distinctiveness and "personality" for a new product or system. It can be used to reinvigorate interest in an old product. It can communicate a sense of the product's value to the consumer and lead to greater satisfaction. Finally, it can reduce operating and maintenance costs.

RESEARCH AND DEVELOPMENT

Product life-cycle engineering has its greatest impact during the early phases of product development, and it is therefore in the area of applied research and development that it can most profitably be employed. An estimated forty-one percent of all scientists and engineers are engaged in applied research and development and their management, so life-cycle engineering is important to the work of a large number of people. The National Science Foundation's Science Indicators for 1985 indicate expenditures for applied research total almost $23 billion, and for development, over $70 billion. The application of life-cycle principles, in even a small degree, to research and development activities would have beneficial results.

ECONOMICS RESEARCH

The ultimate value of the goods produced by engineers is measured in economic terms. Often, however, work done in engineering economics overlooks the product's life cycle. The economic aspects of a design are not usually considered until its final stages. By then it is too late. The Figure shows how little can be done to improve the economic feasibility of a product or manufacturing system once the design is completed.

The central model in engineering economics is the money-flow diagram, which shows estimates of receipts and disbursements over time, the same dimension in which product life cycle is measured. However, as it is currently practiced, engineering economics is not integrated into the process of bringing products and systems into existence.

Over the past few years, the National Science Foundation has shown a growing interest in the problem of engineering for international economic competitiveness. As a result, the foundation has been sponsoring research planning conferences in this field, and the value of traditional methods of economic analysis is being called into question.

At the NSF's Research Planning Conference on Engineering Economics, held in Mountain Lake, Virginia in August of 1984, and attended by representatives from universities and industry, a list of 32 important research topics was drawn up (see the Table). The list reflected the concern of the conference participants in the impact of new technology on design and manufacturing.

After the conference, a further proposal was submitted to the NSF that called for the development of a framework for the field of engineering economics. The proposal defined engineering economics as being "... concerned with the definition and life cycle economic evaluation of technological alternatives in terms of worth and cost."

Technological advances are making possible not only the development of new products and systems, but the ability to anticipate the behavior of the products and systems over their entire life span. Research is needed on how best to integrate technical, economic, environmental, and human criteria into the design process. The NSF has formed a Division for Fundamental Research in Emerging and Critical Engineering Systems, and the following four areas of research have been proposed as showing promise for the future of life-cycle engineering.

- Life-cycle engineering design: research is needed on how to balance the competing nontechnical considerations within a design. This research must focus simultaneously on the development of computer-aided design and of estimating models and simulations that will provide life-cycle evaluations of alternative design, manufacturing, and maintenance ideas.
- Extending CAD/CAM: research in the life-cycle extension of CAD/CAM, focusing on conceptual/preliminary design, should be concentrated on relating consumer needs to product features and on providing real-time communication and modeling to link conceptual design with operational requirements. This *macrocad* concept should be useful in examining the anticipated operational characteristics of various design alternatives from a life-cycle perspective.
- Compressing development time: the product-development phase of the life cycle could be shortened by treating product design, manufacturing engineering, manufacturing itself, and product introduction into the market as parallel aspects of the same phase. The use of expert systems and extended computer-aided methods should be examined in this context.
- Life-cycle engineering economics: further research should help integrate both economic and noneconomic design decisions at the appropriate stage of the product's life cycle. Designs should be considered in terms of their impact on competitiveness.

•

The principles of life-cycle engineering can help integrate technological advances into the design of competitive products. These principles can also be used to evaluate products, processes, and services. Product life-cycle engineering can play an important role in restoring the United States to its competitive position in the world's economy. ME

Reprinted from *Mechanical Engineering*, May 1986

TOWARD A SCIENCE OF MANUFACTURING

To restore America's competitiveness in manufacturing, fundamental and interrelated issues of strategy, integration, and production must be addressed.

PHILIP H. FRANCIS
DIRECTOR
FLEXIBLE INSPECTION AND ASSEMBLY LABORATORY
INDUSTRIAL TECHNOLOGY INSTITUTE
ANN ARBOR, MICHIGAN

That the United States' leadership in the design and production of manufactured goods is under siege from foreign competition is a matter of common knowledge. Although our gross national product is the largest in the world, our nation's productivity, measured by manufacturing output per worker hour, now lags behind that of every other industrialized nation. Since 1971, the United States has incurred an overall and growing annual trade deficit. This rose 20 percent in 1985 to over $148 billion, nearly 30 percent of which was with Japan.

Clearly, profound changes will have to be made if American industry is to maintain a significant role in international manufacturing. Among the goals that must be established are:

- *Better Management Tools.* New management systems must be devised to help companies adapt to shorter product-life cycles and extreme market cycles, distribute decision-making, and develop shared goals with labor.
- *Higher Product Quality.* Improved product designs and better manufacturing engineering should be emphasized in the interest of quality.
- *Adaptive Labor Force.* Creative and effective ways must be found to help our maturing labor force adapt to change. As the aging of the population accelerates, this problem will demand increasing attention.
- *Stronger University Role.* American colleges and universities have been notably slow in developing curricula in

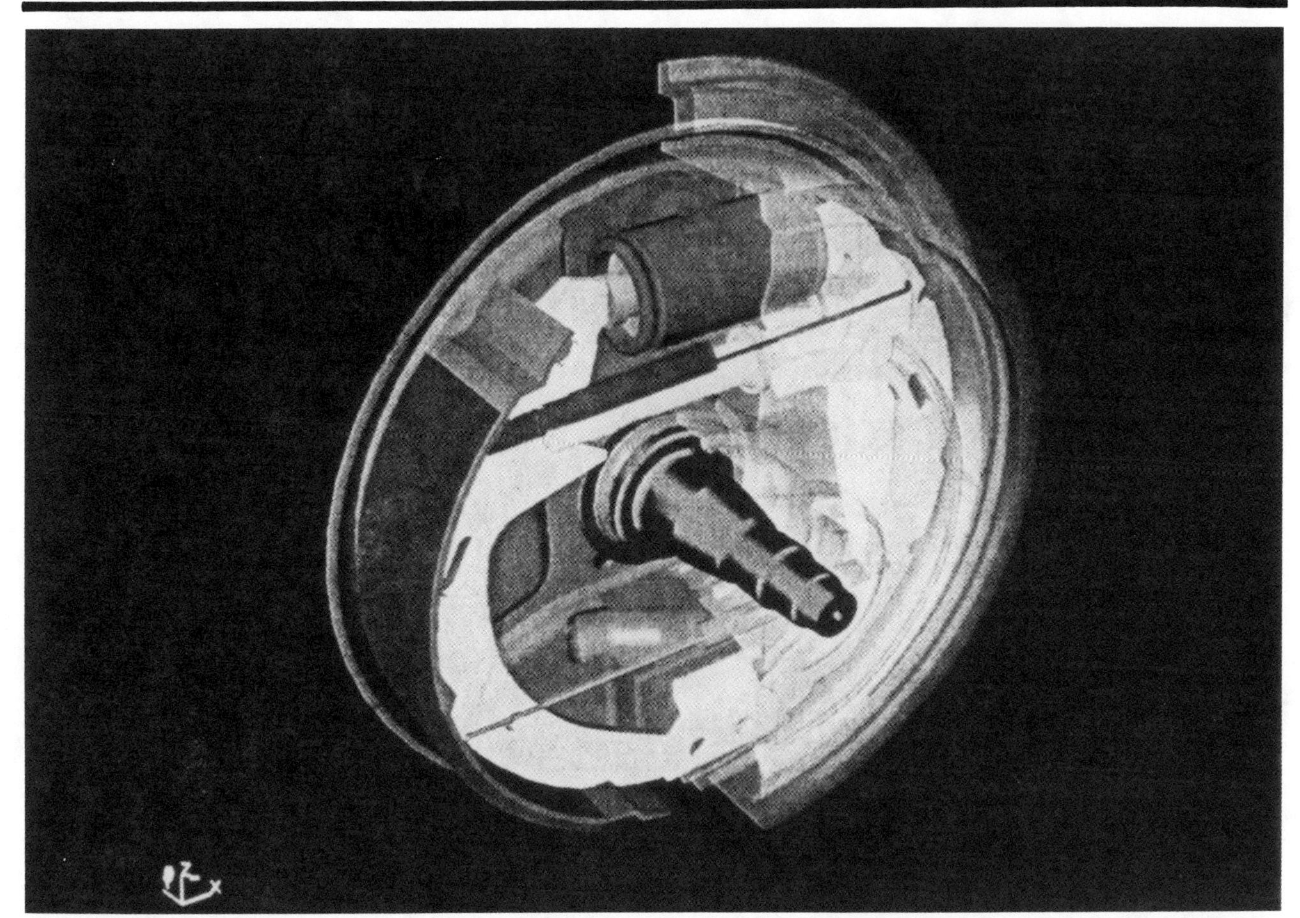

Ease of manufacturing will have to become a dominant design criterion if efficiency is to be improved. The car brake assembly shown here and in the following photographs was designed using the I-deas system from SDRC.

manufacturing engineering. Higher education should encourage applied research, reexamine tenure standards, and seek new relationships with industry to keep abreast of new technologies.

• *Government Support.* The federal and state governments must provide industry with more tax incentives and other inducements to invest in modern manufacturing technology and equipment.

Our country's average hourly manufacturing wage ($12.59 in 1984) is the highest in the world. The corresponding rate in West Germany is 76 percent of that amount; in Japan, 50 percent; in Britain, 46 percent; in South Korea, 11 percent; and in Brazil, 10 percent. It is obvious that manual, labor-intensive manufacturing will not allow the United States to compete successfully with these countries. More productive, automated systems are therefore essential. The Rand Corporation projects that by the year 2000, only two percent of America's workforce will be engaged in manufacturing, down from the present level of about 16 percent. This would represent a reduction in our manufacturing employment pool of 15 percent each year.

Meanwhile, the United States' steel and semiconductor industries are beleaguered. The foreign content of the few existing "American-made" computers is so high as to challenge the term. The U.S. camera industry has virtually disappeared. Nearly all the radio and small home appliance manufacturers left for overseas years ago, and only one American maker of televisions remains. Imported automobiles increased by 24 percent in 1985; this penetration of the industry is expected to continue, due in part to a flood of new cars anticipated for 1990 from Czechoslovakia, Greece, Korea, Spain, Taiwan, Britain, and Yugoslavia.

If we continue to do business as usual, we can expect these trends to continue. Although accelerating the application of sophisticated automation systems is by itself not the solution, it is certainly an essential ingredient. The terms associated with productive automation—computer-integrated manufacturing (CIM), flexible manufacturing systems (FMS), and computer-aided design and manufacturing (CAD/CAM)—are heard so often that they have become almost interchangeable. Implementing the technology that underlies all of them requires that fundamental and interrelated issues of strategy, integration, and production be addressed.

STRATEGY

Strategic issues involve a firm's organization and purpose: what to make, for what market, at what price, and how. In the context of modern automation, some of the most important issues are the development of accounting tools appropriate to new manufacturing processes, the decentralization of hierarchical organizations, and the creation of common goals for management and labor. These matters are so interrelated with technical issues that without understanding them engineers cannot hope to provide industry with effective manufacturing tools.

Productivity. Simply stated, productivity is the result of the optimal use of labor, management, government, material, capital, energy, and technology. The problem is to

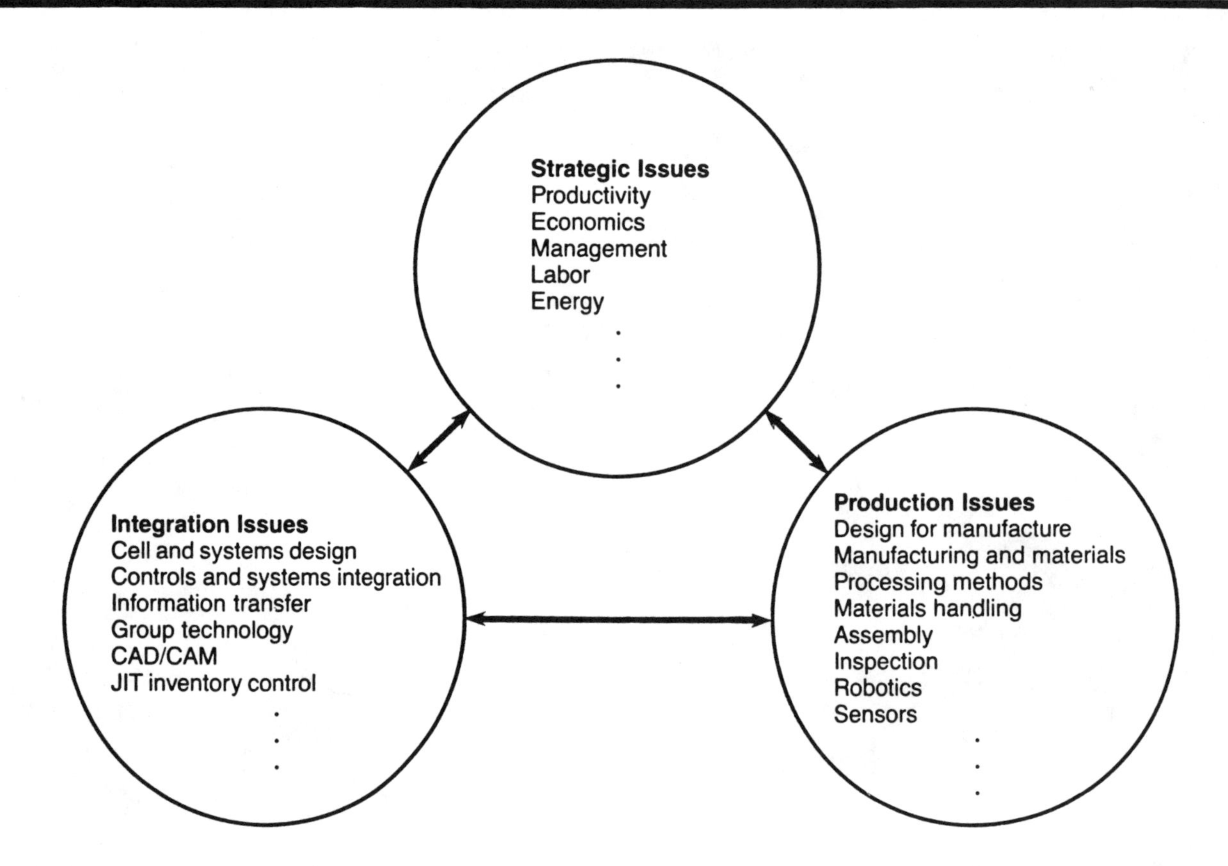

Interrelated issues of strategy, integration, and production.

determine the best combination of these resources, some of which (labor, government, and energy) are not completely under a firm's control. Certainly, more adaptability on the part of management and more responsive information systems are essential in an increasingly dynamic environment.

Economics. From national policy down to the concerns of a particular firm, economic issues influence strategic planning and decision-making. Of concern at the macroeconomic level are the debate on protectionist legislation and foreign imports; job security and retraining; environmental controls; and tax incentives to encourage automation.

At the level of the firm, better tools are needed to evaluate the economics of alternative investments, production systems, software, and data collection systems. Perhaps the most burning issue is to justify the cost of automation. Conventional return-on-investment and payback calculations often do not provide relevant measurements. Instead, it is necessary to regard automation equipment as an investment rather than an expense. Companies also need to find ways of communicating their plans to investment analysts. Accurate indices of a firm's performance (including quality improvements; increased manufacturing flexibility; and reductions in inventory, overhead, throughput, and lead and design-cycle times) must be developed.

Management. The organization of management is changing from rigid hierarchies to more flexible structures. This is in response to a number of pressures: to respond more quickly to organizational and customer demands; to shorten design-cycle times and put the emphasis on customized products and shorter product-life cycles; to allow a more sophisticated labor force to make more decisions; and to give responsibility for quality to those who deal directly with production. In contrast to management's task in the past, which essentially was to maintain the status quo, its primary function now is dealing with continual change. Methods of training managers and engineers to work effectively under these conditions must therefore be found.

Labor. Corporate survival, without which there will be no jobs, should be the immediate overall goal of both management and labor. The two should cooperate in working towards staged programs of job declassification, salaries determined by individual productivity rather than cost-of-living adjustments, and job retraining programs.

Energy. In the past, the cheap labor available in underdeveloped countries was offset by cheap energy and technology in developed countries. The pattern in recent years has been to shift to developing countries the manufacture of goods that require minimal skill to produce, substituting cheap labor for expensive labor and energy. The effects of this trend on the U.S economy and unemployment rate are well known. To reverse the drift toward deindustrialization, we must turn our technical resources to the development of cheaper and continuing supplies of energy.

INTEGRATION

The effect of highly flexible manufacturing systems on the manufacturing process as a whole is to integrate all its aspects. Integration occurs vertically, from the machine-tool controller to the controller's office and back, and laterally, from product design through process planning, component manufacture, assembly, and

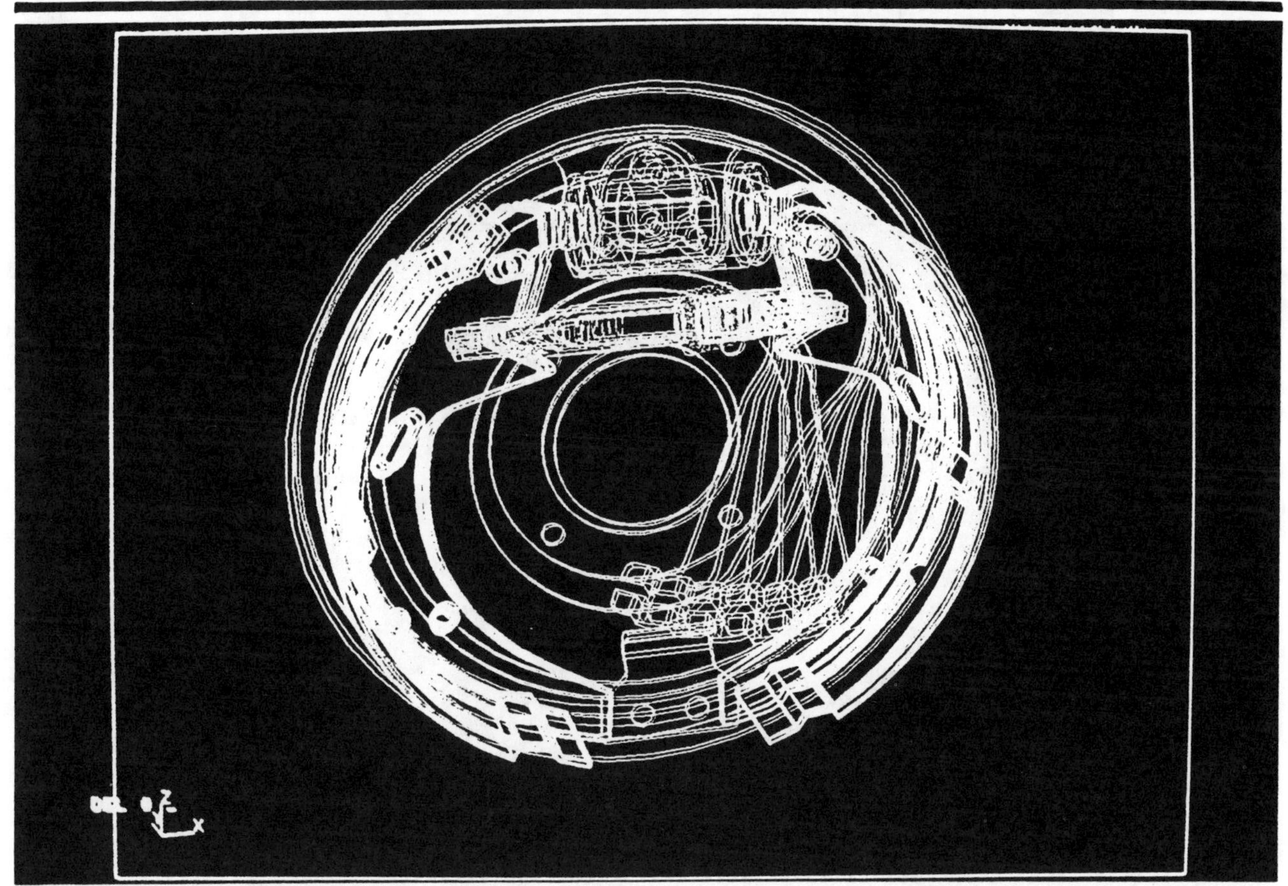

Analysis of the mechanisms in the brake assembly.

shipping. The issues involved in this process include communication protocols, the relationship between businesses and vendors, hierarchical systems control, and CAD/CAM. Artificial intelligence will undoubtedly contribute greatly to the integrated factory of the future.

Cell and System Design. Combined flexible and dedicated manufacturing systems, comprised of manufacturing cells, are different from the fixed automation systems of the past in being both integrated and intelligent. They are integrated because they are capable of transferring information via command hierarchies and communications interfaces. They are intelligent because they are programmed to handle unexpected events (rescheduling and system failures, for example) by making the right decisions. The principal aims of cell and system design are flexibility, workflow-path redundancy, and maximum use of costly equipment.

System Integration and Process Control. At this point, full integration of automation systems is contingent on providing components with a way of communicating with each other. The Manufacturing Automation Protocol (MAP), the result of a development effort led by General Motors, is likely to be *the* factory protocol over the next few years. Manufacturers should now be working out ways to implement MAP to best advantage in the near future. Once the system has been integrated, automated process control can eliminate expensive and unreliable product inspection.

Information Transfer. Improved information systems could provide enormous benefits in terms of increased productivity and reduced development costs and product-development-cycle time. Traditional subject-oriented indexing systems should be replaced by systems oriented to problem solving. Information systems capable of handling graphic data also are sorely needed.

Group Technology. Industry can also reduce costs by designing products that have as much in common as possible. Group technology (GT) eliminates redundancy by identifying similar parts and grouping them according to processing requirements. There are many approaches to implementing GT, and the form it takes must be compatible with a particular firm's manufacturing, resource planning, and inventory systems. Whichever approach is used, GT can result in better coordination between a manufacturer's design and engineering functions, and can reduce parts inventory, process planning, and scheduling requirements.

CAD/CAM. The goal of computer-aided design and manufacturing is for a designer, aided by a relational data base comprising materials, product, and manufacturing process information, to design the product on a CAD system. The design is then downloaded to a program that develops a bill of materials and manufacturing process plans and the CAM system initiates the automatic production of the part. In fact, very little truly integrated CAD/CAM has been implemented. However, the technology is rapidly developing, and companies must assess how best to take advantage of it.

JIT Inventory Control. Just-in-time (JIT) inventory systems depend on receiving small batches of supplies as the need arises. JIT systems put the responsibility for accurate deliveries on the supplier, relieving the manufacturer of the need to inspect each shipment.

JIT inventory systems need not apply only to the relationship between manufacturers and suppliers. Toyota, for example, has applied a JIT system

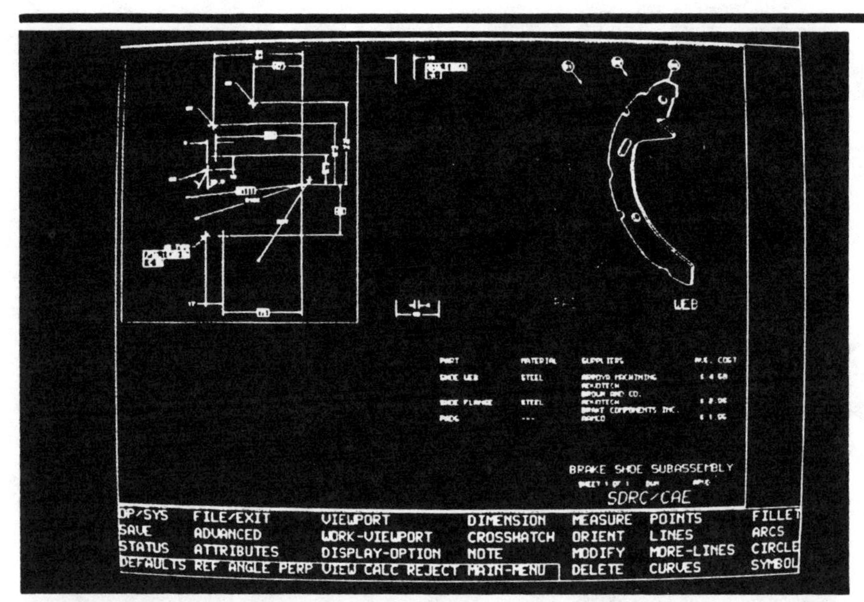

Once design criteria are met, I-deas interfaces with a CAD/CAM system to generate production drawings (top) and NC paths (bottom).

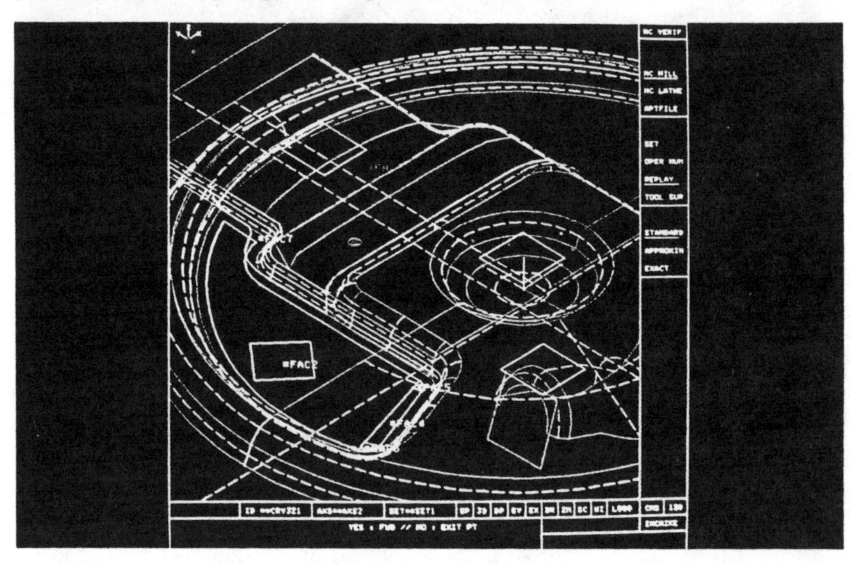

within its manufacturing organization. Instead of the department upstream pushing a product all along the manufacturing stream, each department orders up only what it needs to produce the parts required by the department immediately downstream. In addition to directly reducing costs, JIT systems can reduce scrap rates and improve line balancing.

PRODUCTION

Issues of production are concerned with the technology of manufacturing. They range from product design for manufacturability, materials handling systems, and methods of metal forming/materials transformation to actual assembly, inspection, and process control procedures.

Design for Manufacture. As automation increases, it is important that designs be compatible with and take advantage of the new technology. Design for manufacture (DFM) is a method of addressing this need by seeking new ways of representing and combining design data; new methods of coordinating design and manufacturing departments; improvements in engineering education that will relate design to manufacturing; and a better understanding of the relationships among process flexibility, production volume, types and levels of automation, and the influence of product design on these factors. DFM also makes the designer aware of opportunities to use nontraditional materials in new products and the importance of considering materials handling at the beginning of the design process.

Manufacturing and Materials Processing. Nontraditional materials—HSLA alloys, plastics, composites, and ceramics, for example—open up tremendous possibilities for the manufacture of new products. Currently, however, materials processing and inspection systems adequate to these materials have not been developed. Process control strategies that take into account problems arising from variations in material specifications must also be found, along with new sensor technology, especially improved signal processing and multiple sensor integration.

Materials Handling. A large part of the cost of a product is often consumed by materials handling. This is because of problems associated with the storage and flow of materials: work in process, bulk materials and discrete parts, fixtures and tooling, and containers. A variety of automated products for improved materials handling, including autonomously guided vehicles for handling and docking operations, are now available. JIT inventory control systems and plant layouts that facilitate more efficient storage and distribution of materials and work in process can also recover a significant part of handling costs.

Assembly. For many companies, automated assembly will be the key to productivity. About 85 percent of all manual labor performed worldwide is in assembly work; in the United States, assembly accounts for most of the cost of manufactured products. Successful automated assembly systems not only can reduce labor, but can provide process-control feedback and improve quality as well. Low-volume production is compatible with robotic assembly, while high-volume production calls for dedicated (fixed) automation methods. Robotics opens up the possibility of flexible assembly, which, in turn, will require new approaches to fixturing.

Inspection. Ensuring quality in automated manufacturing will rely less on human inspection and more on process control and on-line automated inspection at critical points. This is already possible thanks to machine vision, acoustic sensors, and high-speed image processing. Automated inspection systems are used to check critical dimensions and tolerance stacking, the quality of welds and painted surfaces, and the completeness of work in process.

Robotics. Already common in the automobile industry, robots are now spreading to such industries as electronics and aerospace. Current trends point to more off-line programming (so that it will not be necessary to shut down a robot while it is being reprogrammed for another operation), more applications in assembly via integrated sensors and improved repeatable accuracies, and fail-safe software.

Obstacles to the widespread adoption of robots include the cost of implementation and maintenance, a lack of trained engineers and technicians, and the tendency to forfeit their flexibility by using them in dedicated automation systems. Moreover, a lack of performance standards often has resulted in inconsistent, misleading, and erroneous performance claims. As the technology matures, however, these problems are disappearing.

Sensors. Advanced sensor technology is in many respects the key to automated manufacturing. Sensors are essential in process control, adaptive machining, automated inspection, systems integration, and applications of robotics and artificial intelligence. As the technology moves in the direction of advanced optical, tactile, and phased-array acoustic processing, improvements in transducer-based sensors will be required. Progress in signal processing is already making it possible to use sensor data in on-line inspection and in the reworking of outmoded machine tools.

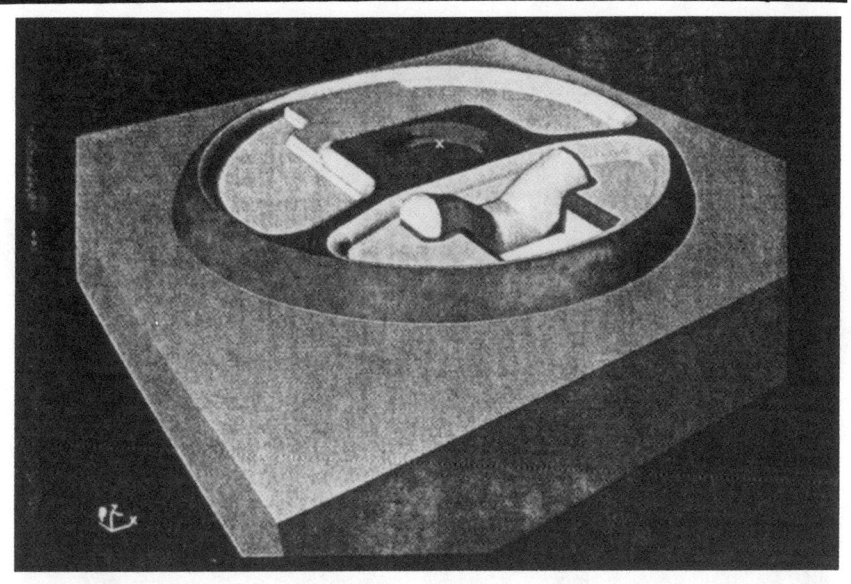

Solid model of machined backing plate (top) and actual NC machining of the plate (bottom).

ASME'S EFFORT

Manufacturing is rapidly becoming more scientific—relying less on subjective methods and more on analysis, simulation, and experiment. A recently published report of the President's Commission on Industrial Competitiveness points out that the single overriding reason for the loss of U.S. manufacturing leadership is that we have failed to apply our own technologies to manufacturing. According to Erich Bloch, Director of the National Science Foundation, the challenge is not merely to apply technology in a systematic way, but to develop a genuine *science* of manufacturing.

Just as changes in industry are essential to maintaining America's position in the international marketplace, ASME will have to innovate to meet the new needs of engineering and manufacturing. The Society has a distinguished history of leadership in industry, from the seminal contributions of F.W. Taylor (the father of scientific management) at the turn of the century to those of Lillian and Frank Gilbreth and of succeeding authorities working through ASME's divisions and committees.

In response to these problems, ASME established the Manufacturing Science and Technology Program (MSTP) at its Summer Annual Meeting last year. The fundamental goal of the program, which is under the direction of the author and reports directly to the Council on Engineering, is to make important contributions to the scientific and technological base of advanced manufacturing systems. The goals established in support of this aim are to improve the science and technology of durable goods manufacturing, to foster the development of an engineering economics appropriate to modern integrated manufacturing, and to communicate all new technological developments via the appropriate channels. Several significant initiatives are already under way, among them the establishment of a major biannual conference—Manufacturing International—to be held for the first time in Atlanta, Georgia, October 12–21, 1987. With the help of members from the various groups and divisions, ASME should be able to make a difference. ME

The author wishes to recognize the opinions and suggestions of the following colleagues: David A. Dornfeld of the University of California at Berkeley; Joseph A. Falcon of Bechtel Corp.; Keith M. Gardiner and Bruce J. Haupt of IBM Corp.; Joel Goldbar of the Illinois Institute of Technology; Nathan H. Hurt of Goodyear Atomic Corp.; Robert Kaplan of Harvard University; Daniel T. Koenig of General Electric Corp.; Keith McKee of IIT Research Institute; and Paul G. Ranky; Henry Stoll, and Nancy Vaupel of the Industrial Technology Institute.

Reprinted from *Electronic Engineering Times*, January 1989

SIMULTANEOUS ENGINEERING

For all practical purposes, American manufacturers traditionally approach the product development cycle in much the same way: sequentially. Each function, from design through manufacturing and all the way to distribution, is taken one step at a time, in virtual isolation from its neighbors.

Recently, this restrictive tradition has come under close scrutiny by overseas competitors. Recognizing that product development can be a powerful driver of competitive advantage, they have shaken off the sequential legacy and are now beating American companies to the marketplace thanks to much shorter cycles.

Over the past few years, United Research has been working with several U.S.-based manufacturers to bolster their competitive abilities vis a vis product development. The process we used is known as simultaneous engineering. It ensures that the appropriate people within a corporation are brought into the development process at the appropriate time, regardless of a business' internal organization.

By Fred Gordon
and Robert Isenhour
United Research, Morristown, N.J.

Simultaneous engineering, as an approach to product development, holds out the hope of consolidating competitive advantage by reducing the time lag between product development and marketing and sales; minimizing development and product costs; increasing quality; and augmenting market share.

As an approach, it involves all of the disciplines that are themselves aspects of product development, starting from market research and working through design engineering, manufacturing and sales. But from the first, all are viewed holistically in order to bring a quality product from concept to market in less time and at competitive cost. This sometimes requires sacrificing short-term profits to gain longer-term competitive advantage.

Because both hard and soft products pass through the same six stages of development, the benefits of this engineering philosophy apply just as well to each. These steps are:

1. Customer need: Identify a marketplace need;
2. Advanced development of concept: Create a saleable product based on that need;
3. Design and development (or product engineering): Translate the concept into reality (i.e., product design). This includes testing prototypes and subsequent design iterations until agreement is reached that the product is truly responsive to the needs of the marketplace.
4. Pilot production: Determine whether the product can be made in existing manufacturing facilities.
5. Field test: Observe how the product performs in the marketplace.
6. Production: Manufacture the product.

Since the essence of simultaneous engineering encourages—even demands—the blurring of conventional boundaries, the six procedures themselves are not seen as distinct sequences. Any and all are called into play whenever they can serve the stated end (see fig. 1).

Simultaneous engineering differs from sequential engineering—and from its cousin, overlapping engineering—not so much in what is done but in when it is done. In the sequential approach, each function is called into play as part of a series whose members are cloistered from one another. Each has its purpose, but each each works autonomously. In overlapping engineering, each function interacts only with its nearest neighbors, those steps that immediately precede and succeed it.

In neither approach, however, does anyone performing a function fully understand or appreciate the issues important to downstream disciplines. Hence, both approaches demand frequent redesigns of a particular product to solve unanticipated problems. Moreover, because the development and support functions do not coordinate their efforts, problems tend to peak just before the start of product sales. This increases costs and lengthens product development time.

In simultaneous engineering, both engineering and support teams work concurrently on all phases of product development, fully sharing information and decision making.

Among the participants on the first side are those involved with product, manufacturing and tool engineering, as well as those concerned with plant and industrial engineering. Estimating costs and supplying the necessary support also demands the talents of those in scheduling, quality/product assurance, purchasing, marketing/sales and manufacturing.

The net effect of having these groups work side by side is to shift the "problem curve" further back along the product development cycle (see fig. 2).

BENEFITS

The benefits of such simultaneous thinking can be seen in the early resolution of problems and conflicts that inevitably surface in the development cycle. Cooperation increases the incidence of product designs that match available resources, thereby reducing multiple redesigns. The results are a better-quality product, produced in a shorter time with fewer changes and, therefore, lower overall development cost (see fig. 3).

United Research recently completed a simultaneous engineering project for a multibillion-dollar manufacturer of automotive components. The division designed and produced automobile hardware and lighting systems for its parent, one of Detroit's Big Three automobile corporations, as well as other companies.

With competitive pressures increasing in the worldwide automotive market, the corporation gave its car divisions the freedom to select their own suppliers. Without a captive customer for 90 percent of its output, the division had to become more responsive to its customers or it would go out of business.

A major reason for the division's poor competitive position was sequential engineering, which lengthened the product development cycle and generated high costs in the area of engineering development.

The division addressed these roadblocks in two ways. First, the business was divided into product teams, each focusing on a specific product family. In effect, the division was carved into a set of smaller businesses, each with a defined market niche.

Second, simultaneous engineering was adopted as the methodology for designing and developing products.

Simultaneous engineering was defined, developed and implemented by a 15-person team. It included representatives from design/product engineering, tool engineering, advanced development engineering, manufacturing engineering, test engineering, product reliability, accounting/estimating, schedule and release, sales, purchasing and two manufacturing facilities.

Simultaneous engineering creates an environment in which all of the disciplines that contribute to and support the product development cycle perform their functions cooperatively.

The team met twice a week for 10 weeks, first defining all of the engineering steps of the design and development phase and determining when each should be performed. The team then decided to what degree each area of expertise would be involved in each step and how functions would interact.

To manage and coordinate the engineering function, engineering teams were formed for each product grouping, headed by a team leader accountable for the total engineering effort from concept through manufacturing. The overall responsibility of the engineering team leader was to assure that product objectives (including quality, cost and delivery) took precedence through all phases of the engineering process.

Aside from clarifying and charting the roles and responsibilities of the teams and their members, all participants were trained in the specific skills needed in the simultaneous engineering environment. Team building and problem-solving skills formed the core of the training, with auxiliary workshops focusing on management skills and conducting effective meetings.

For this division, the steps taken represented a fundamental reordering of its corporate culture. For the first time in the division's history:

- One person is responsible—and accountable—for coordinating the contributions of all of the engineering disciplines that contribute to the product development process.
- All of the functional and support disciplines—manufacturing, marketing, accounting, sales and distribution—now take an active role in product design and development, up front.
- Job descriptions and departmental objectives have been rewritten to link them directly to division objectives regarding productivity, quality and cost control.
- The members of each product team have now adopted the same product-related objectives.

The effects on the division's operations and competitive advantage were dramatic in the critical areas of cost, quality and timely delivery.

- Virtually all critical milestones in the product development cycle are now being achieved on schedule and much faster than ever before.
- Potential problems relating to manufacturability are identified early in the design process. Not only does this improve product quality, but also with less-frequent delays and setbacks, product development time is curtailed and costs reduced.
- Products are now designed and manufactured 20 percent faster than the previous typical time and at substantially lower cost.
- Not just time but money was saved as well. The division realized an annual savings of $750,000 through a 25 percent reduction in the number of engineering changes.
- The quality of the division's finished products now surpasses that of any previous period in its history.

The most dramatic result of the initiatives taken in the area of simultaneous engineering, however, was to restore the competitive life and vitality of the division itself, along with its 18,000 jobs and multimillion-dollar payroll. This human impact is, in the end, the most gratifying and important aspect of the project.

FIG. 1 FUNCTIONS IN THE PRODUCT DEVELOPMENT CYCLE

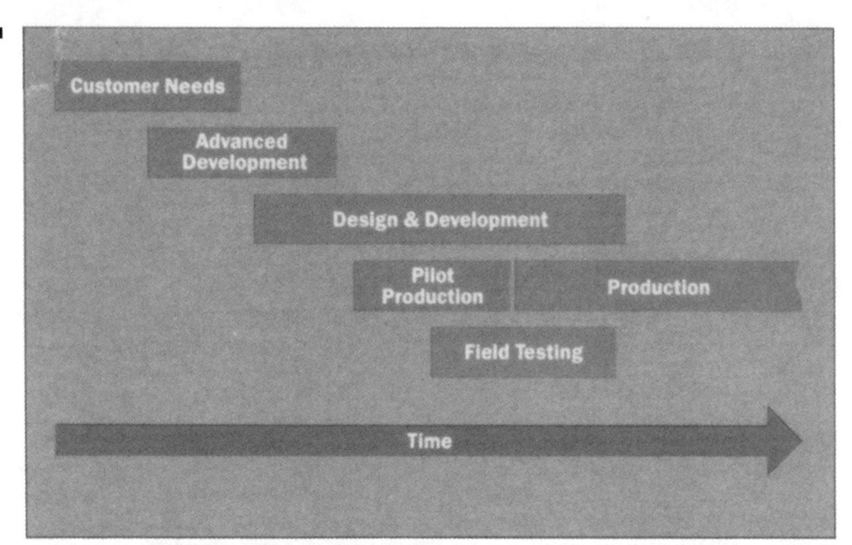

FIG. 2 PROBLEM CURVE: SIMULTANEOUS VS. SEQUENTIAL ENGINEERING

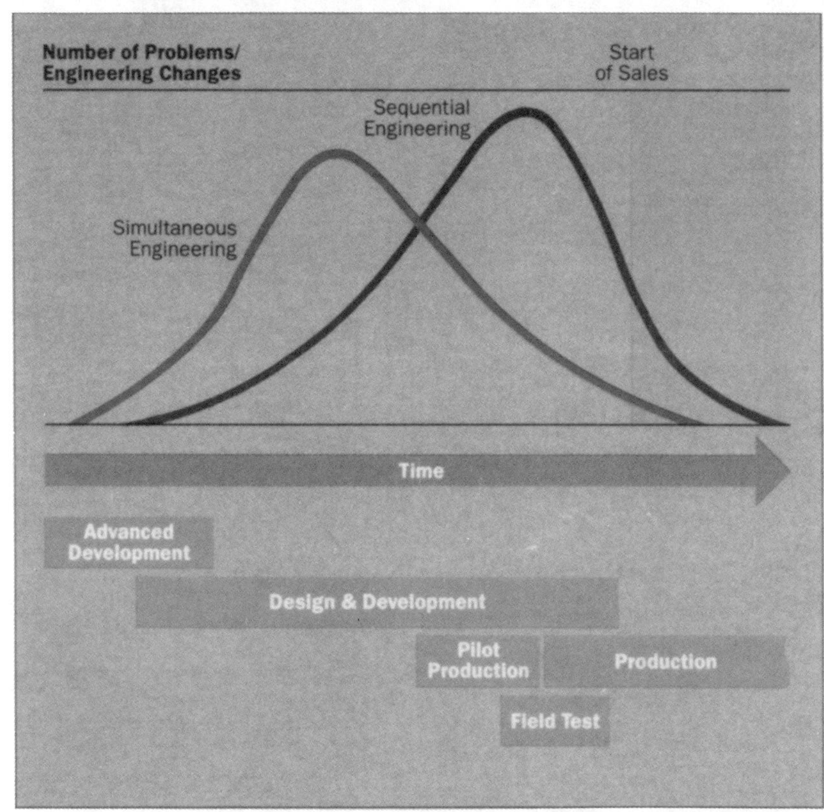

FIG. 3 SHORTENING THE PRODUCT DEVELOPMENT CYCLE

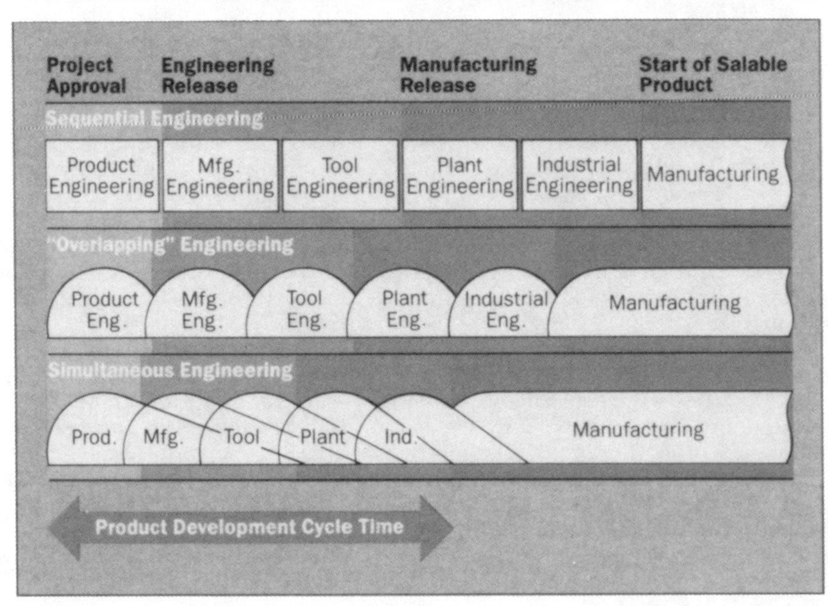

PROVEN METHODOLOGY

Simultaneous engineering is a proven methodology that, properly implemented, can help insure the timely delivery of such products—competitive in cost, quality and innovation—to the marketplace. It creates an environment in which all of the disciplines that contribute to and support the product development cycle perform their functions cooperatively. Such efforts, adapted to each corporate culture, can lower product development costs, ensure much greater built-in quality and help American businesses capture and sustain competitive advantage.

Reprinted from *Manufacturing Engineering*, September 1988

The Final Piece to the Puzzle

By John M. Martin
Endicott, NY
Writer and Consultant

You would think it was the most obvious thing in the world. When you are developing a new product, consider, along with all that product's functions, features, and cosmetics, how on earth you are going to make it.

Yet, shockingly, until perhaps about a year ago, only a relatively small number of forward-thinking companies were bringing their manufacturing organizations into the product development process at an early enough stage to significantly impact the concept, design, and subsequent manufacturability of the product.

Most others were proceeding on their merry way—with lots of headstones marking the trail—finalizing concepts and designs without consulting manufacturing. The result? Manufacturing would get the detailed specifications too late to advise development and design of the process and manufacturability issues involved in the project, and would end up adjusting and customizing the production process to meet the product requirements.

Almost invariably, that meant "a more costly, risk-prone and time-consuming manufacturing operation," according to Henry Stoll, manager of the design for manufacturing section at the Industrial Technology Institute (Ann Arbor, MI).

In retrospect, it's not too difficult to see how this situation evolved. "Years ago," says Wes Allen, a visiting professor at the University of Cincinnati's mechanical engineering department with 30 years' experience in industry, "somebody did both engineering and the business in general a great disservice by fragmenting the product cycle: one designs, one produces. It's an unnatural division of the process, yet we have continued to live with this due to the general pace of technology."

There is certainly also a difference between the design and manufacturing function that has contributed to that division. "You have to recognize the design mentality," says Randy Johnson, manager of manufacturing engineering at Apple Computer's Fremont, CA, facility. "Designers are sitting there waiting for the opportunity to design a new product. Manufacturing shines when it has commonality and mundane-looking things to produce."

Other factors inhibit cooperation as well. "Manufacturing engineers are too busy putting out fires," states Richard Bradyhouse, whose title is technical manager of produceability at Black & Decker (Towson, MD). "Because of that, the interface with development and design suffers. Sometimes, too, they're not even at the same location, and there's travel involved."

The reasons why a more cooperative approach has not worked properly are no longer good enough, not in today's business climate. "These last few years, product as well as manufacturing technology is changing faster. Along with the worldwide competitive environment, we can no longer put up with this lack of communication during product development," Allen says.

Probably echoing a number of other manufacturers who have increasingly admitted the same thing these past few years, Doug Barron, manager of product and process development at Delco-Remy (Anderson, IA), states: "We were a strong manufacturing group, but we kind of went to sleep, we didn't have to compete. Now we're in a global marketplace, and there are entirely new demands."

Adds Allen, "We thought we were on top of the heap, we didn't want things to change, we wanted to freeze it and keep it right there. Well, we did. . . and progress passed us by."

One Good Apple...

"All of our products have both hard and floppy drives," states Randy Johnson, manager of manufacturing engineering at Apple Computer's Fremont, CA, plant. "And commonality is imperative across our product line, especially in the subassemblies where, unless you have common mounting techniques, there will be problems on the line in the areas of handling, fixturing, inventory, and production control."

The MAC II and SE products were being designed at the same time, and Johnson made sure that "the two design groups—which each included design-for-manufacturing teams—talked together. It was tough; each were sensitive to their own development cycles and schedules, each felt they had a way that worked, and each felt their own way was the best. I was in the unenviable position of being the arbitrator."

The resulting cooperation, though, standardized drive configurations across the two product lines and led to a more straightforward manufacturing operation that was in a better position to serve the business requirements of the overall corporation. ■

Tracy Cox

No use crying over spilt milk. In fact, today a veritable tidal wave of interest and activity is swelling in the area of increasing the role of manufacturing engineering in new product development. Whether you call it simultaneous engineering, concurrent engineering, early manufacturing involvement, design for manufacture, design for assembly, or just plain old common sense, an enormous number of companies are starting to realize the benefits of upgrading manufacturing's place in the development process.

"I have seen it in more and more companies," says Stoll of ITI. "I was in a 'missionary mode' for the past three years, but in the past year this issue has really gotten the serious attention of top management, and people really want the tools now to make it happen."

How, in fact, do you make it happen in your company? "Manufacturing engineering can take the lead here," Stoll says, "bringing forward manufacturing and produceability guidelines so design is aware of what we're good at: what our process capabilities are, our tooling, NC and fixturing configurations, and so forth."

To get to that stage, however, first takes getting the ear of top

> Manufacturing engineering must take its place in product development if projects are to succeed

Final Piece to the Puzzle

corporate management. "This is more of a cultural change than anything else," says Allen. "The primary requirement is that top management recognize and comprehend the changes you are trying to make, fully support them, and be active in their implementation. Without top management involvement, the chances of success drop off by an order of magnitude, at least. Top management has to say, 'This is our strategy, we have to do this day in and day out,' or it will fall apart."

You have to "champion" the idea to get that commitment. And once you get it, you've got to set up a structure that works for your company. That doesn't necessarily mean traumatic reorganizations, Allen says. "Most companies that have moved in this direction have made an organizational adjustment, but I don't think that's an absolutely necessary requirement."

He goes on to say that "the initial reaction within a company is normally not too positive. It is obvious right off the bat that this will involve large changes in everybody's way of doing things. And the fact is, nobody really wants to change. We want improvements, better money, but we don't really want to change."

Allen feels that "one of the most beneficial things is to put the manufacturing and design people in proximity to each other. Have them share the same coffee machine so they get to know each other and realize the other people don't have three heads and four arms. Give them an environment where they can approach each other." Apple's Johnson is also high on "the informal aspect, being 'buddies,' part of a group. We sometimes get a lot more accomplished after working hours than during them."

Whatever combination of formal and informal devices you develop to make cooperation happen, the important thing is that you set up a permanent part of the organization to drive it. At Apple, Johnson says, two years ago the company established a design for manufacturing section within manufacturing engineering. Tom Mooney, chief engineer for Xerox's 5028 family of copier products, reports that manufacturing technology works within the development group and that his organization has had a design for assembly group active for the past five years.

The initial reaction within a company is normally not too positive

These kinds of groups have to get into the product development process early and stay on the case. "Before we had the DFM group," Johnson says, "the product development guys thought that early manufacturing involvement meant coming down periodically to make presentations. But that's a 'done deal'—you're put in the position where you have to criticize, and all you get are all the reasons why things can't change.

"What's really worked for us," Johnson says, "is getting in at the conceptual and design phase. I tell our DFM people that their job is to roam the halls of product design. Formal design reviews are out—don't count on them to call you. This is a day-to-day job; you've got to catch them hour-by-hour, minute-by-minute. If a week or a month goes by, and an idea gets out of hand, then it's too late. They'll already have 'ownership,' then they'll want to check it out with manufacturing, and all you can be is the bad guy."

Of course, there are organizations for which the day-to-day approach is impossible. At Black & Decker, for example, Bradyhouse reports that "We are both a centralized and decentralized company. The corporate engineering group designs all the products and we have a number of

Boosting Delco-Remy

Prior to assuming his current position as manager of product and process development at Delco-Remy (Anderson, IA), with responsibilities for long-range product and process development for alternators, starter motors, lead acid batteries, ignition systems, and miscellaneous switches, Doug Barron was superintendent of battery manufacture process development.

"There were five US manufacturing locations, one in Canada, and one in France," Barron recalls. "Development was at another location. I got both product engineering and process engineering together in a joint development effort. The specification and design people sat down with the people responsible for making the products a reality. We did this so well that a decision was made to reorganize the company."

In the fall of 1986, Delco-Remy was reorganized around four separate business units: Power Systems, Control Systems, Battery, and Heavy Duty Systems. "The focus is now on product," Barron says, "and the driving reason for the reorganization was to get the product and manufacturing guys physically located in proximity to each other to facilitate both formal and informal communications."

The results? "When you mix people up like this in a large division, there is going to be some emotional trauma. Some will adapt, some will resist. We're still struggling, but things are going fairly smoothly. We're doing the early-on planning better, decisions are better and being made more quickly, and we're not having to do things like prototypes and tooling over because intent was not interpreted correctly."

Barron feels that the results of the switch to simultaneous engineering are difficult to quantify at this point because "it's only been two years. In another two years, if our products are more competitive in world markets, then we'll know we're doing better."

Barron does know that things worked better when he made these same kinds of changes in the battery area. "One of the keys there is finding ways to improve the utilization of lead in the grid without sacrificing performance. The grid structure of the plates is critical. We brought product and process together in a series of weekly meetings, computer-modeled the grids, came up with ways to achieve very significant cost and performance improvements, and we did it quickly and at reasonable cost." ■

manufacturing plants that build them. It presents the age-old problem of what is the best way to do product transfer."

At Black & Decker, Bradyhouse says, "We are using the simultaneous engineering approach. Manufacturing engineering has to build the product, and they are brought to corporate engineering on a periodic basis—every five to six weeks—throughout, say, a one-and-a-half-year development program. The manufacturing engineers start at the concept phase, review the status of new product development, and are given the opportunity to tell design how they would manufacture the parts. At first manufacturing engineering said that the staff didn't have the time to spare for extra travel to the design center, but now that it's doing it, it sees the payback because it knows what it'll need to put into manufacturing."

Dan Challgren, director of product industrialization at Allen-Bradley's Industrial Controls Group (Milwaukee), states: "We request target cost information from marketing very early in the stages of product development. We want to see how realistic these costs are. We estimate the manufacturing costs, and if they're way off base, that alerts marketing that the plan is either unachievable or involves high costs. Marketing can then make plans to modify its specifications, cancel the project, or reevaluate the target costs and see what they do to the margins. This is additionally done periodically throughout the project as well, so we can wave a flag if targets are not being met."

However you choose to do it, don't take the attitude of "All right, manufacturing is finally going to get the upper hand," says Allen. Adds Stoll, "You can't tell product design, 'You've got to.' You need to say, 'I would prefer such-and-such, and if you can't design it that way, this is what your design will cost.' Manufacturing has to give design more standardized information, information that tells design such things as if a fillet radius is 0.125″ (3.17 mm), the tooling will cost five cents, but if it's 0.12773″ (3.2443 mm), then the tooling will cost a thousand dollars, and it will be the only tool like it in the company."

One way to lay the groundwork for this type of cooperation is cross-training. At Xerox, for example,

Manufacturing engineering on the product development team means improvements at GMC Truck Operation (Pontiac, MI).

Xerox Original

At Xerox's Decentralized Business Unit (Webster, NY), Chief Engineer Tom Mooney reports that "The way we're structured, the entire manufacturing organization reports to me. That includes procurement, configuration management, piece part engineering (the vendor interface), the design organization, and planning."

Mooney's responsibility is the 5028 family of copier products. "This means high-volume manufacture of the lower-speed machines in Xerox's market," he explains. "Manufacturing these products is the heart of our ability to compete. We design for manufacturability and support manufacturing with automation. The advice we get from manufacturing engineering is what makes the difference."

He goes on to say, "Years ago, we were missing the standardized build hours. Our yields were low, as were those of our vendors. We measure defects in machines per hundred, and now we're seeing the payoff of manufacturing engineering's involvement in quality, cost, and time. We're averaging 0.1 defect per machine, and these are things like a mark on the cover, nonuniform intensity of the LED on the panel, that sort of thing."

Mooney makes sure that everyone reports to him early in product development. "The manufacturing resource manager reports to me on volumes and market locations so we can pick our manufacturing sites, the availability of locally sourced materials, and so forth. The operations manager comes in to develop the type of line that is best suited to the job. Manufacturing engineering is brought in with the design team. As we get the floor operations going, supervisors, quality people, and the hourly assemblers also get involved. Their participation begins with the start of the second prototype build and before the pilot plant, staging, and process sheet creation is initiated. One of the hourlies recently said, 'I feel like I died and went to heaven. I'm part of the design team.' "

Mooney reports that he has "forced the design engineers onto the floor" (the design team is housed in the pilot plant). He's also "refereed a lot of small battles. But there are no major divisions any more. We've developed a more homogeneous thought process. We now realize how important manufacturing is to us." ■

Final Piece to the Puzzle

Mooney says that "for an engineering job hire, it is mandatory that he or she spend six weeks out in the field with field engineering and nine months in the manufacturing environment. Conversely, manufacturing hires must also spend six weeks in the field and nine months in the design function."

Johnson would like to see something similar at Apple. "Design engineers, before starting in design engineering, should be hired through manufacturing engineering first, spend six months there, and then go to their regular assignments. If you plan far enough ahead, you can do that. Besides, many of these people are just out of school; they'd waste that time anyway going through five iterations of the design before they hit it right. This would give them the time to get experience, exchange ideas, and meet the manufacturing people."

Johnson provides an interesting angle on how he makes the design-for-manufacture/product development collaboration work at Apple. "The DFM people are mechanical engineering designers, not process engineers. I attracted them to manufacturing by telling them that when you work on this end of things, you get to work not only on one project, but on the design of all the products. Plus, you don't have to grind out the details. I also tell them that manufacturing is where the action is. We work out the design of every new product coming in."

Defining the project is usually the number one problem

That strategy has done wonders for the interface between manufacturing and design, he reports. "Once you've got these savvy designers on your side, they can communicate—and do occasional battle—with the product designers. In manufacturing, we used to lose every battle. We could have a great idea, but if we didn't have the credentials and respect, it was no use. Now the designers know what we want, and they have put it into their mind-set."

Apple, for example, has what Johnson refers to as "typical guidelines we submit to designers. We want minimum parts counts; design to assembly and process capabilities; parts with tolerances and specifications easily met by normal vendor processes and equipment; top-down assembly in one *Z* axis so we don't have to bring in things from the side; parts shaped for easy packaging; reduction of cables and utilization of direct connections; and standardization across product lines."

Stoll feels that, ultimately, "designers are much happier that way. They don't have to spend so much time on what otherwise seem to them to be arbitrary decisions; they have a chance to modularize for fixturing and tooling; costs go down; and, the more repetitive you get, the more quality results."

At ITI, Stoll is currently working to develop just such tools for design for manufacturing. He feels that "the focus has to be on the CAD arena; that's where the manufacturing information has to be available to the designers. We're working on a generic but 'tailorable' software module—we've got a shell right now—of standard primitives that will be easy to mold, machine, or stamp, have all the right dimensions, and follow best manufacturing practice for a particular company."

Bringing about such changes at a company is by no means an easy task. There is a historical inequality between manufacturing and design that has been fostered by corporate culture. Manufacturing people, as a result, "are reluctant, they're used to taking a back seat," says Stoll.

Also, the significant amounts of automation that have been implemented these past few years have compounded the crisis. "Designers aren't familiar with the new manufacturing capabilities; they're still designing the old way. It's difficult to use a new manufacturing process efficiently if the product hasn't been designed for that process. People are still inherently designing for human

Truck and Bus Steps Out

Ernie Vahala, director of manufacturing engineering of GM's Truck and Bus Group (Pontiac, MI), states that when GM's June 1982 reorganization created his group, he decided to use a little common sense.

"Product has to fit process, process has to fit product, and process has to fit the plant," he says. "Our whole thrust is bringing on new products, and all of the manufacturing engineering disciplines are required in order to carry out a product program. We give every department full-time membership on the team—engineering, materials, finance, marketing, and sales–so there is constant communication to develop a successful program that answers all the customer's needs."

Manufacturing engineering "lives side by side with product engineering during design, development, and test," he says, "so they can input to the designer what can be manufactured and what can be assembled to the quality tolerances that the customer expects."

Vahala gives as an example the GMC Sierra and Chevy CK full-sized pickup truck program. "We will be turning out about 670,000 units annually when the second shift is fully operational at our third assembly plant," he says. "There was a big debate concerning a ± 0.0005 mm tolerance that fabrication felt it could not hold because coils of steel tend to get rolled off different billets and the degree of cold compression and springback varies."

Cooperation between the various departments resulted in a cab and box assembly that "lets the fab plant get its tolerances while the final assembly comes out closer than any of the individual components," Vahala reports. "It's the best example we have of how effective it is when these groups live and work together."

He concludes: "The only way to meet the needs of the customer for world-class quality is to get manufacturing engineering and product engineering working together up front so that each meets the other's needs and the desires of the customer. It can't happen separately—that's not good enough. By doing this you'll eliminate rework, and rework used to be a fact of life. But it cannot be now." ■

assembly," Allen says.

Stoll spells out some other difficulties. "People don't want to tamper with 'evolutionary' design, where there's a lot of art involved, a long life cycle, and failures might not show up for 20 years. So they stick to what works. Also, highly constrained, performance-driven designs are a problem. People are reluctant to do something that has never been done before: put it in a smaller package, use new materials, make it more produceable."

There is inertia to be overcome, and political and turf battles to be fought. But they must be fought. "Excellence of product is the goal, not 'this is my job, this is yours.' Manufacturing must take a leadership role in planning at an early stage, getting the information to designers in the form they need it, and working together with them in a team approach," says Stoll.

Says Challgren, "Our role is to be extra proactive with the marketing and product development people throughout the entire project. We have to focus on the key success factors: achieving target production costs, meeting product quality targets, maintaining the introduction schedule, and controlling the implementation costs."

A company will save time in a business climate that heavily rewards quick response to market changes

Challgren adds, "Defining the project is usually the number one problem. If the product specifications are not detailed and clearly written, everyone will go off on different tangents. This also forces marketing, design, and, to some degree, the other functions to sit down and work out these specifications, exchange information and ideas, and reach final agreement."

The results can be nothing short of spectacular. A company will save time in a business climate that heavily rewards quick response to market changes; will reduce costs by reducing surprises and rework, by getting things done right the first time, by standardizing product features and manufacturing processes, and by reducing the charges associated with tooling, fixturing, procurement, and the storage of parts; and very likely will create a degree of cooperation and level of enthusiasm that has been missing from the manufacturing business for a long time.

"It builds on itself," says Xerox's Mooney. "When you've gotten into the practice, you learn to design better, market better, manufacture better. All the separate components start to function as one." Barron says, "You develop more confidence in each other when you work together earlier. These are 'soft side' issues, but they are critical to an effective organization."

Adds Challgren, "You develop a real overall understanding of the business. The team takes on its own identity; everyone wants it to succeed, and there's a lot of pitching in. No one person can have 180 good days in a row. You need a lot of help from your coworkers, and as you get to know people better and work out these problems with them, it's amazing what you learn."

What you're learning is critical to your survival. Allen: "The people I talk to and work with are of the opinion that if a company is going to stay competitive, the old way of doing things has to change or it won't exist down the road."

Indeed, the ultimate arbiter of every company's survival or failure is the customer. Says Barron of Delco-Remy, "Design by itself will not ensure that a product is useful to a customer. Cost, repeatability, and quality are critical, too."

And competitive cost, repeatability, and quality can only come about through what Stoll calls "coordinated product and process design, where the product, tools, fixtures, and material flow are all designed as a coordinated system so there is no mismatch between the design and the best way to make the product." ■

Romeo, Romeo

At Wilson Automation (Warren, MI), Chief Engineer George Fisher has spent the last year, along with other Wilson engineers, working closely with Ford Motor Co. on that firm's new 4.6-liter engine. Wilson, a division of Newcor, Inc. (Bay City, MI), just got the purchase order for the assembly line in January 1988.

"Before, we'd get the line concept that was structured by Ford as an inquiry, and then they'd give us a PO for their line," Fisher states. "This time a team was formed with the Ford people, weekly meetings were held, every component we had access to was analyzed, and numerous different layouts were discussed with Ford before the team decided on the 15-section combination synchronous/nonsynchronous line that will assemble their engines in the Romeo, MI, facility."

The benefits? "It keeps the costs down," Fisher says. "It gives us a jump on our design and manufacturing requirements. We can look at areas and strategies that we would not normally have the time or opportunity to examine. And we'll be more familiar with the equipment beforehand, before it goes into production."

As part of the collaboration, Wilson visited every Ford engine plant in the US and Canada. "We looked at our equipment, talked to their people, saw what they liked and didn't like, what the problems were. We also collected efficiency data for the plants while we were reviewing them."

Wilson engineers chair portions of the weekly meetings between the two companies. The vendor also keeps a room at its facility that is completely at the disposal of Ford people. And they've given Ford a five-year warranty on the line, commencing with job one.

"This has created a lot of enthusiasm and a great deal of trust in each other," says Fisher. "We've done this collectively as a team, a true partnership. We feel it's a competitive edge both for us and for Ford."

Fisher states that, in the process, "We and Ford have reconfigured the way product development is implemented." ■

Reprinted from *Manufacturing Engineering*, January 1988

Design for Manufacture

MANUFACTURING ENGINEERING Technical Report series...special technical articles detailing the latest manufacturing breakthroughs adapted from material presented in the Tool and Manufacturing Engineers Handbook *series, fourth edition.*

What is design for manufacture, and why is it important? Here are 10 different approaches; one—or more—may be appropriate for you

By Henry W. Stoll
Industrial Technology Institute

Design for manufacture (DFM) represents a new awareness of the importance of design as the first manufacturing step. It recognizes that a company cannot meet quality and cost objectives with isolated design and manufacturing engineering operations. To be competitive in today's marketplace requires a single engineering effort from concept to production. The essence of the DFM approach is, therefore, the integration of product design and process planning into one common activity.

The DFM approach embodies certain underlying imperatives that help maintain communication between all components of the manufacturing system and permit flexibility to adapt and to modify the design during each stage of the product's realization. Chief among these is the team approach or *simultaneous engineering*, in which all relevant components of the manufacturing system including outside suppliers are made active participants in the design effort from the start. The team approach helps ensure that total product knowledge is as complete as possible at the time each design decision is made. Other imperatives include a general attitude that resists making irreversible design decisions before they absolutely must be made and a commitment to continuous optimization of product and process.

The objectives of the design for manufacture approach are to identify product concepts that are inherently easy to manufacture, to focus on component design for ease of manufacture and assembly, and to integrate manufacturing process design and product design to ensure the best matching of needs and requirements.

Meeting these objectives requires

the integration of an immense amount of diverse and complex information. This information includes not only considerations of product form, function, and fabrication, but also the organizational and administrative procedures that underlie the design process and the human psychology and cognitive processes that make it possible.

Because of the complexity of the issues involved, it is convenient to divide the subject of DFM into two considerations: (1) the DFM approach or process by which a product can be effectively designed for manufacture; and (2) the methodologies and tools that can be used to help enable the DFM approach and to help ensure that the physical design meets the DFM objectives.

Many different versions of the DFM process have been proposed. Each version is likely to be similar in the issues addressed and the concepts embodied. Differences would likely reflect idiosyncrasies imposed by the organization in which a particular version originated and the type of design problem it was meant to address.

One proposed version of the DFM process is shown in *Figure 1*.[1] The four activities comprising this process are arranged in a circular fashion to emphasize the iterative nature of the process. Traditionally, many products have been designed by starting with functional optimization of the product design itself, followed by detail design of each part to be made by a particular process, then simplification, and finally design of a process to manufacture and assemble the product. As shown by the arrows, the progression of steps in the proposed DFM process is just the reverse of the more traditional design approach.

The DFM process begins with a proposed product concept, a proposed process concept, and a set of design goals. All three of these inputs would be generated by a thorough product plan developed using the team approach. Design goals would include both manufacturing and product goals.

Each of the activities within the DFM process addresses a particular aspect of the design. Optimization of the product/process concept is concerned with integrating the proposed product and process plan to ensure inherent ease of manufacture. The simplification activity focuses on component design for ease of assembly and handling. This activity can often be rapidly effective because the integrated product and process requirements and constraints help identify problem areas. The third activity ensures conformance of the design to processing needs. For example, if an assembly is to be built on a particular flexible assembly system (FAS), it is important that the assembly be designed in such a way that it can be assembled using the programmable gripper engines, flexible fixtures, and assembly operations available within the FAS. Finally, functional optimization considers appropriateness of material selection and parameter specification that maximize the design objectives.

By reversing the process, this DFM approach helps ensure that all of the design constraints, including assembly, material transformation processes, and material handling requirements are included as part of the functional optimization of the design. In this way, the DFM process enables the designer or design team to consider all aspects of the product's design and manufacture in the early stages of the design cycle, so that design iteration and accompanying engineering changes can be made easily and cost effectively. Finally, by integrating the product and process design, it is possible to include manufacturing recommendations and a process plan as part of the engineering release package.

This has great advantages because it leads to few, or no, manufacturing surprises. Also, both manufacturing and engineering share equally in ownership of and ultimate commitment to the design.

The development and use of design methodologies that help the design team achieve an optimized design solution is an important part of the DFM approach. *Figure 2* provides a selected list of DFM methodologies and indicates where they might fit into the proposed DFM process. Use of these design methodologies helps promote the objectives of DFM by guiding the design team in making better informed design decisions and providing systematic procedures that help ensure that all aspects of product function, manufacture, and operational support are considered from the start.

An axiomatic approach to design is based on the belief that fundamental principles or axioms of good design exist and that use of the axioms to guide and to evaluate design decisions leads to good design. By definition, an axiom must be applicable to the full range of design decisions and to all stages, phases, and levels of the design process. Design axioms cannot be proven, but rather must be accepted as general truths because no violation or counterexample has ever been observed.

A study of many successful designs by several individuals in 1977 led them to propose a set of hypothetical axioms

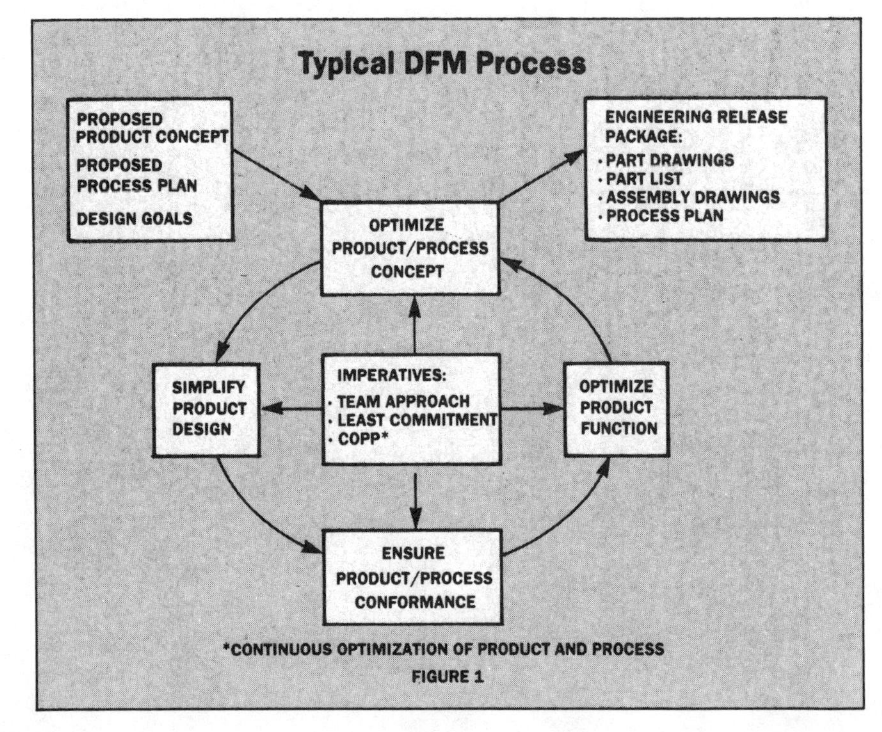

FIGURE 1

for design and manufacturing.[2] Analysis and refinement of the initial axioms has shown that good design embodies two basic concepts. The first is that each functional requirement of a product should be satisfied independently by some aspect, feature, or component within the design. The second is that good designs maximize simplicity; in other words, they provide the required functions with minimal complexity.

Use of design axioms in design is a two-step process. The first is to identify the functional requirements (FRs) and constraints. Each FR should be specified such that the FRs are neither redundant nor inconsistent. It is also useful in this step to order the FRs in a hierarchical structure, starting with the primary FR and proceeding to the FR of least importance. Once the functional requirements and constraints are specified for a given product or design problem, the second step is to proceed with the design, applying the axioms to each design decision; each decision should be guided by the axioms and must not violate them.

Application of the design axioms to the analysis and design of products and manufacturing systems is not always easy or straightforward. Because the axioms are quite abstract, their use requires considerable practice as well as extensive on-the-job design and manufacturing experience and judgment.

DFM guidelines are systematic and codified statements of good design practice that have been empirically derived from years of design and manufacturing experience. Typically, the guidelines are stated as directives that act to both stimulate creativity and show the way to good design for manufacture. If correctly followed, they should result in a product that is inherently easier to manufacture. Various forms of the design guidelines have been stated by different authors, a sampling of which follows:

1) Design for a minimum number of parts
2) Develop a modular design
3) Minimize part variations
4) Design parts to be multifunctional
5) Design parts for multiuse
6) Design parts for ease of fabrication
7) Avoid separate fasteners
8) Minimize assembly directions; design for top-down assembly
9) Maximize compliance; design for ease of assembly
10) Minimize handling; design for handling and presentation
11) Evaluate assembly methods
12) Eliminate or simplify adjustments
13) Avoid flexible components.

DFM Tools

DFM TOOLS	OPTIMIZE CONCEPT	SIMPLIFY	ENSURE PROCESS CONFORMANCE	OPTIMIZE PRODUCT FUNCTION
	DFM ACTIVITY			
DESIGN AXIOMS	●	●	●	●
DFM GUIDELINES	●	●		
DESIGN FOR ASSEMBLY METHOD		●		
TAGUCHI METHOD	●			●
MANUFACTURING PROCESS DESIGN RULES		●	●	
DESIGNER'S TOOLKIT			●	
COMPUTER-AIDED DFM	●	●	●	●
GROUP TECHNOLOGY	●	●	●	●
FMEA	●			●
VALUE ANALYSIS				●

FIGURE 2

DFM guidelines show the way, but do not replace the talent, innovation, and experience of the product development team. They must also be applied in a manner that maintains and, if possible, enhances product performance and marketing goals. Design guidelines should be thought of as "optimal suggestions," which, if successfully followed, will result in a high-quality, low-cost, and manufacture-friendly design. If a product performance or marketing requirement prevents full compliance with a particular guideline, then the next best alternative should be selected.

The design for assembly (DFA) method was developed by G. Boothroyd and P. Dewhurst while at the University of Massachusetts (Amherst). Details of the methodology are presented in *Design for Assembly—A Designer's Handbook*.[3]

Based largely on industrial engineering time study methods, the DFA method developed by Boothroyd and Dewhurst seeks to minimize cost of assembly within constraints imposed by other design requirements. This is done by first reducing the number of parts and then ensuring that the remaining parts are easy to assemble. Essentially, the method is a systematic, step-by-step implementation of the DFM guideline numbers 1, 7, 8, 9, and 10.

The Taguchi method addresses the problems associated with determining robust design by using statistical design of experiment theory. Robust design implies a product designed to perform its intended function no matter what the circumstances. In particular, the Taguchi method seeks to identify a robust combination of design parameter values by conducting a series of factorial experiments and/or using other statistical methods. Termed parameter design by Taguchi, this step establishes the mid-values for robust regions of the design factors that influence system output. The next step, called tolerance (allowance) design, determines the tolerances or allowable range of variation for each factor. The mid-values and varying ranges of these factors and conditions are considered as noise factors and are arranged in orthogonal tables to determine the magnitude of their influences on the final output characteristics of the system. A narrower allowance will be given to noise factors imparting a large influence on the output.

In establishing the tolerance or allowance range for a particular parameter, Taguchi uses a unique concept defined as a loss function. In this approach, loss is expressed as a cost to either society (the customer) or the company that is produced by deviation of the parameter value from design intent. Because any deviation from design intent produces a loss, allowance or permissible deviation should be determined based on the magnitude of the cost associated with this loss. The concept of loss and other Taguchi concepts provide valuable insight into quality and the role design plays in determining the quality of a product or system.

Process-driven design seeks to ensure that parts and products are correctly designed to be produced using a particular production process or method. Design requirements for a given process are often stated in the form of design guidelines and rules of thumb. Typically, these guidelines are highly specialized for a particular industry, process implementation, plant, or equipment installation within a particular plant. Making the designer aware of these process requirements and constraints early in the design process, before concepts are finalized and lines are put irreversibly on paper, is a key goal of design for manufacture. Design tools that help ensure product/process conformance and enable process-driven design can generally be

Guide to DFM Methodologies Comparison

1. **Implementation cost and effort.** A rating of "better" indicates that the methodology can be effectively implemented simply by creating awareness through seminars and/or brief training and by providing management expectation that it be used. A "worse" rating indicates that implementation may require extensive company-wide commitment, purchase or development of expensive software or hardware, extensive and/or costly training, and possibly extensive preparation for methodology use. An "average" rating indicates that implementation requirements are relatively uncertain and may involve varying degrees of software expense, training expense, and organizational and procedural change.

2. **Training and/or practice.** A rating of "better" indicates that relatively little training or practice is required for effective use. A "worse" rating indicates that extensive training requirements and/or user experience is needed or that effective use is directly dependent on effective training. An "average" rating indicates that a significant commitment to training and/or extensive practice in using the method is required.

3. **Designer effort.** A rating of "better" indicates that little or no additional designer time and/or effort is required to make effective use of the methodology. A "worse" rating indicates that significant additional design time must be allocated for use of the methodology. An "average" rating indicates that the designer must make a commitment to using the methodology, that perseverance may be required, and that some additional design time must be allocated.

4. **Management effort.** A rating of "better" indicates that little or no management effort or expectation is required. A "worse" rating indicates that significant management effort and commitment are required and/or that effective use is directly dependent on management expectation and support. An "average" rating indicates that successful use of the method requires management expectation that the method be used, coupled with good support in using the method.

5. **Product planning/team approach.** A rating of "better" indicates that effective use of the methodology requires good product planning and/or the team approach. A "worse" rating indicates that the methodology neither depends on nor encourages good product planning and/or the team approach. An "average" rating indicates that the methodology *can or may* require and/or foster good planning and the team approach, depending on circumstances.

6. **Rapidly effective.** A rating of "better" indicates that the method is likely to be rapidly effective in producing beneficial results. A "worse" rating indicates that benefits will likely be a long time in coming and use of the methodology therefore requires a long-term view. An "average" rating indicates the methodology *can or may* be rapidly effective, depending on circumstances.

7. **Stimulates creativity.** A rating of "better" indicates that effective use of the methodology tends to require design innovation and creativity. A "worse" rating indicates that use of the methodology in itself is not likely to require or to stimulate design creativity. An "average" rating indicates that there is a good potential that design creativity will be stimulated and possibly required to successfully apply the methodology.

8. **Systematic.** A rating of "better" indicates that the methodology involves a systematic, step-by-step procedure that helps to ensure that all relevant issues are considered. A "worse" rating indicates that there is relatively little or no step-by-step procedure involved. An "average" rating indicates that there are aspects of the methodology that involve systematic, step-by-step procedures.

9. **Quantitative.** A rating of "better" indicates that the methodology is primarily quantitative in nature and that one or more quantified design ratings are generated. A "worse" rating indicates that the method is primarily subjective and qualitative in nature and that there are no quantitative ratings generated. An "average" rating indicates that the method has both qualitative and quantitative aspects and that one or more useful quantified design ratings may be generated.

10. **Teaches good practice.** A rating of "better" indicates that use of the methodology teaches good design for manufacture practice and that formal reliance on the method may diminish with use. A "worse" rating indicates that use of the methodology does not in itself teach good practice and that the benefits produced by the methodology depend directly on formal use of the methodology. An "average" rating indicates that the methodology teaches good practice, but it must still be formally applied to achieve intended benefits. ■

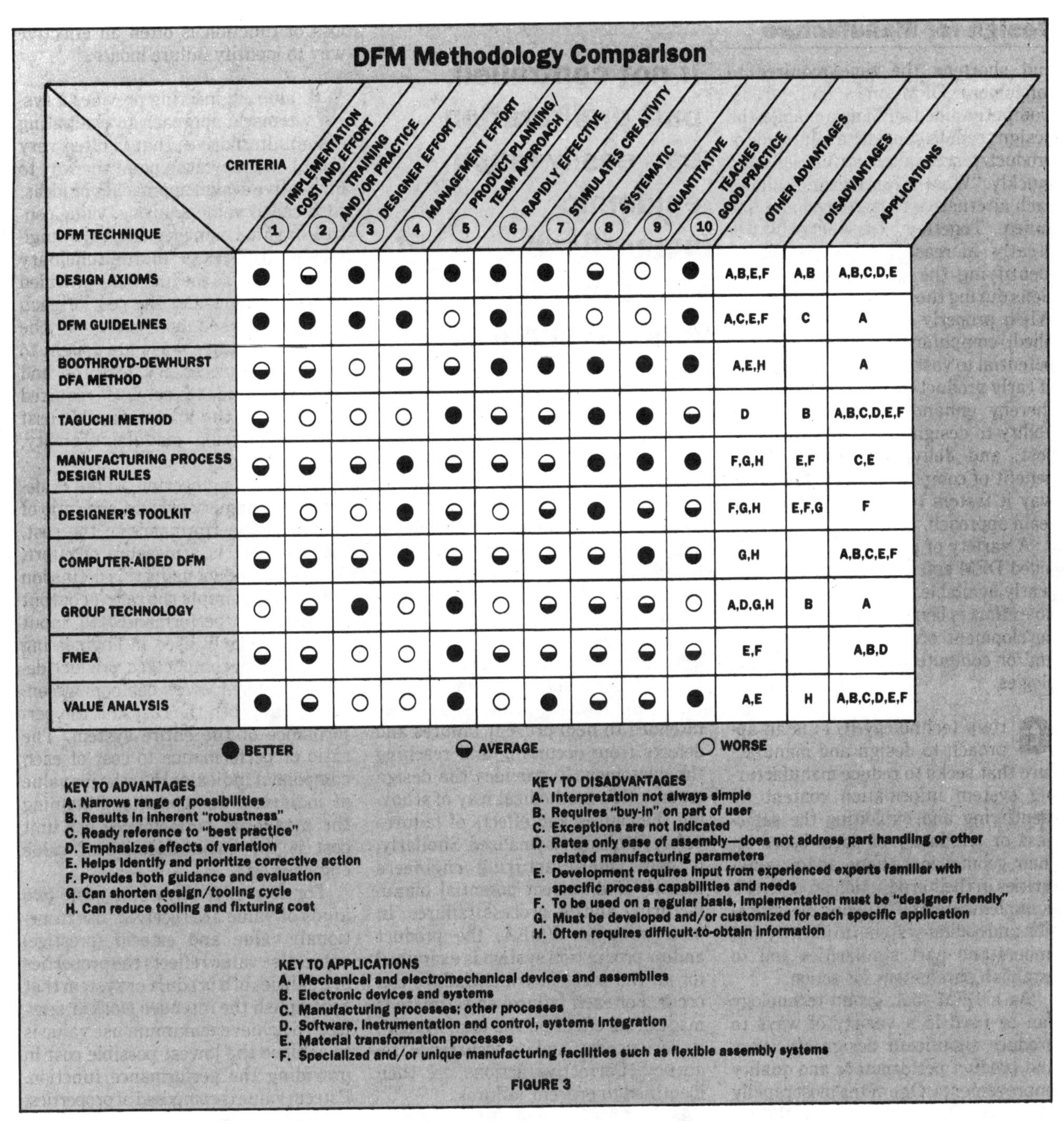

DFM Methodology Comparison

DFM TECHNIQUE / CRITERIA	(1) IMPLEMENTATION COST AND EFFORT	(2) TRAINING AND/OR PRACTICE	(3) DESIGNER EFFORT	(4) MANAGEMENT EFFORT	(5) PRODUCT PLANNING/ TEAM APPROACH	(6) RAPIDLY EFFECTIVE	(7) STIMULATES CREATIVITY	(8) SYSTEMATIC	(9) QUANTITATIVE	(10) TEACHES GOOD PRACTICE	OTHER ADVANTAGES	DISADVANTAGES	APPLICATIONS
DESIGN AXIOMS	●	◒	●	●	●	●	●	◒	○	●	A,B,E,F	A,B	A,B,C,D,E
DFM GUIDELINES	●	●	●	●	○	●	●	○	○	●	A,C,E,F	C	A
BOOTHROYD-DEWHURST DFA METHOD	◒	◒	○	◒	◒	●	●	●	●	●	A,E,H		A
TAGUCHI METHOD	◒	○	○	○	●	◒	◒	●	●	◒	D	B	A,B,C,D,E,F
MANUFACTURING PROCESS DESIGN RULES	◒	◒	◒	●	◒	◒	◒	●	●	●	F,G,H	E,F	C,E
DESIGNER'S TOOLKIT	◒	○	○	●	◒	○	◒	●	◒	◒	F,G,H	E,F,G	F
COMPUTER-AIDED DFM	◒	◒	◒	●	◒	◒	◒	◒	●	◒	G,H		A,B,C,E,F
GROUP TECHNOLOGY	○	◒	●	○	●	○	◒	◒	◒	○	A,D,G,H	B	A
FMEA	◒	◒	○	○	●	◒	◒	◒	◒	◒	E,F		A,B,D
VALUE ANALYSIS	●	◒	○	○	●	○	●	◒	◒	●	A,E	H	A,B,C,D,E,F

● BETTER ◒ AVERAGE ○ WORSE

KEY TO ADVANTAGES
A. Narrows range of possibilities
B. Results in inherent "robustness"
C. Ready reference to "best practice"
D. Emphasizes effects of variation
E. Helps identify and prioritize corrective action
F. Provides both guidance and evaluation
G. Can shorten design/tooling cycle
H. Can reduce tooling and fixturing cost

KEY TO DISADVANTAGES
A. Interpretation not always simple
B. Requires "buy-in" on part of user
C. Exceptions are not indicated
D. Rates only ease of assembly—does not address part handling or other related manufacturing parameters
E. Development requires input from experienced experts familiar with specific process capabilities and needs
F. To be used on a regular basis, implementation must be "designer friendly"
G. Must be developed and/or customized for each specific application
H. Often requires difficult-to-obtain information

KEY TO APPLICATIONS
A. Mechanical and electromechanical devices and assemblies
B. Electronic devices and systems
C. Manufacturing processes; other processes
D. Software, instrumentation and control, systems integration
E. Material transformation processes
F. Specialized and/or unique manufacturing facilities such as flexible assembly systems

FIGURE 3

classified as either process specific or facility specific.

Process-specific DFM involves the design of parts to be manufactured using particular methods or processes such as casting, forging, injection molding, and stamping. Typically, these tools facilitate systematic application of specialized process knowledge in the form of codified statements of design guidelines and rules to the design of parts to be made using a particular manufacturing process or method. Examples include design for casting, design for injection molding, and design for metal stamping.

Facility-specific DFM tools facilitate correct design of products intended to be manufactured using highly specialized or unique advanced manufacturing facilities. Such tools, which could be aptly described as "designer toolkits," provide design rules, physical examples and models, various CAD design aids, and other specific information about a specialized manufacturing facility in a readily usable form to the designer.

Development of manufacturing facility-specific DFM is, at present, in its infancy and is likely to advance very quickly as the relevance of this approach becomes more widely recognized. Typical applications that could benefit greatly from the designer toolkit approach include such diverse situations as flexible assembly and manufacturing system concepts, design of stampings for production on certain classes of triaxis transfer press lines, and design of weldments for production on special flexible welding fixtures or lines.

A major barrier to DFM is usually time. Design and manufacturing engineers are typically operating under very tight schedules and are, therefore, reluctant to spend time learning and using DFM approaches. Computer-aided DFM helps simplify the effort and shortens the time required to implement DFM on a daily basis. Computer-aided DFM also enables the design team to consider a multitude of product/process alternatives easily and quickly. "What-if" optimization allows each alternative to be refined and fine tuned. Together, these capabilities greatly increase the probability of identifying the most desirable solutions during the early stages of design. When properly implemented and applied, computer-aided DFM has the potential to vastly improve the quality of early product/process decisions and thereby enhance the design team's ability to design for effective quality, cost, and delivery. Another major benefit of computer-aided DFM is the way it fosters team building and the team approach.

A variety of proprietary computer-aided DFM software packages is currently available. In addition, considerable effort is being directed toward the development of new computer-based and/or computer-aided DFM methodologies.

Group technology (GT) is an approach to design and manufacture that seeks to reduce manufacturing system information content by identifying and exploiting the sameness or similarity of parts based on their geometrical shape and/or similarities in their production process. GT is implemented by utilizing classification and coding systems to identify and understand part similarities and to establish parameters for action.

As a DFM tool, group technology can be used in a variety of ways to produce significant design efficiency and product performance and quality improvements. One of the most rapidly effective of these is the use of GT to help facilitate significant reductions in design time and effort. In using a GT system, the design engineer needs only to identify the code that describes the desired part. A search of the GT database reveals whether a similar part already exists. If a similar part is found to exist, and this is most often the case, then the designer can simply modify the existing design to design the new part. In essence, GT enables the designer to literally start the design process with a nearly complete design.

Group technology can also be effectively used to help control part proliferation and eliminate redundant part designs by facilitating standardization and rationalization approaches. If not controlled, part proliferation can easily reach epidemic proportions, especially in large companies that manufacture many different products and product models. By noting similarities between parts, it is often possible to create standardized parts that can be used interchangeably in a variety of applications and products.

If not controlled, part proliferation can easily reach epidemic proportions, especially in large companies

Failure mode and effects analysis (FMEA) is an important design and manufacturing engineering tool intended to help prevent failures and defects from occurring and reaching the customer. It provides the design team with a methodical way of studying the causes and effects of failures before the design is finalized. Similarly, it helps manufacturing engineers identify and correct potential manufacturing and/or process failures. In performing an FMEA, the product and/or production system is examined for all the ways in which failure can occur. For each failure, an estimate is made of its effect on the total system, its seriousness, and its occurrence frequency. Corrective actions are then identified to prevent failures.

In FMEA, function is defined as the task that a component, subsystem, or product must perform, stated in a way that is concise, exact, and easy to understand for all users. Functions are typically actions such as position, support, seal, retain, and lubricate. Failure is defined as the inability of a component/subsystem/system to perform the intended function (design intent).

Failure modes are the ways in which a component/subsystem/system could fail to perform its intended functions. Typical failure modes would be fatigue, fracture, excessive deformation, buckling, leakage, fails to open, fails to close, and requires excessive force. Asking what could happen to cause loss of function is often an effective way to identify failure modes.

Value engineering provides a systematic approach to evaluating design alternatives that is often very useful and may even point the way to innovative design approaches or ideas. Also called value analysis, value control, or value management, value engineering utilizes a multidisciplinary team to analyze the functions provided by the product and the cost of each function. Based on results of the analysis, creative ways are sought to eliminate unnecessary features and functions and to achieve required functions at the lowest possible cost while optimizing manufacturability, quality, and delivery.

In value engineering, value is defined as a numerical ratio, the ratio of function or performance to the cost. Because cost is a measure of effort, value of a product using this definition is seen to be simply the ratio of output (function or performance) to input (cost) commonly used in engineering studies. In a complicated product design or system, every component contributes to both the cost and the performance of the entire system. The ratio of performance to cost of each component indicates the relative value of individual components. Obtaining the maximum performance per unit cost is the basic objective of value engineering.

For any expenditure or cost, two kinds of value are received: use (functional) value and esteem (prestige) value. Use value reflects the properties or qualities of a product or system that accomplish the intended work or service. To achieve maximum use value is to achieve the lowest possible cost in providing the performance function. Esteem value is composed of properties, features, or attractiveness that makes ownership of the product desirable. To achieve maximum esteem value is to achieve the lowest possible cost in providing the necessary appearance, attractiveness, and features that the customer wants.

A value analysis is generally carried out in two phases, the analytical phase and the creative phase. In the analytical phase, the use value and esteem value offered by the product are systematically investigated by a team of experts representing all relevant components of the manufacturing system. Findings generated in the analytical phase are then used by the team in the creative phase to define innovative design solutions that maintain the desired balance between use value and esteem value, maximize these values by providing required functions for the lowest cost, and eliminate identified waste.

A number of different DFM methodologies and tools have been discussed. All of these techniques are effective and, if properly applied, can produce significant improvements in product quality and performance, manufacturing system productivity, and life-cycle cost. Ideally, these methodologies, as well as others that are just beginning to be developed (AI/expert systems), should all be implemented and applied to effectively address design for automation needs. The question that arises for many managers is how to begin to do this in the most effective way possible. Which methodologies can be implemented most easily and quickly? Which are most rapidly effective? What would be a good long-range implementation plan?

To help provide insight into these questions, a relative comparison of the various DFM methodologies and tools listed in *Figure 2* is presented in *Figure 3*. In this comparison, each methodology is rated with respect to a variety of different criteria. Also included in *Figure 3* is a listing of specific advantages, disadvantages, and appropriate applications for each methodology. Rating of the methodologies is based on the assumption that no DFM tools are currently being used. This means that the ratings given could change depending on the actual DFM capabilities and level of DFM experience that exist within a particular company. For example, value analysis is rated "worse" with respect to rapid effectiveness. One of the reasons why this rating was assigned is the difficulty involved in obtaining accurate cost estimates early in a design project. This rating could change dramatically if a group technology database was available for use in estimating cost of new parts based on known cost of existing parts.

Figure 3 can be used in a variety of ways. For example, by consulting column 1, it is apparent that efforts to begin using the design axioms and DFM guidelines can probably be initiated fairly quickly. Also, because design for assembly can be rapidly effective and requires less formal training, it might be selected for implementation before the Taguchi method.

The design situation is a complex array of diverse and often contradictory human activities and technological issues that differ with each problem and continually change and evolve during the realization of the design. Design is open-ended, with many solutions possible and the final result determined largely by the way and extent to which the design problem is understood and by the process with which it is solved. The increased complexity of the modern information age, the continual need for change, and the constant emergence of new materials and technology are placing ever-increasing demands on proper and complete understanding of the design problem and on the broad spectrum of needs to be met by the design process. The DFM process and associated DFM methodologies help the design team deal more effectively with these demands.

In so doing, DFM helps improve the quality of early design decisions. Design decisions, especially those made early in a design project, have a tremendous impact on the life-cycle cost of the product. The price paid for poor decisions can be devastating. Conversely, early quality design decisions can ensure business success by enabling the production of better performing, more robust and reliable, and lighter weight products with greatly improved cost, quality, and productivity.

DFM, and the concept of simultaneous engineering that underlies it, is recognized as key to minimizing life-cycle cost and design time, assuring product quality, eliminating "over-the-wall-to-manufacturing" mentality, and realizing the productivity increase promised by advanced manufacturing technology. For many companies, the DFM approach can ultimately lead to the innovative product and process solutions needed for a measurable competitive edge and a healthy balance sheet.

REFERENCES

1. H.W. Stoll, "A Design Backwards Approach to Product Optimization," presented at the SME Simultaneous Engineering Conference held June 1, 1987 (Dearborn, MI: Society of Manufacturing Engineers).

2. N.P. Suh, A.C. Bell, and D.C. Gossard, "On an Axiomatic Approach to Manufacturing and Manufacturing Systems," *ASME Journal of Engineering for Industry*, vol. 100, no. 2 (May 1978).

3. G. Boothroyd and P. Dewhurst, *Design for Assembly—A Designer's Handbook*, Department of Mechanical Engineering, University of Massachusetts-Amherst (1983). ■

Reprinted from *Automation*, May 1989

Product Manufacturability

By John P. Tanner, PE

The design engineering/manufacturing engineering interface is the essential element in designing for quality, reliability, automated assembly, and manufacturability. Here are ways to establish and improve that interface.

"Design for quality." "Design for reliability." "Design for automated assembly." "Design for manufacturability." These are popular expressions in the product design and development environment. The manufacturing engineer who works with design groups that expound these goals is fortunate. Even more fortunate is the ME who has the opportunity to participate in the early stages of product design, influencing the use of design features favorable to good manufacturing processes. Unfortunately, a close working relationship between design and manufacturing engineering is not a naturally occurring phenomenon in all companies.

The design/manufacturing engineering interface is too important to be left to chance, because an effective interface always results in enhanced product designs and improved manufacturing processes. The interface can be developed and maintained through the same good human relations practices that work in other relationships, and must include courtesy, mutual trust, and respect. When achieved, an effective interface will prove that design functionality and manufacturability are compatible features of a good product design.

Manufacturability defined. Manufacturability can be thought of as the art and science of designing a product that is easy to manufacture. The ME must be involved at the earliest stages of the design phase, discussing design concepts and reviewing preliminary designs. He may provide cost estimates of the various design alternatives to help the design engineer narrow down the solution range. He can often point out design approaches that have caused problems in the past. By MEs becoming involved in the design process, potential manufacturing problems can be found and corrected before the design is frozen and drawings are released.

As a minimum, before the design is released, there should be a worst-case dimensional analysis to determine whether there are any tolerance stacks at assembly. If testing is required, the assembly should lend itself to stage-by-stage testing. Any test points or adjustments should be readily accessible. Product design should lend itself to mechanized and automated assembly techniques.

Cost trade-offs should be made between such design considerations as castings versus forgings, sheet metal structures versus machined hogouts and composite structures, and chemical process finish versus paint. Manufacturability also includes materials. Standard materials, standard parts, and standard hardware means standard costs and procurement lead times. Special materials, nonstandard parts, and nonstandard hardware means special manufacturing processes and special costs and lead times. In the final analysis, it is the product design that truly determines the cost of manufacture. Improved methods, processes, and tooling have minimal impact when the product is not a readily producible design.

Manufacturability by design. Some high-technology manufacturing companies, where new and untried designs are often put into production before they are ready, invoke manufacturing design standards to ensure producible hardware. Such standards are prepared by MEs and, essentially, define the key manufacturability requirements that must be met by design engineers for castings, forgings, machined parts, form parts, nonmetallic parts, structural assembly, circuit card fabrication and assembly, equipment assembly, paint and finish, etc.

When the design is frozen, but before drawings are released to manufacturing, the ME should have the opportunity for a final sign off-on all drawings. This review will help assure that the item is producible in manufacturing. Of course, unless the ME has been involved in the initial concept stage, through the design layout stage, having a final sign-off may not be worth much.

The interface. The product knowledge obtained from direct contact with the designer becomes the cornerstone of the information data base which the ME must build to successfully perform the design function he is responsible for --the manufacturing process. At the same time the ME is obtaining the product knowledge he will use to develop the manufacturing process, he should assist the design engineer by providing inputs to make the product compatible with the most economical and effective manufacturing techniques. This knowledge of and experience with manufacturing processes and practices provides the basis for the ME's contribution to the team effort, complementing the design engineers expertise in the areas of aesthetic and functional product design. The success of this team in creating a product design which is functional and adaptable to cost-effective production is critical to the long term success of the company.

Reducing costs. Checklists are convenient tools when doing manufacturability analysis. Checklists help assure that opportunities for design improvement are not overlooked, along with opportunities to reduce manufacturing costs and improve product quality. The check lists that accompany this article are not all-inclusive, but are intended as a guide or starting point for tailoring lists to virtually any manufacturing operation. In addition, the following items apply specifically to automatic assembly applications:

- Minimize the number of parts.
- Ensure that the product has a suitable base part on which to build the assembly.
- Ensure that the base part has features that will enable it to be readily located in a stable position in the horizontal plane.
- If possible, design the product to be built up in layer fashion, each part being assembled from above and positively located. This will eliminate the effect of horizontal forces during the machine index period.
- Try to provide chamfers or tapers to help guide and position parts correctly during assembly.
- Avoid fastening operations that are time consuming, expensive, and frequently labor intensive, such as soldering and screw fasteners.

Manufacturability checklist

Materials, parts, and components

--Can material be obtained in standard stock configurations (bar stock, sheet, standard extrusion, etc.)?
--Is material compatible with the most desirable manufacturing process (i.e. ease of forming, casting, machining, etc.)?
--Is the material available from reliable sources?
--Is the material subject to wide price fluctuations over time?
--Is the material compatible with the anticipated operating environment of the end product?
--Special alloys and exotic materials should only be used as required for special environmental or functional demands.

Fabricated parts

--Are specific tolerances reasonable for functional requirements?
--Are tolerances attainable within the normal capability of the manufacturing process to be used?
--Are data points, surfaces, and tooling points clear, and are they accessible?
--Does part configuration minimize the need for special processes and special tooling?

Product assembly

The following checklist for reviewing manufacturability of an electronic assembly is representative of the kind of questions the manufacturing engineers should be asking while reviewing design concepts and layouts:

--Are tolerance dimensions realistic?
--Is marking and stenciling defined and visible?
--Are assembly notes complete and definitive?
--Is internal wiring layout critical? If so, is the location and routing specified?
--Are test points and adjustments accessible?
--Is harness development required. If so, can the harness be fabricated outside the unit and installed as a subassembly?
--Does the design lend itself to automated assembly?
--Are component parts accessible for assembly?
--Can testing be performed without disassembling the unit?
--Are standard connectors and assembly hardware used?
--If circuit cards are used, are they designed to plug in?
--Has the assembly been analyzed to meet electrical, thermal, vibration, and shock specifications?
--Can printed circuit flex cable or molded ribbon be used in place of hard wiring?
--Can plastic tie-wraps be used in place of lacing or spot ties?

About our author

Mr. John Tanner is a registered professional engineer with over 20 years of industrial management experience with companies such as McDonnell Douglas Astronautics, Martin-Marietta Aerospace, and Applied Devices Corp.

He received an MBA in Management (Rollins College, 1965) a BS in Industrial Engineering (University of Miami, 1954) and a BBA in Economics (University of Miami 1951).

Author of Manufacturing Engineering, published in 1985 by Marcel Dekker, Mr. Tanner has been published in numerous technical and trade magazines. He is a senior member of IIE and SME.

Now president of John P. Tanner & Associates, an industrial consulting firm, Mr. Tanner shares his manufacturing knowledge and experience with AUTOMATION readers in an exclusive five-part series on manufacturability.

This month, in Part I, Mr. Tanner introduces the manufacturability concept and describes methods by which design and manufacturing engineers can work together towards achieving it.

In succeeding parts, Mr. Tanner will describe how the manufacturability concept can be applied to metal forming and fabrication, finishes and tolerances for machined parts, assembly and wiring, and automatic assembly.

Vol. 35, No. 4, Page 232-237, November 1988

Assessing the Development/Production Transition

By R.C. Thurmond and D.V. Kunak

I. Introduction

It has been widely recognized that most products go through life phases of introduction, growth, maturity, and decline [5]. The introduction phase typically includes both product development and initial production. During this phase, product redesign is generally required, since development designs stress concept verification whereas a production design must be manufacturable at minimum cost in quantity. Several authors point out that this transition of a product from development to production is critical to its evolution [1], [7], [15], [16], [18], [21], [22], but most literature has focused on the overall new product development process. Authors who have primarily addressed the development to production interface include Abita [1], Glaser [7], and Moore [15]; Patterson [18] has addressed the interface in military equipment development.

This article proposes the concept that total life-cycle profitability is dependent on the quality of the development-production transition (DPT), making it strategically important in the new product development process. Factors thought to be critical to the DPT are discussed, and a paradigm which may be useful in assessing product transition difficulties is introduced. The focus is primarily on high technology industries, both commercial and military, but many of the conclusions may be applicable to their firms as well.

II. Strategic Importance

As existing product lines age and market demand declines, development of new products is vital to every manufacturing business in order to survive and grow. Some authors stress that effective new-product development is virtually synonymous with success in such high-technology industries as electronics and computers [13]. At the same time, it has been recognized that development of new products is not only costly, but also uncertain [5]. According to a recent study, only about 65 percent of all new product introductions are considered successful [4], while another study suggests that only one product out of every 10 in the research and development stage will ever reach market [21].

In addition to the uncertainty of new product success, there is evidence that the amount of time available for product development and transition is decreasing. For example, the introductory life stage of household appliances shrank from 12.5 to 2 years over a 50-year period starting in 1922 [20]. With respect to high-tech products, several authors indicate that short product life cycles, rapid competitive response, and rapid obsolescence are particularly common [8], [21], [24].

A shorter product life typically has several effects. First, it reduces the time available to develop the product and transition it to a manufacturable design (DPT), and second, it limits the time available to make design changes during protection due to factory problems or customer feedback. A third effect of decreasing lifetimes is to increase the number of new products per year which a firm must develop in order to stay competitive, since the average product is profitable for a shorter time than just in the past. The situation thus exists that a continually increasing number of product developments must be done while maintaining a high level of development and transition quality, since there is also little time for design changes in production.

Rapid competitive reaction also places great urgency on the product development/transition process, as explained by the experience curve (popularized by the Boston Consulting Group [9]). This concept teaches that in production, cost of value added decreases by 25 to 30 percent for each doubling of production volume. Consequently, the first product to reach market in a particular market segment will capture market share and provide higher profitability for some length of time [2], since new competitors cannot match the production cost of the "experienced" manufacturer. The experience curve is thought to be applicable to the early growth of technological markets [2], making it particularly critical that product transition occurs smoothly and quickly so that market share is not lost, perhaps irretrievably.

Unfortunately, it appears that without a good knowledge of critical success factors, the product transition is often inefficient [1], taking longer than necessary and resulting in an inferior product design. A literature survey suggests the following costly impacts:

1) Market introduction of the product is delayed, reducing market share and profitability [2], [22].
2) The product is more costly to produce than it should be, requiring more labor and more costly parts/processes [3], [22]. This reduces margin or drives up the required sales price.
3) The life cycle of the product is shortened because the poorer design is less reliable [7] and will not retain customer loyalty.

It is also reasonable to conclude that a less efficient DPT would:

4) Be more costly than it should be due to higher labor costs from longer work effort [22].
5) Result in a product having less customer utility (less time to "fine tune" the design), and therefore more vulnerability to competitive displacement.

Although quantitative measures were not found, it is apparent that a development to production transition of poor quality can have a significant impact on the product life cycle and profitability.

III. Critical Factors of Transitions

Among factors which the literature suggest as critical to the DPT, useful groupings are product-related factors and company internal factors. Product factors may be categorized as market fit (acceptance), and company fit (alignment) [4] which are relatively independent, and the dependent product variable, manufacturability (a result of the DPT process).

The company internal factors are more complex; an augmented version of the Leavitt model [11] of organizational variables will be used to categorize them. This model adds the variable culture [6] to the four interactive variables described by Leavitt: people, structure, technology, and task. Culture represents the dominant values of the firm [6]. Structure includes such things as the management hierarchy and the communications networks of the company; technology represents the technological tools used in performing tasks, such as machines, software, and techniques. Task includes the overall business work of the firm and all associated subtasks [11].

The model of Fig. 1 is proposed as a means to illustrate the position of the product transition task (DPT) between product conceptualization/ development and product manufacturing; culture, technology, people, and structure all play supporting roles in the accomplishment of the task. The market provides a source of many product concepts, and the customers for finished products. The prototype product must fit into an area of company expertise in order to be successfully transitioned, and the manufactured product must have market fit in order to sell. Market information influences the transition process through people (individuals) and structure (marketing and management policy). These factors are discussed in the following sections.

A. Product Factors

Referring to Fig. 1, product factors may be divided into external factors and internal factors. External factors are primarily associated with the market acceptance or fit, as reflected in market demand for the manufactured product. Internal factors are product alignment (fit) to the main business areas of the firm, and closely related, the producibility or manufacturability of the product. Product alignment has a bearing on the difficulty of the transition task, and producibility of the final product design is both a measure of transition success and an indicator of how easily the product can be manufactured.

New product developments range from small incremental changes in an existing line to radical breakthroughs which are unlike any existing product [16]. It would appear that while the market fit for incrementally different products should be fairly predictable, the market reaction to radical products may be very difficult to judge. Some authors even feel that radical innovations require supply-side marketing: "for technological innovations, markets must be created, not surveyed" [24]. Consequently the following comments on market acceptance apply more to the incremental-type innovation than the breakthrough invention.

1. Market Fit: Approximately 87 percent of respondents to a recent survey indicated that product fit to market needs was an important factor in new-product development [4]. Functional and aesthetic characterization (specification) take place at the beginning of product development so that an appropriate structure can be developed. Some companies even place engineers in marketing positions to insure that the functional product design is appropriate to market needs [17]. However, structural and processing trade-offs required during the DPT to make a product manufacturable can easily influence its function and appearance such that market acceptance is compromised. A familiar example of this might be a new car design. The prototype would be attractive and have high performance due to hand assembly and finishing. However, highly automated assembly and finishing would be necessary in production to reduce cost. The assembly process change would require design compromises; resulting lower product quality and performance could significantly reduce appeal and market acceptance.

2. Company Alignment (Fit): According to some authors, new products should be evaluated with respect to other products presently or recently manufactured by the firm and products available on the market [13], [14]. Approximately 60 percent of respondents to a recent survey indicated that product fit to internal functional strength was important to new product development [4]. New products which are radically different from the company's existing product lines are likely to be more difficult to prepare for production, whereas products requiring smaller changes should be more easily accommodated. This can be due to lack of appropriate experience, or

lack of technological tools. This factor must be considered during the selection of products for development. It also must be analyzed after development, at the beginning of DPT, since the development process can cause divergence in product fit relative to the firm's areas of transition and manufacturing strength.

3. Producibility (Manufacturability): The primary purpose of the design to production transition (DPT) is to create a product design which is easily manufactured (producible), therefore producibility of the final product design is a strong measure of DPT success. Boothroyd states that "the most up-to-date manufacturing methods are of little consequence unless the design of the product lends itself to efficient manufacture" [3].

Manufacturing unit cost at the anticipated production volume is an important parameter which must be considered when assessing producibility. It has been found that the contribution margin of a product is important for success in product innovation [13]. Since the contribution margin is partially dependent on manufacturing cost, which is a function of producibility, the producibility is again shown to be a very important factor of the DPT.

Because redesign (DPT) cost is reduced when the initial product design is fairly producible [22], it is useful to have design guidelines and measures of producibility. Design guidelines for military equipment development have been recommended as a method for insuring a better product [18]. Since much assembly is labor intensive, ease of assembly is synonymous with reduced manufacturing cost; analysis systems have therefore been developed which rate equipment's ease of assembly, thus providing useful feedback to the designer [3].

B. Organizational Factors

As illustrated in Fig. 1, organizational factors which influence the DPT can be grouped as culture, people, structure, and technology. These factor groups are considered to be mutually interactive in supporting the DPT.

1. Culture: Concepts which a company values most strongly are the basis of its culture. Some of the reasons suggested for these values are the dominant ideas of founders, the company's history, and the stage of technology. They may shape the organizational culture to such extreme

differences as innovator or follower/copier, high-tech or low-tech firm, high quality/price or low quality/price, etc. [6].

The firm's culture is perhaps best indicated by its goals and values beyond the financial realm; these have been found to make a critical difference between mediocre firms and superior performers [19]. With respect to the engineering to manufacturing interface, it is suggested that a company which has a tradition of close customer relations, shrewd product selection, and smooth well-coordinated product developments will be more likely to experience smooth DPT's than a firm which does not have this culture.

2. People: The second grouping of critical variables from the model relates to people [11]. Many attributes can be ascribed to people, such as knowledge, attitude, power, position, etc. Regarding a smooth and efficient DPT, the literature [1], [3], [12], [15] suggests that knowledge, skills, attitude, and motivation may be the most important attributes, probably because these relate directly to capability and desire to accomplish the task. Power and position are considered to relate more to the structural grouping (discussed later).

a. Knowledge/skills: Adequate knowledge and skills are necessary in order to minimize interface difficulties and facilitate the product transition. Boothroyd says of producibility: "it is hoped that all discrete component parts will be designed so that they are both easy to manufacture individually, and then to assemble. This can occur only after the product designers are provided with the information necessary..." [3].

Development and transition personnel who are not knowledgeable of manufacturing requirements and processes will almost guarantee a difficult and costly transition.

It would appear that knowledge of the following areas is essential in order to design a product which can be manufactured at low cost with good quality:

1) mechanical/process operations normally used to form piece parts;
2) assembly/fastening methods;
3) electrical connection/wiring operations;
4) finishing operations/processes; and
5) testing and inspection methods.

Increasing the manufacturing knowledge of development personnel (and upgrading the technical knowledge of production personnel) by seminar or interdisciplinary assignment are valuable techniques recommended for better product transition [15].

b. Attitude/motivation: People's attitudes and values are critical to product translation [15]. Engineering and manufacturing people tend to have quite different goals, attitudes, and operating modes [1], and conflict can develop between these groups. A typical scenario might involve engineering feeling that communications is a major problem: that production personnel cannot read drawings or follow instructions. Production might feel that there was a lack of competent engineers, too many mistakes and changes on drawings, and a lack of support for the production viewpoint [12].

Several factors can be responsible for dysfunctional group conflicts in organizations [23]:

- o group preferences and personality differences;
- o difference in skills, training, and job activities;
- o competition for authority, power, and influence.

Group preferences are, unfortunately, intrinsic to functional groups [1]; however, cross training and education could reduce problems caused by the second area (skill differences). Competition for power is discussed later under management.

Application of Herzberg's motivation-hygiene theory [10] to the dilemma would indicate that certain factors, called hygiene, such as favorable company policy/administration, good supervision, and adequate salary are required to minimize organizational frictions and job dissatisfaction. Meanwhile, another set of qualitatively distinct factors, including opportunity for achievement, recognition, responsibility, and personal growth provide intrinsic motivation for high performance [10].

Any combination of deficiencies in these two sets of factors can cause severe problems for the transition effort. Inadequate hygiene factors can directly hamper the DPT by creating disgruntled personnel, while an absence of motivators will reduce the transition team's capability to respond to challenges.

3. Structure: The literature indicates that critical transition factors which would fall under the definition of structure are marketing, management, and company organization/communications.

a. Marketing: It would appear that a significant amount of marketing support is required during the DPT phase of new products, since the final production design occurs at this time [22]. Market-critical attributes such as function and appearance must be continually monitored by marketing during transition so that the production-ready design retains adequate market acceptance or alignment (fit). At this stage, market knowledge via customer interaction has been found critical to success in product innovation [13]. Application engineering has been suggested as an important source of such market knowledge in high-tech firms [22].

b. Management: While management performs many functions in a firm, the literature suggests that the following factors are most critical to the design transition phase: strategy, planning/coordination, support, and review/control.

Some management consultants point out that competition forces firms to excel at developing new products. This requires a new product strategy which identifies the roles new products must play to fulfill corporate objectives, and assesses internal capabilities to "identify relevant company strengths and weaknesses" [4]. It is proposed that assessment of factors critical to the DPT should be part of the overall new product assessment, and correction of weak transition factors should be part of the larger strategy.

Planning and coordination are not only basic management functions, they have been found critical to product innovation [13]. They are also required in order to insure that various parts of the company interact constructively in the transition process. Consequently, planning and coordination must be assumed to be critical for successful product transition to manufacturing.

Some authors cite management support, and the disposition of key organizational actors (such as top management) as critical to new product developments; it would therefore appear that these are also extremely important factors in the product transition area [4], [13], [21].

Traditional management control is needed during the DPT in order to keep the creative process on track. Marketing and management must undertake adequate design reviews to insure that the product being transitioned to manufacturing takes on the desired characteristics, including manufacturability. Reviews are most effective when they are conducted by personnel with adequate technical knowledge, when design deficiencies and trade-offs are identified, and when formal reports are prepared [18].

Because design transition requires the close cooperation of several different departments, it has been recommended that top management construct an authority hierachy which parallels the informal prestige hierarchy. This type of structure is said to minimize dysfunctional internal power competition [23].

c. Company organization/communications: The organizational form typically has an important effect on the efficiency of the product innovation process, including the DPT. At least two authors suggest the value of a group committed to the transition process, headed by a product transition manager [1], [15]. Depending on the type of product, companies which are most successful in new product development use one of three basic organization/management styles [4].

The first style, called entrepreneural, is very autonomous, loosely structured, creative, and is well suited to the development and transition of radically new products. It usually takes on the form of an independent venture team. The second style, called managerial, is highly structured, controlled, and usually part of a functional department. It is best suited to incremental innovations which are closely linked to existing products. The third approach, called collegial, lies between the other two in ability to handle innovation, being best suited to enter new business areas or add products to existing lines. The collegial structure is probably best reflected by an interdisciplinary product team [4].

It has been said that the DPT tests cooperation and communications between engineering and production like no other activity [22]. Within this context, "the greatest problem of communications is the illusion that it has been achieved" [15]. Fortunately, several techniques can be used to improve communications. First, manufacturing personnel can be assigned to R&D programs, or visa versa, engineering personnel can transition to production along with their product. Another technique involves mutual engineering/ manufacturing project reviews during the R&D cycle along with reciprocal technical seminars [15]. A final means of improving communication is to reduce the required data flow between individuals and departments by establishing standards and guidelines.

4. Technology: Technology includes the tools used to perform tasks, such as software, machines, and techniques. Two such tools are particularly pertinent to the DPT: CAD and integrated CAD/CAM (computer aided design/computer aided manufacturing) [18].

According to Patterson, "the use of adequate CAD technology is a significant factor in reducing the risk in development projects, particularly as those projects make the transition from development into production" [18]. The introduction of discipline to the design process is greatly facilitated by CAD since three-dimensional modeling and many types of technical analysis (such as stress, vibration, and thermal) can be done very easily, disclosing potential design problems. Data can also be built into the CAD software to automatically control the design: design standards, preferred parts libraries, preferred materials, and available manufacturing processes/rules [18].

The common database implicit in an integrated CAD/CAM system makes data transfer between engineering and manufacturing much easier and more accurate. This could allow more attention to the overall product considerations which have sometimes been neglected in the past. It has been stated that "integrated CAD/CAM software architecture for multiple access and control, coupled with a common data base, greatly improves system effectiveness and facilitates the design-to-production transition" [18].

IV. Predicting the Transition

Based on the DPT model of Fig. 1, and subsequent discussion of its components, the quality of product transition from engineering to manufacturing can be analyzed and predicted in term of two primary dimensions. These dimensions are product alignment including market and company fit, and organizational capability including the factors culture, people, structure, and technology as major inputs to the transition task. This concept is further illustrated in Fig. 2 as a transition quality matrix.

The transition quality matrix of Fig. 2 portrays four extreme DPT situations, as illustrated by the captions. Most product transitions would fall between these extremes, but the matrix may help to distinguish major forces influencing the transition.

A. The Last Great Product

"This is really different from our normal products, and we usually have a lot of trouble getting them into production!" This may indeed be the last great product idea for that company where the product fits poorly into the company business areas, does not enjoy good market acceptance, or both, and the firm is weak in the organizational factors important to transition. This can occur with a radical "new to the world" product, or a product very unfamiliar to the firm. The product may have to go through many cycles of DPT as the transition team attempts to produce a design which can be manufactured, while marketing works at the creation of a demand for this radical product. Such transitions often fail or achieve only modest success at best. This is an example of a product which should never have been developed by that particular company. A product which offers much promise but is particularly resistant to successful transition can absorb enough attention and resources to destroy a weak company.

B. Another Marginal Product From a Market Follower

"This is almost like Product XYZ, but we have trouble building that too." This quadrant corresponds to a product which has good fit to the company business and the market, but the firm is weak in transition capability. Because the firm is poor at transitioning, it is probably late to market with new products, and also cannot produce at minimum cost due to poorly optimized designs and a low position on the experience curve. This could easily relegate the company to being a market follower rather than a market leader in its market niche.

C. IBM PC's Edsels

"This is really different from our normal products, but we can figure out how to build it if marketing can sell it!" If the product is significantly unlike other company products (IBM PC), or does not have good market acceptance (the Edsel), or both, but the organization has strong transitioning capability, then the lower right quadrant of the matrix applies. If the transition team is confronted with an unfamiliar product, the transition will likely be a long one. Strong organizational capability will eventually result in a manufacturable product at minimal unit cost, but it will be successful only if marketing is able to create demand.

D. Another Razor From Gillette

"This is almost like product XYZ, and we really know how to build those things!" This is the comfortable situation where the product is similar to existing company lines, enjoys good market acceptance, and the organization has strong transitioning capability. Another example would be the many incremental model changes to the Volkswagen Beetle. Product transition under these conditions is straightforward. Other business factors being good, it would appear that a company in this situation should have a large market share.

DPT difficulties arising from weak organizational capability (the left half of the matrix) can be analyzed in terms of the firm's culture, people, structure, and technology. Such an organizational audit can highlight weak points, help define corrective actions, and set the stage for much more efficient transition tasks.

Difficulties associated with the DPT of unfamiliar products (lower half of the matrix) can be minimized by assessing the flexibility of the transitioning and manufacturing personnel, and the size of available transitioning funds. Radically new products should stimulate a thorough assessment of potential markets and the firm's marketing capability. A weak evaluation in these areas should be reason to question further investment of resources in such a product. A stronger evaluation could result in consideration of limited transition and test marketing.

Using this strategy, products which have evolved too far from the firm's areas of technical and business strength can be detected before lengthy and costly DPT's occur. Additionally, this strategy will more accurately predict the scope of resources required to accomplish a successful DPT.

V. Conclusion

There is limited literature on the transition of a product design from engineering to manufacturing. However, meaningful insights are also available from studies of product innovation, indicating that product transition is a sensitive and critical point in the product life cycle. Many product developments die at this stage, and of those which survive, the profitability of resulting production can be related to the quality or effectiveness of the DPT.

A proposed transition model developed from the literature suggests factors critical to the quality of the transition process: alignment of the product design with respect to the market and existing product lines, and organizational capability in terms of culture, people, structure, and technology. Discussion of these factors included methods by which some companies have been able to improve their transition capability, and the rationale for this improvement. Based on the model and discussion, a transition quality matrix was presented; this provides examples of extreme DPT situations and likely related market positions. While the model can be of help in understanding those forces which influence the transition process, the matrix should promote a more strategic view of the DPT and clarify the effect of transition policies.

Several areas could benefit from research and further study. Quantitative measurement of life cycle profitability as a function of DPT quality would verify the importance of the DPT. Additionally, further analysis of transition critical factors in various industries would allow the model to be improved. Finally, the transition matrix concept should be validated by researching the relationship between product transition and market position.

References

[1] J.L. Abita, "Technology: development to production," IEEE Trans. Eng. Manag., vol. EM-32, pp. 129-131, Aug. 1985.

[2] P.H. Birnbaum, "Strategic management of industrial technology: A review of the issues," IEEE Trans. Eng. Manag., vol. EM-31, no. 4, pp. 188-189, Nov. 1984.

[3] G. Boothroyd, "Design for producibility--the road to higher productivity," Assembly Eng. pp. 42-45, Mar. 1982.

[4] Booz-Allen & Hamilton, New Products Management for the 1980's, New York: Booz-Allen & Hamilton, 1982, pp. 7, 6, 19, 22.

[5] D.J. Dalrymple and L.J. Parsons, Marketing Management, Strategy, and Cases. New York: Wiley, 1983, pp. 293, 308, 309.

[6] G.B. Davis and M.H. Olson, Management Information Systems: Conceptual Foundations, Structure, and Development, 2nd ed. New York: McGraw-Hill, 1985, p. 343.

[7] M.A. Glaser and A.H. Rubenstein, "Barriers to innovation--the R&D/production interface; Some early findings," Prog. Organic Coatings, vol. 13, no. 3-4, pp. 237-252, Sept. 18, 1985.

[8] A.G. Goldman, "Short product life cycles: Implications for the marketing activities of small high-technology companies," R&D Manag. vol. 12, no. 2, pp. 81-89, Apr. 1982.

[9] B. Henderson, The Logic of Business Strategy. Cambridge, MA: Ballinger Pub., 1984, pp. 47-49.

[10] F. Herzberg, The Managerial Choice: To Be Efficient and To Be Human. Salt Lake City, UT: Olympus Pub., 1982, pp. 58-60, p. 71.

[11] H.J. Leavitt, "Applying organizational change in industry: structural, technological, and humanistic approaches," Handbook of Organizations, J.G. March, Ed. Chicago, IL: Rand McNally, 1965, pp. 1144-1145.

[12] L.V. Leonard, "Forces that reshape production engineering...better communications unsnarl production tie-ups," Prod. Eng., pp. 61-65, Jan 1977.

[13] M.A. Maidique and B.J. Zirger, "A study of success and failure in product innovation: The Case of the U.S. electronics industry," IEEE Trans. Eng. Manag., vol. EM-31, no. 4, pp. 192-203, Nov. 1984.

[14] G. Maioulis and P.J. LaPlaca, "A systems approach for developing high technology products," Ind. Marketing Manag., vol. 11, no. 4, pp. 253-262, Oct. 1982.

[15] R.F. Moore, "Five ways to bridge the gap between R&D and production," Res. Manag., pp. 367-373, Sept. 1970.

[16] R. Mueser, "Identifying technical innovations," IEEE Trans. Eng. Manag., vol. EM-32, no. 4, pp. 158-176, Nov. 1985.

[17] K. Ohmae, "Managing innovation and new products in key Japanese industries," Res. Manag., pp. 11-18, July-Aug. 1985.

[18] D.O. Patterson, Best Practices: How to Avoid Surprises In the World's Most Complicated Technical Process, the Transition From Development to Production, (Series: NAVSO P-6071 Ser). Washington, DC: GPO, 1986, pp. 4-23, 4-49, 4-75, 6-45.

[19] T.J. Peters and R.H. Waterman, Jr., In Search of Excellence. New York: Harper & Row, 1982, p. 103.

[20] W. Qualls, R.W. Olshavsky, and R.E. Michaels, "Shortening of the PLC-- and empirical test," J. Marketing, vol. 45, pp. 76-80, Fall 1981.

[21] J.A. Raelin and R. Balachandra, "R&D project termination in high-tech industries," IEEE Trans. Eng. Manag., vol. EM-32, no. 1, pp. 16-23, Feb. 1985.

[22] H.E. Riggs, Managing High Technology Companies. Belmont, CA: Lifetime Learning Pub., 1983, pp. 50-51, 123-131.

[23] J.A. Seiler, "Diagnosing interdepartmental conflict," Harvard Bus. Rev., vol. 41, pp. 121-132, Sept.-Oct. 1961.

[24] W.L. Shanklin and J.K. Ryans, Jr., Marketing High Technology. Lexington, MA: Lexington Books, 1984, p. 64.

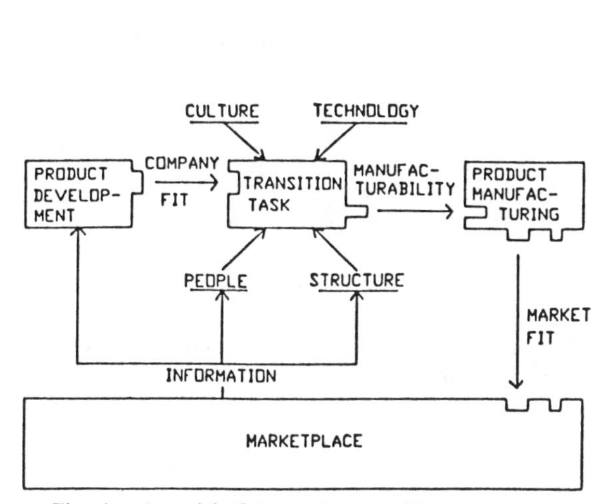

Fig. 1. A model of the product transition process.

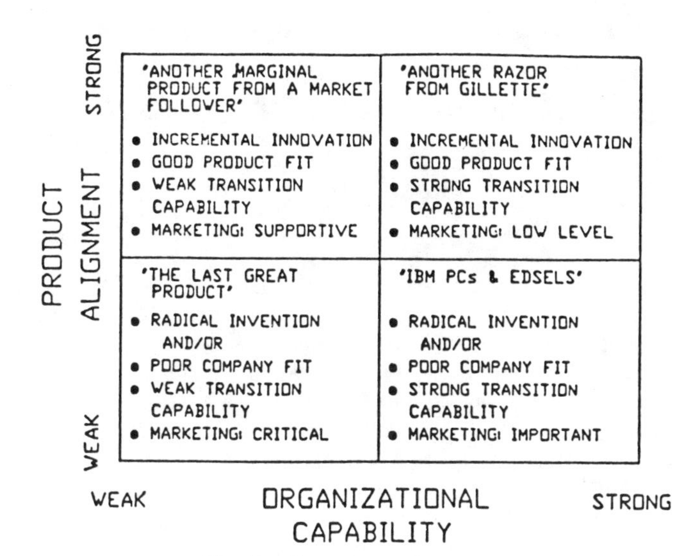

Fig. 2. Transition quality matrix.

Manuscript received April 27, 1987; revised January 25, 1988. The review of this paper was arranged by Editor D. Kocaoglu.

R.C. Thurmond is with Hercules Defense Electronics Systems, Inc., Largo, FL.

D.V. Kunak is with the University of South Florida, Tampa, FL.

IEEE Log Number 8822894.

Reprinted from *Production*, July 1987

Simultaneous Engineering

Management's New Competitiveness Tool

The concept is as simple as it is powerful: obtain as much information as possible as early as possible, then go to work creating world-class products and the processes that make them cost effective. But it isn't as easy as it may sound or seem.

By Gary S. Vasilash
Senior Editor

THERE'S A NEW MANAGERIAL STRATEGY that's sweeping through American industry with a furor not seen since the days of quality circles. It's called simultaneous engineering. Or a variation thereof.

In effect, it can be thought of as quality circles increased exponentially: QC^2. More than a simple give-and-take between management and workers about the existing state of affairs, simultaneous engineering brings together groups that have historically had, perhaps, more friction between them than labor and management: design engineers and manufacturing engineers, or product and process people. What's more, they don't talk about givens; they must take concerted action on things that don't exist.

Their combined objective is, quite simply, to develop better products, whether it's a home appliance or an automobile. *Better* is expected to result from the fact that the two groups are working together. This may not seem to be extraordinary. After all, both groups are engineers; both groups work for the same company. It would be only natural that the two work in concert. But with few, relatively recent, exceptions, each group operates in semi-impenetrable isolation within the major organizations.

Dr. Henry W. Stoll of the Advanced Manufacturing Technologies Laboratory, Industrial Technology Institute, Ann Arbor, MI, explains, "The product design is developed, 'thrown over the wall,' and the production system is designed. The 'can-do' boys get to ask some questions, but so much is cast by that time that resultant changes are few. This arrangement forces the design into suboptimal areas because the decisions are made too late."

Not only are the design and manufacturing people brought together, but there are a few more ingredients that really get the pot boiling. For example, the input from marketing and financial people becomes relevant to product and production decisions like never before. And the OEMs no longer select suppliers to work to specs, but actually preselect suppliers who help develop the specs. Customers and vendors work together, sometimes even in the same office.

Another factor that exacerbates the difficulty of coming to grips with simultaneous engineering is the fact that its definition is more slippery than that of FMS. Some call it the "team approach," or "concurrent engineering," or "life-cycle engineering," or "process-driven design," or "design for manufacture," or...

But when all is said and done, what does it give you? On the one hand, it can give you the biggest management headache that you can conceive of. On the other, it can give you a tremendous analgesic in the form of a product that is either bona fide "world-class" (as distinct from lip-service world class) or as near to it as your company will ever get.

It would be convenient to provide some examples of major products manufactured in the U.S. that have been the result of simultaneous engineering. But the approach is still too new. The available domestic examples tend to be an amalgam of the old and the new. (Examples in Japan are, as you might expect, quite numerous. Among them are an automobile by Honda, a plain-paper copier from Fuji-Xerox, a camera by Canon, a personal computer from NEC, and a dot-matrix printer by Epson.)

Americans at work. Which is not to say that American manufacturers aren't making some major strides in this area. Ford Motor Co. attributes the success of its Taurus/Sable line to what was called the "Team Taurus" approach, which included input of relevant personnel throughout the company. Assembly line workers were queried, vendors were qualified, designers and process people were coordinated, and the $3 billion, 4½-year program resulted in automobiles that not only have tremendous customer appeal, but which are produced more efficiently. It wasn't easy. For example, in 1981 it was deemed necessary to scrap the then-current prototype. The reason: the consensus felt that the car was designed too small.

Roger W. Masch, chief engineer at the Flint Engineering Center, Buick-Oldsmobile-Cadillac Group, Flint, MI, explains that since January 1, 1987, his organization has been involved in "integrated engineering," which includes "actual integration of product and process into one group in one location." Additionally, there

Ford's "Team Taurus" approach resulted in high-quality cars that have played no small role in the company's competitive strength in the market. It's proof that cooperation can work.

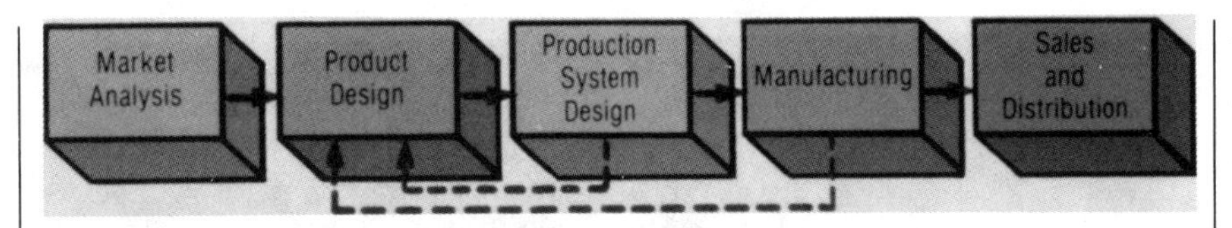

This is the way that manufacturing is normally conducted. It is simple to manage. The dotted lines indicate that while there is some feedback, there tends to be too much fixity of design to result in products that have high manufacturability. Simultaneous engineering hopes to change this situation.

The simultaneous engineering model is nonsequential and consequently more complex. Management can be a real problem. But ideally, the give-and-take between the various groups results in product and process designs that are highly complementary. The goal of world-class, cost-competitive products is, proponents of the process believe, attainable through this approach.

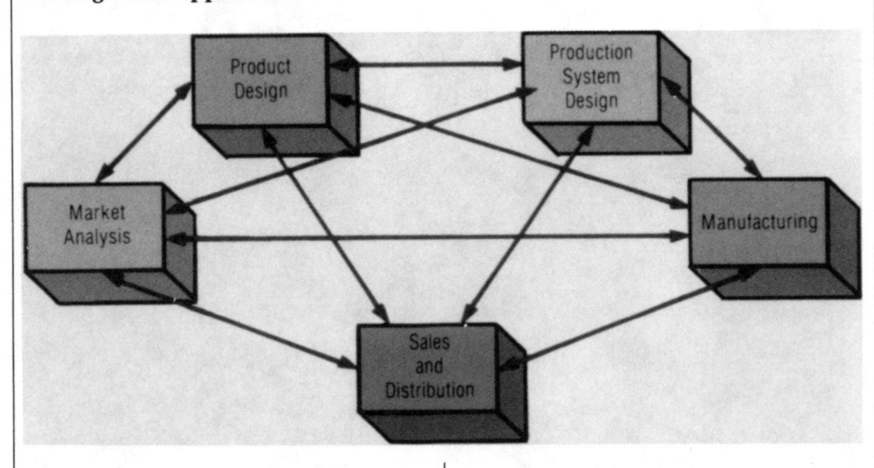

is input from materials management and financial personnel.

There are three assistant chief engineers under Masch, one each for: (1) chassis/electric/interior trim; (2) body and exterior; (3) vehicle validation and systems. The latter is a support organization for the preceding two. Essentially, then, the car is separated into the inside and outside. Under each assistant chief engineer are both product and process people. While Masch admits that the theoretical ideal would be to have simply a single organization, product complexity makes management of such a mega-team virtually impossible.

Masch says that his group, which is responsible for the C- and H-body luxury cars, is now in a proactive mode rather than reactive. He cites the fact that the organization is well ahead of conventional milestones for the 1989 product as evidence that integrated engineering is working.

At the subassembly level. There have, however, been successes on the subassembly level that more nearly meet the notion of simultaneous engineering. Paul Misegades, manager of equipment services engineering and advanced manufacturing services, General Electric Co., Lexington, KY, says that when it's time for a new product, a design and manufacturing team is formed. They then begin to examine existing products, their own and the competitions'. After determining what's right, they begin to do models. Each model is scored on how well its design facilitates assembly. A better design would include assembly steps that have components placed in a top-down fashion, since gravity is a natural assist, features that make mating virtually fool-proof, and other characteristics. GE developed its scoring approach through its analysis of the Hitachi, Ltd., Tokyo, Japan, assemblability evaluation method (AEM), the design for assembly (DFA) method developed by professors Boothroyd and Dewhurst, and other methods.

As an example of where this worked at GE, Misegades cites a self-cleaning oven door. "Everyone thought that the self-cleaning oven door that existed when we began to make the new one was a marvelous work of art," Misegades says. "The more they got into it, they discovered that it was a turkey." There were parts and fasteners which had been believed to be essential that just fell away as the project team started the redesign. The final result: a door that was less expensive to manufacture, yet which exhibited higher quality.

Yet just as there have been successes, there have, according to one highly knowledgeable source, been an equal number of failures.

Over the wall. That there have been failures should not be at all surprising. After all, industry has been enculturated to work separately and sequentially. Separation begins at engineering school and typically continues throughout a person's career. While ITI's Stoll and several others use the phrase "throw it over the wall" to describe the separation of design and manufacturing, the metaphor is actually too kind in many cases. The two groups are typically on a different floor or at the other side of the building—or in another building, perhaps across town. It's much too far to throw anything. What's more, there's great separation between vendors and the OEMs. The opportunities for interaction are minimal. The mindsets are different. Yet simultaneous engineering calls for harmony.

A solution—perhaps *the* solution—is to do what Masch of BOC proposes: put the people in a single location. As one manager put it, "There should be a single coffee area. Say someone is stuck on a problem and goes for a cup of coffee. He bumps into another person at the coffee machine; they begin to share ideas. If they don't see each other, they won't talk."

What's more, it must be made clear that it is not a situation wherein the design people are still loyal to the chief designer and the manufacturing group to the head of manufacturing while all are nodding and saying "teamwork." After all, human nature being what it is, an individual knows who is fundamentally responsible for his or her on-the-job well being. If process and product are under different heads, with each of those heads reporting to the team leader, then it's clear that the focus from below will be on the immediate supervisor. People will protect their turf. And simultaneous engineering won't work.

There must be an intimate meshing of personnel. Which is easier said than done. As one observer put it, "It's one hell of a cultural problem." And that observer is a proponent of simultaneous engineering.

Say good-bye to sequential. Then there is the fact that most products are made in a sequential manner. One name for it is the NASA phased program planning (PPP) system. Here, a project is concepted, feasibility is determined, there is a more narrow definition, the product is designed, then it's processed, manufactured, and shipped. It's a neat step-by-step approach. And because it is a neat step-by-step approach, it is easily managed. But it is also very time consuming because each downstream organization must wait for the preceding to do its job. When group *D* has a problem with what's been done by *C*, chances are *C* is already fairly well set with what it has done; it will be difficult to make it change for the benefit of *D*, especially since they are undoubtedly of two different engineering allegiances. Additionally, *C* will have its own budget and schedule, which it's not likely to want to mess up for the sake of the other group.

The real lever of simultaneous engineering, the thing that makes it work, is information. The real problem in most projects, that which throws off schedules and amplifies costs, are changes. Through simultaneous engineering, information is consolidated, or pushed back from the far end forward. This can minimize changes.

Take design. According to Henry Stoll (ITI), design is an interactive process; change is a part of it. "But we want to minimize the number of changes," Stoll says, "and make them early on, when it's nothing more than erasing a line on a piece of paper, not obsoleting a piece of equipment that has already been contracted."

OEMs no longer select suppliers to work to specs—both work to develop specs.

Changes have always been a major cost driver in any production program. Although designs are imagined to be (or are treated as) complete when they come from the product department, inevitably there are changes. Perhaps it's something minor like a set of holes that can't be put into a surface because a mounting bracket will get in the way. Although it may be seemingly minor, it can have a major negative impact on tooling. Machine tool builders have historically found changes to be a cause of friction between them and their customers. By having all concerned parties examine the relevant parts and products at the front end of the project, the chances are better that there will be fewer hiccups getting to the finish.

Simultaneous engineering applies to suppliers of dashboards as well as to producers of special machines. Although the latter is concentrated upon here, the fact that simultaneous engineering's influence is all-pervasive should be kept in mind. No one group has a corner on the opportunities—or on the obstacles.

Something must change. E. Kidder (Kit) Meade, senior vice president of Lamb Technicon Corp., Warren, MI, is straightforward about why Lamb sees simultaneous engineering as being important. Its areas of concern are machining systems for the powertrain and welding systems for body assembly. Increasingly, he says, the market for these products is decreasing as a direct result of imported cars, imported bodies, imported engines, and imported systems into transplant facilities. So about two years ago, Lamb recognized that something must be done. "We decided up front that our whole industry must change," Meade says. "Our company has bent over backwards to accommodate change, and so have General Motors, Ford, and Chrysler. We all have our jobs to do or we are going to lose our respective competitive positions in the world."

Bob Bowen, who is heading up the simultaneous engineering activities at the Ingersoll Milling Machine Co., Rockford, IL, points out that in business-as-usual arrangements there is a two-track approach. The customer, probably an auto manufacturer, does its work and the machine tool builder its; only rarely do the twain meet. This is one reason why it takes longer than is appropriate for the market for new projects to become launched. And it also contributes to the use of what are obsolete technologies in "brand new" systems: by the time the equipment hits the floor, someone has already found a better way of doing something. Unless the simultaneous engineering approach is fully embraced, the likelihood of any modifications being made to the system early on—or at any other time, for that matter—are slim.

Upfront work, analysis, and imagination can pay big dividends. For example, Bowen cites one result of a study performed on a preliminary assembly layout for an engine valve body. The relocation of one hole by 3/32" would make it possible to combine drilling, boring, and tapping for the entire assembly, which results in a 15 percent reduction in machining costs and setup time. That's just one hole. Since each of the elements in the entire powertrain is considered in turn, there is the possibility of compounded savings throughout the manufacture of the product.

Another difference in the current approach is that no longer does a company receive a complete part description or an entire process sheet (a document that is becoming rare as engineering staffs at OEMs shrink), then develop an appropriate machine, which would undoubtedly be a variation on a transfer line. For the valve body manufacture, for example, Ingersoll examined the viability of four different types of machines: single-spindle CNC machining centers, twin-spindle CNC machining centers, transfer machines, and head indexers. The company determined at which production rate the equipment would be economically viable, and weighed in the factor of flexibility.

Clearly, a consideration that all manufacturers must put right up there with quality and cost effectiveness is the ability to effect changeovers. And simultaneous engineering adds a twist. The OEM doesn't simply select a vendor for a single project. Instead, it picks a partner for, ideally, a long-term relationship. Consequently, flexibility is important because modification to the product design will occur—must occur, if simultaneous engineering is to work.

Traditional Two-Track Approach

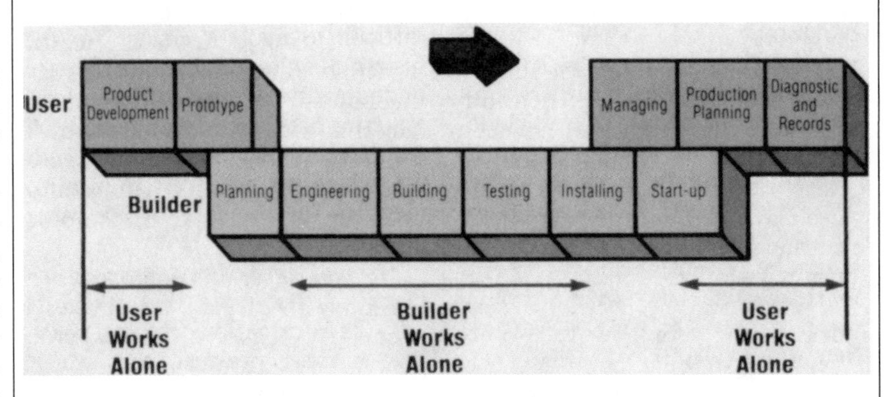

What is the normal pattern of events that leads to the creation of a manufacturing system? It's a disjointed, or two-track, approach. Simultaneous engineering makes the activities more nearly parallel, as these illustrations from Ingersoll indicate.

According to ITI's Stoll, a design change should be effected as soon as possible "if it's a directionally good improvement." He provides an example of how things ought to work: "If you can make a design change that eliminates a couple of stations, and if it leads to a more optimized product and process, it should be done, even though it means obsoleting some tooling. The savings in the long run from making the change will more than pay for it."

He cautions against the attitude of collecting a number of modifications before doing anything. As he succinctly explains, "The Japanese are continuously improving and fine tuning. It's never Miller time for them."

The effects on vendors. The partnership approach has many ramifications. One reason that the customers opt for this working relationship is because it permits them to pare the number of vendors that it must work with to obtain products. Partnership selection, then, is a very critical activity, all the more so since the supplier is brought into decision-making; it doesn't simply serve to fulfill a set bill of goods.

One effect of simultaneous engineering is that there may be fewer special machine builders. The reason why this is so is simple. The personnel at the auto companies who are involved in making the selection of vendors are typically personnel at a higher level than those persons machine builders have historically called on.

Conventionally, there's a competitive bid situation. The OEM has a given part that it wants to have produced; the companies on the bid list then work to develop the piece of equipment that will best do the job. As Richard L. Lund, president of Hueller Hille Corp., Troy, MI, a Thyssen company, puts it, "The main thing that we have to sell in the special machines business is ideas. The bread-and-butter of systems from all the suppliers are basic steel fabrications and iron castings. Where a company can make a big difference is with a super idea about how to handle or fixture a part, or a new unit that can eliminate three."

The best idea—and the best price—usually gets the order.

But with simultaneous engineering, the users are asking lots of questions. They are obtaining information about the special machine builders on subjects ranging from financial strength and technological capabilities to the builders' track records on producing systems for particular parts. In some cases, companies are

Simultaneous Engineering

Product Development | Prototype | Managing | Production Planning | Diagnostics and Records

Planning | Engineering | Building | Testing | Installing | Start-up

- **Collapses Time Frame**
- **Eliminates Repeated Steps**

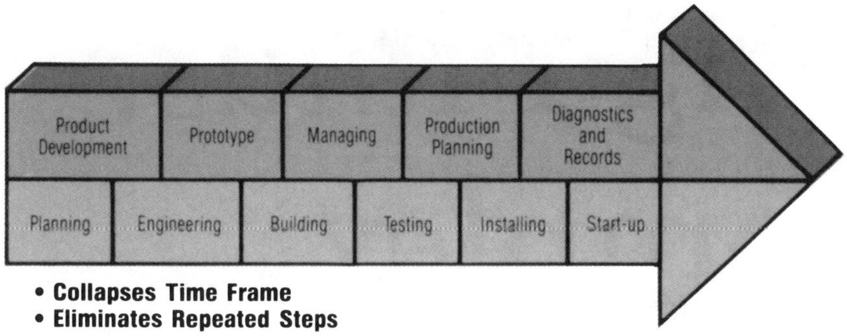

asked to quote on a concept. If their price doesn't fall within a predetermined window, they are undoubtedly removed from consideration. But a key factor is image. Since the product to be made doesn't exist—it is to be produced as a result of the partnership—the OEM must have confidence that the company it selects to be its partner will be capable of doing the job technologically and financially.

What will be a driver in the selection process? Lorne G. Greenwood, Jr., executive vice president at Hueller Hille, answers, "With simultaneous engineering, there's a tendency to go with the traditional supplier. There's less risk."

And there will be fallout from this selection process. David E. Reichard, vice president of sales and marketing, Wilson Automation, Warren, MI, comments, "If this goes to the ultimate, it will promote the big companies with the big resources. The little guys, who are usually more creative, will be eliminated or must become subsuppliers."

However, given the present overcapacity in the machine tool industry, it's possible that the little guys would fall by the wayside, with or without simultaneous engineering.

Edward A. Cooke, vice president of sales at Newcor Bay City, Bay City, MI, sees the upside and the downside of simultaneous engineering for a company: "If you end up a winner, that's great. You're locked into a partnership. If you don't make a mistake, they'll come right back to you with the next order if you did a good job. But if you lose, how do you get back into the loop again?" The answer seems, at this early point, that you may not get back in.

Which is not to say that it can't be done. Despite current successes, Lamb's Kit Meade says that they are aggressively pursuing technologies that will permit them to take on other types of projects. Out doesn't necessarily mean down.

The technology question. There's some question of whether the auto companies, by signing up with a single vendor, may not be actually being as technologically competitive as they would like to be. For all of its faults, the competitive bid process did tend to make the machine builders work very hard as they tried to get the business. As is evident by the small number of companies that have survived, some have truly proven themselves. But with simultaneous engineering, this competition will probably be a thing of the past. And if a company isn't put to the test, isn't there the possibility that it may not be as imaginative and as resourceful as it was when it was doing battle?

Meade retorts that concept by pointing out that since there will be a greater sharing of information by both the OEM and the supplier, there will undoubtedly be better, more appropriate systems devised.

With regard to pricing, it's said that OEM purchasing departments don't like simultaneous engineering. After all, with competitive bidding, it's clear who is charging how much for what. With simultaneous engineering, the price of the project evolves. The purchasing department has estimates, not bidded figures. Therefore, some of the OEMs are asking to look into their suppliers' books. They want to be assured that the costs of doing business are appropriate. Some people don't like that. Yet it is a difficulty that must somehow be resolved. After all, it is necessary that there be an amicable feeling between the two companies if they are to be working together over the long run.

One observer of simultaneous engineering puts it this way: "There is a need to establish a good, fair working relationship. If I was a supplier I would have no difficulty looking a purchasing agent in the eye and saying, 'I'll do this for you for 15 percent profit above my cost.' If he replies, 'I've got a guy down the street who will do it for eight percent,' I'd say, 'Good luck, and you'd better make sure that he counts it the same way that I do; I can't run my business on eight percent.'

"When we get into this type of thing, we do get into the need to look at books and control costs. We are breaking new ground." P

For an additional copy of this article, write on company letterhead to PRODUCTION, 6600 Clough Pike, Cincinnati, Ohio 45244.

Speeding Up

Manufacturers Strive To Slice Time Needed To Develop Products

They Cut Bureaucratic Cycles
And Use Suppliers More To Compete With Japan
Honeywell Inc.,'s 'Tiger Team'

By John Bussey and Douglas R. Sease
Staff Reporters of *The Wall Street Journal*

A few years ago, Xerox Corp. executives were stunned to learn that Japanese competitors were developing new copier models twice as fast as Xerox and at half the cost. Its market share eroding, Xerox faced a painful choice: either slash its traditional four to five-year product development cycle or be overtaken by more nimble competitors.

Today, after a sweeping reorganization and millions of dollars of investment, Xerox can produce a new copier in two years. But that still isn't as fast as some of its Japanese competitors, who have also quickened their pace. "We'll be 30% more efficient by 1990," promises Wayland Hicks, Xerox executive vice president.

Quality in U.S. Industry may be up and costs down, but American companies like Xerox are still getting sideswiped by foreign competitors who get new and improved products to market faster. The edge those competitors get from shorter development cycles is dramatic: Not only can they charge a premium price for their exclusive products but also they can incorporate more up-to-date technology in their goods and respond faster to emerging market niches and changes in taste.

'The Next Battlefield'

"It's the next battlefield," says Vladimir Pucik, a University of Michigan business professor who has studied product cycles in the auto industry. "The game the Japanese (auto makers) are going to play is to leave the Americans (building) well-engineered by boring and obsolete cars."

Now, U.S. manufacturers of all stripes are scrambling to shorten their development cycles and be the first to market at home and abroad. To do that, they are attempting to break out of old, stratified ways of developing new products--methods that, up to now, have left them uncompetitive. The Big Three auto makers, for instance, all recently formed task forces to cut ponderously bureaucratic development cycles that have swollen to nearly five years. The Japanese, by comparison, can design and build a new car in a little over 3 1/2 years.

Some of the "innovations" U.S. industry is using to close the gap are really just common sense by any other name. A couple of years ago, for example, General Motors Corp. management was puzzled by a huge testing backlog at its Milford, Mich., proving grounds. The backlog was delaying entire car projects, so GM called in consultants. They found that drivers were completing a full test cycle to check just a handful of parts. Why not double the number of parts tested on each run? the consultants asked. That done, the backlog quickly shrank.

'Business Discipline'

"We're not talking about rocket science here," acknowledges James Rucker, who heads a GM group searching for ways to trim lead-time. "We're talking about business discipline."

Similarly, Allen-Bradley Co., a unit of Rockwell International Corp., recently abandoned its old, "sequential" method of developing new industrial controls. Under that approach, the marketing department handed off an idea to designers, who drew up concepts they then passed on to product engineers. The engineers, working in virtual isolation like everyone else, built a batch of expensive prototypes and then handed one off to the manufacturing department, which had to find a way to build the new product.

Now, all of Allen-Bradley's departments work together to find--from the start--a design that fits both the customer's demands and the company's manufacturing capability. Results have been striking: Allen-Bradley recently developed a new electrical contactor in just two years. It would have taken six years under the old system.

Honeywell's 'Tiger Team'

It is often desperation, not enlightened planning, that drives U.S. industry to shorten development cycles. Until recently, Honeywell Inc. required four years to design and build a new thermostat. Then a customer, worried about the delay, threatened to take its request for a new climate-control device to a competitor. In response, Honeywell set up a special "tiger team" of marketing, design and engineering employees and gave it carte blanche.

"We told them to break all the rules but get it done in 12 months," says John Bailey, a Honeywell vice president and general manager. The team did.

Such innovations can turn old-line companies upside down. The Big Three auto makers, for example, have all adopted "parallel engineering" programs similar to Allen-Bradley's, scrapping their traditional sequential approach. "It's a whole change in philosophy," says Donald Mullaney, the manager of development of medium and large four-wheel-drive cars at Ford Motor Co. "We've got to teach our engineers to get involved upstream, to learn to be predictive."

Indeed, the auto companies provide a good case study of how U.S. industry is tackling the development-cycle problem. For years, the car companies decreed public taste, often cranking out one million cars of a single design in a year. Management regularly changed designs at the last minute--incurring huge retooling costs--and let stylists run free.

In one moment of excess, the Chrysler Corp. design shop ordered up a fin for the middle of the trunk lid on an early 1960s Plymouth. "It was a magnificent fin, a fin like this," recalls Stephan Sharf, former head of manufacturing, running his hand high into the air. There was just one problem: No stamping machine could press it, so factory workers would have had to weld the fin by hand onto each car. At the last minute, recalls Mr. Sharf, Chrysler scrapped the idea.

Such reverses were common. Until 1981, in fact, Chrysler's manufacturing group wasn't even represented on its product-design committee. That might explain why, in the past, the company regularly raised program budgets by 25% to 40% to cover expected midstream changes.

By contrast, Japanese auto makers have long designed with an eye toward production. At Honda, for example, stylists generally don't craft door panels that require more than four operations by a stamping press to shape them. (In the U.S., intricate designs often require six, seven or eight operations.) And at Honda, the concept of all departments working together is old hat. The company even has a nickname for the resulting discussions--"wai-gaya," loosely translated as "hubbub."

The benefits of this approach: Up to now, Japanese auto makers have needed only half to two-thirds the number of engineering hours that U.S. companies require for comparable car projects, estimates Kim Clark, a professor of business administration at Harvard University. Consumers in the U.S. have seen the difference in Japanese firsts--for instance, 16-valve engines offered as standard equipment.

Now, rapid market segmentation is forcing auto companies to design more cars than ever, making speedy development cycles increasingly critical. Ford, Chrysler and GM are responding with strategies of their own, and all essentially incorporate the same truth: A lot of headaches can be prevented by planning ahead, sticking to decisions and working as a team.

The push for shorter cycles is evident in each phase of the development process. In design and engineering shops, engineers are building early "math models" of vehicles on computer-aided design systems. This reduces the need for laboriously crafted clay mockups and the construction and testing of dozens of prototypes. Now, says Donald Atwood, GM vice chairman, you, "in effect, only build a clay model to validate your aesthetics."

These computer designs--which contain detailed specifications--also help the factory get a head start on tooling production equipment. (Previously, it had to wait for design engineers to take measurements from clay mockups and pass those specifications on to manufacturing.) Simultaneous, or overlapping, engineering like this, for example, recently saved Chrysler two weeks in the tooling stage of a new midsized car program, according to John D. Withrow, executive vice president of product development.

Americanized "wai-gaya" also is helping save time on the factory floor. A few years ago, for example, Chrysler suppliers ran into problems making the complex three-tone bumpers the auto maker wanted on its cars. That quickly backed up manufacturing, so Chrysler management handed down an edict: no more car designs with bumpers in more than two tones. To avoid such problems in the future, the company is increasingly moving manufacturing personnel into offices with the engineers designing the product, instead of having the two groups work in isolation.

At the same time, the auto makers are awarding more contracts to trusted suppliers without a lengthy bidding process. These suppliers, in turn, are doing more design work for the Big Three, speeding information over common computer systems rather than waiting for dog-earred blueprints to arrive in the mail.

In another effort to streamline production, Ford has halved the number of suppliers it uses in North America, keeping only the ones who don't pose quality problems. It, like other auto makers, is also relying more on suppliers to do time-consuming assembly work. A Dana Corp. division, for example, now delivers "sub-assemblies" of an entire power system--including axles, drive shafts and transfer cases--for Ford's Tempo and Topaz four-wheel drive models. Before, it just sent axles.

The auto makers claim such changes have pushed development cycles below five years, and some say that segments of the cycle are now comparable to Japanese efforts. GM's Chevrolet-Pontiac-Canada group, for example, has squeezed 14 weeks out of its cycle by getting the finance and engineering departments to review designs at the same time, instead of sequentially, says Gary Dickinson, the director of engineering. He contends efforts like this already have helped cut engineering costs on an average project 35% since 1986.

Faster isn't always better, of course. GM got its Fiero to market ahead of other low-cost, two-seat sporty cars, but with an engine prone to bursting into flames. And the new gospel isn't yet fully reflected in products now on the road. At Chrysler, Richard Dauch, the head of manufacturing, taps a pointer impatiently against a photo of one of the company's 1988 luxury cars on his office wall. It takes 10 separate stamping operations to form the car's intricate fender, which complicates the manufacturing part of the development cycle. "That design is beautiful, but it's inefficient," Mr. Dauch says. "We can't live with that in the future."

Clearly, the pressure is on the auto industry in general to fix this sort of problem quickly--and find other ways to catch up with the Japanese. Donald Smith, an auto industry expert at the University of Michigan, says the world's inefficient car companies will have to cut current development-cycle time by a further 25% to 33% and development costs by 50% to keep pace with the top auto makers five years from now.

"Whenever the last bell rings and we've fought the last round," he says, "this is going to determine who won and who lost."

CHAPTER 2

IMPLEMENTATION OF SIMULTANEOUS ENGINEERING

Presented at SME Simultaneous Engineering Conference, June 1989

Simultaneous Engineering: What? Why? How?

By C. Wesley Allen
University Of Cincinnati

ONCE UPON A TIME

Once upon a time there was a world of engineering that changed only slowly. Product models remained unchanged for ten to fifteen years. Engineering was simpler, everyone understood the manufacturing process (it changed slowly), standard parts were purchased from the supplier across town, and the materials choices were the standard steels. The engineering force was small and everyone knew and trusted everyone else. Communications freely flowed between design and manufacturing.

But days were hectic; keeping the parts list on three by five cards, being sure there was a ninety day supply of raw materials and purchased parts on hand, process sheets lettered by hand, trying to find a batch of small parts in the warehouse, and blueprints to be make from inked drawings.

Were those the "good ole' days"? Some say they were, others say they were just hard work, and young engineers ask why we put up without modern devices.

CURRENT STATUS

Whatever those days were, they are long gone. We now have a host of new manufacturing technologies, an abundance of new materials, parts lists sorted and printed by computers, drawings stored as 0's and 1's, and reports sent to the other side of the world in seconds.

Along with these advances have come new ways of doing business; monthly profit reviews, a choice of assembly locations from around the globe, customer markets that appear around new products and grow to a peak in months - only to be replaced the next year with a higher function and lower cost model. New technologies that have wiped entire old product lines from the market and, in many instances, wiped out the companies also.

CHANGES IN ENGINEERING

During this period the tools of the engineer have changed - from the sliderule to 3-D color graphics. At the engineers fingertips is a calculating power more than that available to even the best computing centers of a few years ago. The input to engineering operations has also changed. Costs are assigned to each step of the automated product production, using accounting procedures that track costs daily, or hourly if you desire.

In a few years manufacturing engineering has seen many plants change more than they changed in the last generation. Research words of just a few years ago now appear on the manufacturing floor - lasers, local area networks, vision systems, on-line data, and area controllers.

Design engineering may have moved to a new building across town. Design engineers no longer stop by the assembly line to talk to manufacturing engineers. Manufacturing now has monthly quotas, and sometimes weekly, daily, and hourly quotas. Parts lists are printed in seconds, without anyone thinking about each entry. Everyone becomes a specialist; at computer software for materials planning, using the CAD system for finite element analysis, or programming the new robot to assemble the next product.

There are significant new tools for the engineer. But in a quest for competitive position we have isolated each group, physically and mentally, from their engineering colleagues and from the overall engineering picture. We calculate answers with great precision, to problems we do not understand. We sometimes get lost in the forest of technology, but we must stay on schedule, there is no time to consider where we are going.

Everyone has an assigned responsibility and is measured by the progress within that responsibility. Every once in a while we assign a coordinator to assure that everything within a project is communicated adequately. But no engineer has time for the coordinator. Besides, the new coordinator was wandering around without an assignment and is planning to retire next year.

Current engineering organization and management is well behind both the magnitude of the change in manufacturing processes and the new power of the computer/information tools we use daily.

We have not arrived at our current position due to the misguidance of a few, nor as the victims of any plot. The reasons lie more with the high rate of change in technologies associated with engineering and the inertia of humans to change from the known, familiar, and proven. What changes were made within the engineering structure and procedures were probably more to accommodate the changing technologies than to develop a new structure that better capitalized on the new technologies.

Simultaneous engineering is one change that addresses the need to change our way of using technical resources, both personnel and technology, in a manner to better accomplish our current objectives. Recognize that what was the best just a few years ago is not competitive today - neither in consumer products, nor in effective organization of our engineering force and their bag of technical tools.

WHAT IS SIMULTANEOUS ENGINEERING?

Simultaneous engineering is not something physical, like a calculator; nor is it a set process, like making coffee in the morning; nor is it a solution for this months profit, like delaying an expenditure. Simultaneous engineering is not the magic elixir that many have sought in the past - and many still do.

But neither does it conflict with good business practices. Nor does it require a large investment in new equipment. Nor does changing to simultaneous engineering within the organization preclude also adopting other current techniques, such as JIT, 3-D graphics, SQC, or Project Planning.

Simultaneous engineering is an approach that has lasting effects and the potential to cut costs by fifty percent. Simultaneous engineering is adopting some of the operational characteristics we have inadvertently left behind over the years and fitting them into todays' environment. Simultaneous engineering is an operational mode for engineering that allows the total engineering effort to be more effective in its mission.

But Simultaneous engineering is not without its price, a price that many people and organizations are not willing to pay. That price is the change to a new way for engineering to operate day-to-day. For most organizations, changes in the basic mode of operations is not welcome; people desire a change in results without a change in the way they work. Managers do not want to change their ways, nor do engineers. The known is preferred to the unknown, and any change presents elements of the unknown. We want to continue to do things just as we have, but to reap the benefits of improved output. Such action sounds like magic, or the old time elixir we all really wish was available.

As a formal definition, I like this following:

> Simultaneous engineering is designing a product and designing the process to manufacture that product at the same time.

Simultaneous engineering is a way of putting the engineering operation together again. Operating as a single entity that attacks engineering problems and does not limit its focus to the detailed assignment and measurement of individual tasks.

WHY CHANGE TO SIMULTANEOUS ENGINEERING?

As mentioned earlier, there is a price to change. Changing to simultaneous engineering is only worthwhile if the results of design and manufacturing working more closely have benefits for the business. Social comradely within engineering is a insufficient reason.

There are benefits to design and manufacturing working side-by-side and those benefits have been significant in many instances:

One company that adopted the simultaneous engineering methodology now designs and develops products 20 percent faster than previously. They also have an annual savings of $750,000 through a 25 percent reduction in the number of engineering changes and the highest product quality in their history. Simultaneous engineering (vs. sequential engineering) has allowed potential problems to be identified earlier in the product cycle, where the time and cost of modifications as significantly less.

Honeywell developed a new thermostat in 12 months that previously would have taken four years. A change that not only reduced product development costs, but provided a competitive much earlier.

One American company discovered several years ago that their Japanese competitors were developing products twice as fast and at one-half the cost. They have since realigned their engineering organization and reduced their product development time from four or five years to two years.

Another company has reduced by one-third the time to put a product into the market place.

A team at Black and Decker reviewed their list of "standard washers" and found they had 448. The team worked across functional and product lines to reduce the number to seven. Not only a benefit to engineering, but to many parts of the business such as purchasing and stocking of repair parts.

The "mouse" for the Apple computer was redesigned through a combined effort involving the total engineering operation. The result was an increase in manufacturing yield from 40 percent to 99.9 percent, a materials cost reduction of 45 percent, and a factor of 20 reduction in the rework cost.

For one product Black and Decker reduced the parts count from 30 to 10, and the assembly time by a factor of 2.5. But the largest significant improvement was in product quality!

Boeing is using "design-build" teams that evaluate the design from the build viewpoint as the design is progressing. The build team has access at any time to the latest level design through the CAD system.

The automotive companies, and their suppliers, were some of the first to support simultaneous engineering as a significant business benefit. These efforts, not surprisingly, represent a range of success.

Hewlett-Packard successfully merged the work of the design and manufacturing functions for the development of a new plotter. Benefits included reduced parts count, increased reliability, lower manufacturing costs, and lower development costs.

General Electric, who has had internal corporate education programs to give more consideration to the manufacturing process earlier in the development cycle, has 5000 engineers who have attended classes in design for manufacturability.

All of these examples involved the cooperative effort of both design and manufacturing engineering to achieve the results. After these successes, a new standard for engineering has been set. These companies will continue to improve on the process and search for ways to assure they retain their willingness to change and maintain their competitive advantage.

Other companies are satisfied with their past modes of operation and are not willing to face the trauma of change. Such companies are becoming an endangered species.

ACHIEVING THE CHANGE TO SIMULTANEOUS ENGINEERING?

To obtain a different result, we must change the way we perform. Part of the difficulty in achieving change is the reluctance to give up the way things are being done now. We are all comfortable with todays routines, yet we would like a better result.

Engineering is inherently change; every engineering project produces a new and different product. Yet engineers ar people and also have the reluctance to change. But engineers are used to change and should be leading their management toward better operational methods for their profession.

Studies have shown that there are common factors for project success and that one or more of these factors are lacking in projects classified as failures. The top success factors include the following:

First: management must understanding and support the change. This support must reflect not just support by memo, but by active participation through their own daily management of the engineering function.

Second: the need for all involved to understand the reason for the change and the intended results so they become a part of the effort to achieve the change.

Third: every part of engineering must have the same common goal. Thus each person and department becomes committed to an overall goal and understands how their effort is a part of something larger.

While these three elements sound very basic and simplistic, in the past we have not followed them. To effectively achieve simultaneous engineering, some set of such actions are required. We have sometimes been too rushed or management centered to follow these basic guidelines.

I guess most of us have been given an assignment that started out something like, "On this new, important program your part will be (___________) and you have until (________) to complete your assignment." If those weren't the words used, then maybe those words reflect the essence of what was said.

If you were in product design, then the emphasis was on schedule, function, and a satisfactory model to test. If you were in manufacturing, then the emphasis was on tooling cost, product cost, and product ship date.

One result of this "parceling out" of assignments was that the separate groups were actually discouraged from working together - time away from your task was time lost on your assignment and your department might not make the schedule. There was no factor that accounted for the need for the pieces to be accomplished such that they all fit into place in the end.

It is time for engineering management and engineers to recognize the need and value of design engineering and manufacturing engineering working more closely. To accomplish this, simultaneous engineering has proven to be a significant factor in improving engineering productivity and providing competitive benefits to the corporation.

DESIGN INTEGRATED MANUFACTURING

Plastics processors are sharpening their competitive edge with a new spirit of partnership between designers, manufacturers, and suppliers.

Welcome to the 1990s. The future is already taking shape at many companies where one of the most sweeping changes affecting plastics processing in recent years is underway. Product designers and manufacturing engineers, together with materials and equipment suppliers, are forming integrated teams to make design-integrated manufacturing a reality.

While still in its infancy in some places, the trend is taking dramatic form in others:

- General Motors ended the traditional barriers between design and manufacturing with a whole new concept called simultaneous engineering.
- "Black-box" molders have emerged as key players. They perform all design and manufacturing functions for specific parts based on end-use performance requirements.
- Major partnerships of food processors, plastics converters, equipment manufacturers and resin producers are driving change in the rigid packaging markets. Alliances like the one forged at Campbell Soup (cover story) are revolutionizing the prepared foods business.

"The time has come for design engineers to be more cognizant of manufacturing problems and failures to eliminate the costly 'error' from trial-and-error engineering," says Terry Peterson, engineering project manager at Marley Pump Co., Mission, Kan.

Nick Zissimopoulos, engineering manager for Motorola in Schaumburg, Ill., says: "The relations of the design and manufacturing engineers had better be much closer than it is today if that company wants to be in business five years from now."

There are many reasons behind the push toward design-integrated manufacturing, simultaneous engineering, or whatever you want to call the trend. A study conducted by Cahners Research for this special issue of *Plastics World* shows seven factors are driving this trend.

Number one is cost and quality improvement. A prime example is IBM's Proprinter, the winner of *PW's* Better Way award two years ago. Designed specifically for totally automated assembly, the Proprinter is held together by snap fits and self-locking features between mating parts. Not a single screw is needed. The printer contains just 60 parts, about one-third the number in previous designs by other original equipment manufacturers (OEMs). As a result of the improved design, an IBM analysis showed that the Proprinter could be manufactured less expensively in the U.S. than in Asia.

The "design for manufacture and assembly" methodologies developed by consultants Geoffrey Boothroyd and Peter Dewhurst were instrumental in the Proprinter development. (See story on p. 44)

The second biggest factor is changing attitudes: American industry was excessively secretive and as a result the U.S. was not gaining adequately from the experience and insight of others.

Third is the proliferation of new resins. Few profound new chemistries have been introduced in the last 12 to 15 years, but thousands of upgrades, blends, and alloys have been developed. And don't think this trend applies only to engineering materials. Look at the significant advances just this year in the stress crack resistance in blow-molding grades of high-density polyethylene. Add the infinite design possibilities provided by new additive combinations and you quickly see why designers need to be better informed about processing.

Fourth is the evolution of processing. Consider the explosion in coextrusion since the first appearance of the Heinz ketchup bottle and the advances in Detroit with structural reaction injection molding. Even plastics veterans can take nothing for granted. One of the oldest plastics-making processes—resin-transfer molding (RTM)—is receiving new life because of improved resins and equipment.

These innovations made RTM the choice for 4,000 seat shells in the new United Airlines terminal at Chicago's O'Hare Field. The shells went from production startup to shipment in 50 days, an impossible accomplishment in RTM just a few years ago. The achievement was possible because the seat designer was

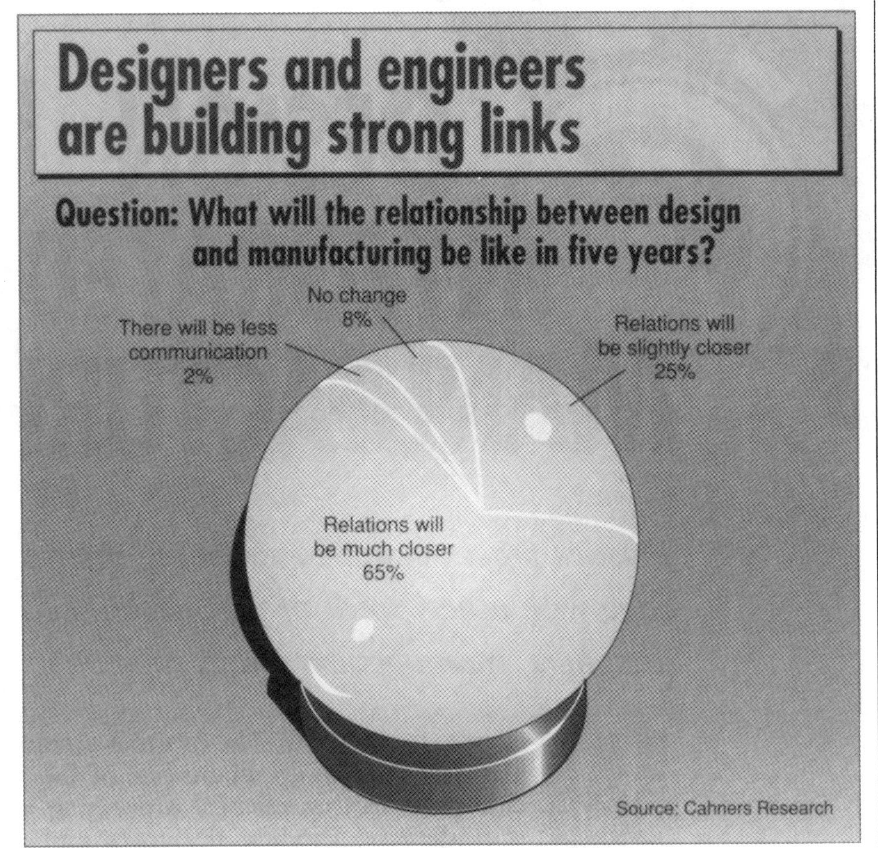

Design-integrated manufacturing is fueled by a number of developments, ranging from cost reduction to design staff reductions at a few manufacturers.

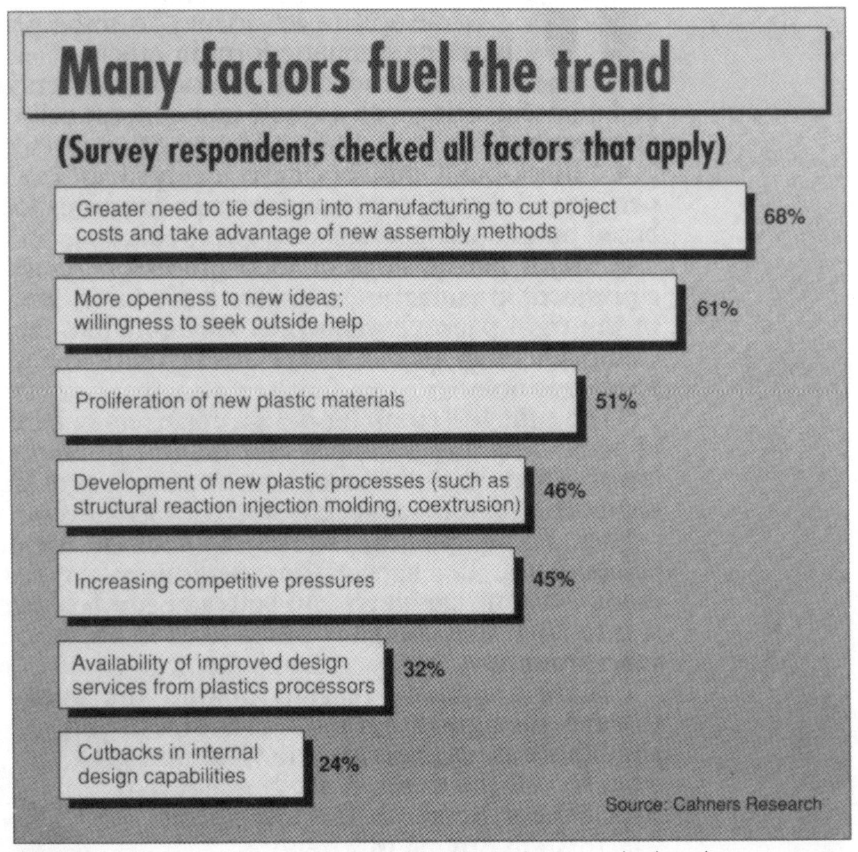

Nine in 10 expect more coordination with manufacturing, both in their own company and with upstream, or vendor, companies.

aware of process improvements.

Other reasons for the design-integrated manufacturing trend include increasing competitive pressures, availability of improved design services from processors and resin suppliers and cutbacks in internal design capabilities at OEMs.

Xerox pinpoints design

One of the best studies of the issue was made by Xerox Corp.

The copier giant suffered a significant decline on its return on assets in the early 1980s and plastics played a big role. Plastics represent 10-30% of the costs of Xerox office machines. The company found that its plastics costs were 15-60% above its foreign competitors.

It launched a detailed investigation of the reasons. The factors studied were resin prices, manufacturing productivity, tool costs, and design practices. At the time of the study, Xerox (or more accurately its molders) suffered some disadvantage in each of those categories. Interestingly, however, differences in design practices acccounted for 50% of the competitive cost gap!

There were two problem areas: immature designs reached manufacturing, causing expensive changes, and there was "a failure to bring the manufacturing technologies and process tradeoffs to the product design community."

The result was significant changes in how Xerox designers interact with molders. Xerox now identifies "model molders" who enter into team relationships with Xerox designers.

Such improvements in communication are becoming commonplace throughout plastics design and processing. In GM's "simultaneous engineering," manufacturing considerations are merged into product design, not only at GM, but ultimately

How the major players fit in now...

Plastics processors

72% of designers seek assistance from processors for help on: (in order of importance)

1. Materials selection
2. Feasibility exploration of original product concept or application.
3. Manufacturability
4. Discussion of competitive processes e.g. structural foam versus pressure forming)

Plastics machinery manufacturers

15% of designers seek assistance from machinery manufacturers for help on:

1. Feasibility exploration of original product concept or application
2. Materials selection
3. Product prototyping
4. Discussion of competitive pressures

Resin suppliers

47% of designers seek assistance from resin suppliers for help on:

1. Materials selection
2. Feasibility exploration of original product concept or application
3. Manufacturability
4. Discussion of competitive pressures

...and in the future

Question: How do you feel the roles of these sources will change your design decisions in the future?

Plastics processors

Equipment manufacturers

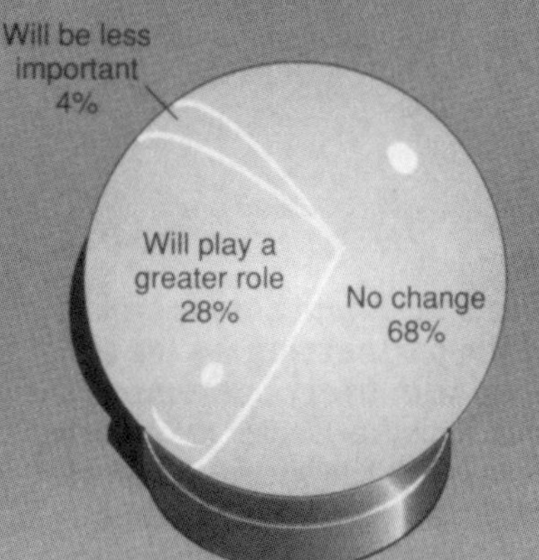

Resin suppliers

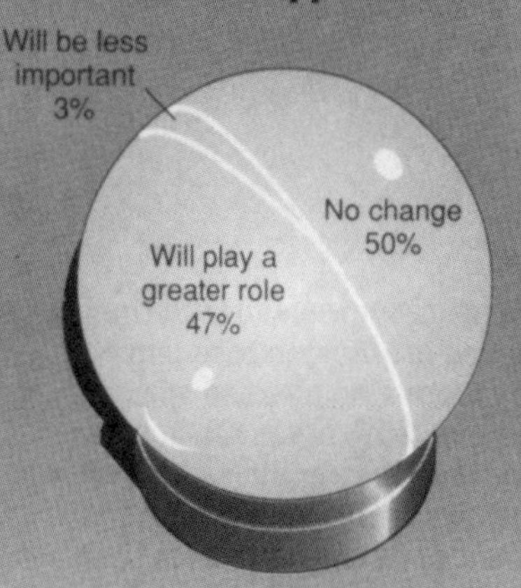

Source: Cahners Research

at suppliers as well. GM, for example, gives broad product concepts and performance parameters to suppliers who are then expected to develop blueprints and preliminary prototypes.

One of the most dramatic successes was the development of a new automotive knee bolster. Pontiac needed help in meeting the federally mandated safety standards for restraints used to protect drivers' knees during collisions. A crude model was sent to Diversified Plastics Corp., which studied the design, available materials and processes. The answer: mold the piece from an expandable styrenic maleic anhydride copolymer bead. As a result, Diversified pioneered a whole new industry, close-tolerance expandable copolymer molding.

GM designers, removed from the latest developments in resin and process technology, could never have designed the same piece at the same cost.

Room to improve

The Cahners study showed that almost three in four product designers regularly consult with processors about materials, product manufacturability, the feasibility of the original product idea, and process to be used. More than half of the respondents expect communication with processors to grow stronger.

Designers said they would like to see better communication from processors in these areas:

- Suggest changes in part design to simplify tooling *before* the tools are made;
- Suggest, other less-expensive materials. In general, broaden range of material knowledge;
- Develop the ability to prototype in several different processes;
- Link digital design data bases to eliminate drawings for mold building; and
- Expedite prototype production process.

A design engineering manager for a company that makes chemical processing equipment comments: "Be honest in appraising the moldability of a part. Too often molders say they can do it with no problem or concern just so they can get a foot in the door. Later, they charge an arm and a leg for tooling changes."

Nearly half of the survey group said they consult with resin suppliers when making design decisions. Not surprisingly, the great majority are seeking more data on resin properties. Although resin companies are improving the quality of these services, many designers are still not satisfied.

Richard L. Panicci, senior design manager at Kiddie Products in Avon, Mass., says: "It takes too much time to have 'simple' questions answered by major materials suppliers

Another complaint: Resin suppliers need to "cater their design service to the 'little people' who use less-than-railcar volumes", says Steven Pyshos, design manager for Ametek/Panalarm in Skokie, Ill. George Kluzak, design manager for Mankato Corp., Mankato, Minn., would like to get past the sales force to the lab people.

There also needs to be more objectivity in the performance data provided. Less puff and more on problems to look out for. Another suggestion: Robert Baur, vice-president of engineering at Tri-Star Electronics in Venice, Calif. would like to see more uniformity in data sheets from different suppliers to allow for easier comparisons. Several designers would like to see more breadth of performance data in areas such as chemical resistance and weatherability.

While there are still plenty of rough edges in the growing plastics partnerships, there are also plenty of spectacular success stories. A packaging revolution at Campbell Soup. A reinvigoration of the color copier business at Kodak. Whole new opportunities in automobile design triggered by the blow-molded gas tank. *See page 127 for more details on the concept of design-integrated-manufacturing by two pioneers, Boothroyd and Dewhurst.*

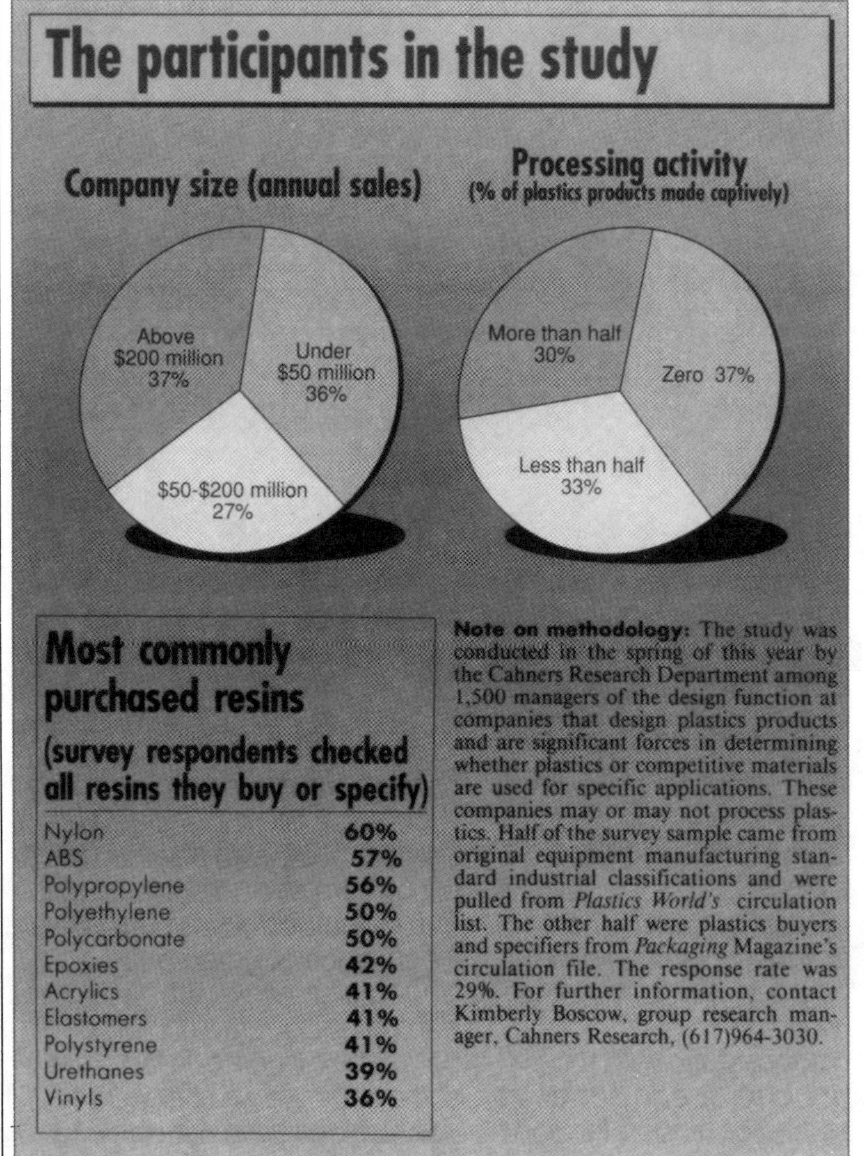

Most commonly purchased resins (survey respondents checked all resins they buy or specify)

Resin	%
Nylon	60%
ABS	57%
Polypropylene	56%
Polyethylene	50%
Polycarbonate	50%
Epoxies	42%
Acrylics	41%
Elastomers	41%
Polystyrene	41%
Urethanes	39%
Vinyls	36%

Note on methodology: The study was conducted in the spring of this year by the Cahners Research Department among 1,500 managers of the design function at companies that design plastics products and are significant forces in determining whether plastics or competitive materials are used for specific applications. These companies may or may not process plastics. Half of the survey sample came from original equipment manufacturing standard industrial classifications and were pulled from *Plastics World's* circulation list. The other half were plastics buyers and specifiers from *Packaging* Magazine's circulation file. The response rate was 29%. For further information, contact Kimberly Boscow, group research manager, Cahners Research, (617)964-3030.

Presented at SME Simultaneous Engineering Conference, June 1989

How Process Logistics Planning Can Enhance The Effectiveness Of Simultaneous Engineering

By Edward J. Budill
Tompkins Associates

HOW PROCESS LOGISTICS PLANNING CAN ENHANCE THE EFFECTIVENESS OF SIMULTANEOUS ENGINEERING

In simpler times a manufacturing company would assemble its product design team, and in a reasonably short time be able to deliver a new product to manufacturing who would then produce it. Our memories may have dimmed but it seemed as if that process took place without the need for extensive engineering lead time and it did not seem to result in numerous recalls of the product. This was the traditional, sequential approach to product design and manufacture. All this may indeed be ill-placed nostalgia and maybe the sequential engineering process really did generate major headaches for manufacturing and maybe it didn't produce products which were as defect free as we recall. At any rate, the good old days are gone forever, and we are in an era where governmental requirements and intense foreign and domestic competition place a great premium on a company's ability to rapidly bring new products or major modifications to existing products to the marketplace in the minimum amount of time.

Speed is an essential ingredient, to not only remain competitive, but is absolutely necessary if a company is to beat its competitors to the market. Speed itself is not the only answer. The products must come to the market as defect-free as possible, which means that manufacturing is an important consideration in the design of the products. The current buzz word is Simultaneous Engineering, which is either something very new or it adds some innovations to the bromide of going "back to basics". Where it has been used, it has achieved quantum improvements in product design, manufacture and delivery of quality products to the marketplace.

DELIVER TOTAL PACKAGE IN SHORTEST TIME

Simultaneous engineering's purpose is to deliver a total package in the shortest time possible and at the highest quality level. It is not a route to reducing product development and engineering cost. Its power lies in its demonstrated ability to cut traditional product development time in half. For example; using SE, AT&T product development time for telephones has been reduced from two years to one; at Navistar, product development for trucks has been reduced from five years to two and one-half; and Hewlett Packard brought Desk Jet printers to market in twenty-two months. In fact, SE will probably increase some of those costs because it is an iterative process and some portions of the work will have to be redone as better ways are found to configure the product and as manufacturing discovers the need for changes in product design to enhance manufacturability.

If the product development cycle can be shortened, than those personnel can be made available for the next project sooner, thus improving a company's use of scarce technical resources. In fact, a new term has come into our lexicon, which is called, "Break-even time". This is the time required from the inception of the project to the beginning of the new product's profitability. The shorter this time the sooner the company will start to make money and get a return on its investment.

If SE can produce so many desirable results, why is it not embraced overnight like so many management fads i.e. Theory Z, Quality Circles, etc. Probably because it runs afoul of the organizational "Chimney's" in the companies. In an automobile company, SE will have designers, stylists, planners and manufacturing all intruding on each others "turf". These turf struggles are not merely reactions to the other guy telling you a better way, but involve some real cultural problems where these groups have traditionally spoken different technical languages. The new computerized tools are making communication easier and are speeding the analytical process. However, the creaky management bureaucracy has not been able to handle the rate of change of these high performance communications systems.

Even in manufacturing, there are "chimneys" which are major obstacles to progress. Process engineering may successfully negotiate with product engineering on the technological approach to automatically assemble some components to a vehicle. But they need an automation friendly conveyance, special line-feed mechanisms, parts packaged in special dunnage and delivered to the plant in special side loading trucks which interface to an automated plant delivery system. This negotiation would require Facilities Engineering, Material Handling Engineering, Materials Management, Purchasing and Traffic to agree and to support the program. Who gets credit for the savings and who gets charged for the investment or additional cost even if the whole proposition produces net savings to the company? Clearly, more of these barriers need to be broken down.

How do we do this? Simultaneous Engineering is significantly more than just engineering. Simultaneous Engineering requires the focusing of the total effort involved to bring the product from conception to delivery to the user, be it a consumer or an industrial customer. The truth then, is that we are not focusing on simultaneous engineering as much as the total functions of simultaneous product delivery. These functions include:

- Concept development,
- Styling,
- Product engineering,
- Process engineering,
- Manufacturing logistics.

MANUFACTURING LOGISTICS SYSTEM CRITICAL TO SE

These functions are performed concurrently and use such techniques as design for manufacture, design for assembly and design for automation. However, most of the writings about SE ignore a critical element - the manufacturing logistics system. This is a combination of material handling, process support and facilities layout and integrates all of the logistics within the factory. In an automotive assembly SE team, the logistics system must be represented from the outset. Why is it important that SE include manufacturing logistics? In the usual manufacturing development approach, manufacturing logistics, which includes material handling plant layout and process support brings up the rear of the manufacturing development cycle. It usually comes into focus after the product has been engineered and bills of material are used to develop the logistic requirements for manufacturing the product. This, in fact, is too late, because by this time a significant portion of the total product lead time has been used and there is little time for developing the proper strategy for manufacturing the product.

In the automotive industry the introduction of a new vehicle into one or more assembly plants is a major event which may require major additions to the plant and/or gutting an existing plant either of which would require extensive downtime of an existing product line. A new plant is a major expenditure and there are several possible strategic configurations for the new plant and the systems inside it. Management needs to see the impact on a plant's design caused by alternative configurations of its manufacturing systems early enough to make decisions on those systems. The alternatives have to be reviewed as far upstream as possible, so that whichever alternative is selected there will be adequate time for rational development of the manufacturing facilities plan.

SURROGATE DATABASE AND SE

How can this be done until all the data is available and all the parts designed and all the bills of material are established? The answer is, it probably can't be done if manufacturing logistics must wait for the development of the actual data. However, surrogate data can be used to develop realistic alternatives which can then be analyzed. This data when used as the basis for these strategic studies can provide meaningful forecasts concerning the effect on manufacturing facilities and operations of the implementation of alternative manufacturing strategies.

One logical approach which has proved effective is to quantify these issues, using a manufacturing database derived from an existing vehicle line. The validity of this data will be accepted by the various factions within the company, which must participate in developing the advanced manufacturing concepts. It is more important that the data elements be accurate in relation to each other than that they precisely represent the new product. Therefore, data derived from the surrogate product can serve as the database for these strategic facility studies. When actual product data becomes available it can be substituted for the surrogate data on a modular basis and the manufacturing model updated dynamically.

DATABASE ELEMENTS

The database should describe the manufacturing process in terms which would include:

- The manufacturing process description
- The tooling
- Estimated time to perform the work process
- Production parts used, including all levels of selectivity (color, style, option - left, right)

- Fasteners
- Applicability of parts to product based on option use

  - On what percent of vehicles is an optional part used

- Containerization of parts

  - Size - quantity of parts in a container

- Freight mode

  - How is it shipped, how many containers per conveyance load

- Freight costs
- Material source(s)
- Material cost

DATABASE SOURCES

This data generally comes from two or three sources within the company. The first source is the manufacturing process sheets which have most of the technical description of the process and tooling. The second would be an assembly line balance report from the appropriate assembly plant. This will establish the number of line stations, their sequence and what processes are performed at a specific line station. Additional data comes from the plant logistic or inventory report which contains all the specific logistics date for each individual part number, product cost and shipping information and the containerization report which specifies applicable types of containers. This data can be brought together to form an integrated parts database. This now describes the total logistic effort required to supply material to any given station in the plant. These listings are not the end in themselves, but merely a means to an end to remind the product/process engineering team of the total number of parts required to build the baseline vehicle. The SE team can be supplied with lists representing specific areas of the vehicle, such as major assemblies which are being studied specifically for automation or design for assembly. The SE team then works with the surrogate bill of material making changes to it when parts are combined, eliminated, simplified or added. The modified bills of material then become the basis for planning the physical aspects of manufacturing the new product.

These efforts to simplify the product may result in reducing the number of parts by 20-25%. This will significantly change assembly plant logistics requirements. The new part configurations will be used to determine appropriate container sizes and quantities of parts per container. The new database reflects the number and sequence of assembly stations required to build the product and quantity of material and rate of supply actions required to build a given production mix at a specific production rate. This is critical to the preliminary layout process because the factors which effect the plant layout include:

- .Process system layouts - number of stations, line lengths for both main line and sub-assembly systems.
- .Amount of material which must be received and by what means.
- .Material storage - back-up and line stock.
- .Material handling throughput and systems used in the material transport system.
- .Line feed methods for manual and automated assembly operations.

SURROGATE DATA PROVIDES BASIS FOR PLANNING

The surrogate database provides a dynamic basis for establishing the key factors which determine the requirements for developing the system layouts and determining:

- .Space
- .Manpower
- .Capital cost

The data can provide a factual basis for determining what the assembly system should look like and what type of logistics support within the factory is necessary to support its line rate. Studies could evaluate the impact on the facility of various production line rates and could evaluate the impact on the facility of various inventory levels for all or specific families of components. A number of other studies can be performed to evaluate such issues as; - What is the potential for installing on-site focused manufacturing lines to produce parts which have high selectivity?

- -What would be the effects on overall product cost?
- -Would these operations be appropriate for incorporation into an assembly plant?
- -Would these be focus factories in the JIT context or under-utilized capital facilities?

RATIONALIZATION OF IN-PLANT PROCESSES

Freight and material costs could be used to evaluate if selected parts should be manufactured in-house or contracted out. If the following JIT strategies were implemented, what would be the effects on plant space, manpower, capital cost and on outside suppliers' costs?

- Inline sequenced delivery of components
- Preparation of parts kits at the component manufacturers plants
- Packaging parts in special containers\dunnage friendly to assembly automation (robots) - off-site or at the point of manufacture
- Moving certain assembly operations upstream from the assembly plant to suppliers plants
- What parts, or groups of related parts could be manufactured in the vicinity of the assembly plant, or by outside suppliers to support JIT strategies.

Other studies could include evaluation of different automation friendly conveyance systems for the main assembly line and sub-assembly lines and developing a material handling strategy for the overall plant. In order to perform the above studies, it is necessary to have all the information concerning the plant, the material and the processes because much of the relevant data which effects plant cost is material oriented.

BASIS FOR SIMULATION STUDIES

Another method for testing the performance of the proposed plant design is through computer modeling and simulation. Individual computer models can be developed for the individual focused factories or sub-assembly systems and a macro model for the main line. The data to substantiate the dynamic material flow will be provided by the surrogate database. High level simulation studies can be performed to quantify material flow relationships between:

- Stamping plant and the body shop
- Body shop and paint shop
- Paint shop and trim chassis and final lines

The simulation studies will look at such issues as:

- What provisions should be made for model changeover?
- How does downtime of automation processes affect production?

The surrogate database provides the basis for developing assembly system computer models and performing simulation studies to determine:

- What provisions should be made in the system design to accommodate reasonable equipment downtime and reasonable cost?
- Should there be buffers in the system - where and how large would they be?
- What systems should be backed up based on criticality to overall system up-time?
- How does the sequence of the vehicle body styles from the body shop and vehicle colors from the paint shop affect the logistics of line supply in relation to specific trim and chassis material which are keyed to body style and vehicle color?
- How the quality issues in body and paint affect the sequence of vehicles starting on the trim line and therefore the ability to use JIT for line supply of selectivity sensitive items?

These are all issues that require strategic studies and require manufacturing logistics data as well as the process data. The studies provide a dynamic look at how the total plant would operate, how much space would be required, and the requirements for both direct and indirect labor.

Of all the many useful outputs of these studies none is possibly more valuable than the overall plant strategy studies. Does the new assembly system fit into an existing plant? If expansion of the plant is necessary how much, where and at what cost? How long would it take to install the new systems in place of the existing systems? What are the implementation strategies which could be used to implement the new systems with a minimal amount of plant downtime? These are top management decisions, but rarely do they have the alternatives presented to them in time for careful deliberation.

INTEGRATED MANUFACTURING DEVELOPMENT CAN REDUCE BREAK-EVEN TIME

Simultaneous engineering or integrated manufacturing development is an important philosophic approach for reducing the break-even time for new products. It is a powerful organizational tool which can provide management with good data upon which to base decisions. This approach won't make the decisions, in fact, it might highlight the difficulty of some of the decisions by providing a comprehensive picture early in the game. It will highlight the alternatives and provide the framework for further specific design action on the part of the various functions which have to develop the new product's manufacturing plan. The development of a surrogate product manufacturing database can facilitate the performance of a variety of realistic advanced strategic manufacturing studies.

These studies can evaluate the cost for new products and the effectiveness of various manufacturing strategies, well in advance of complete development of new product data. This planning methodology is not a data processing technique. Rather it is a powerful method of providing quantative data to the manufacturing facilities and logistics engineering effort which can quantify requirements for system function and performance, which can then be used to evaluate performance of the proposed conceptual systems.

Reprinted from *Automotive Industries*, December 1987

Make Me A Match

Getting design and manufacturing together—simultaneously.

by Cathy Coffman

In today's frantically competitive automotive industry, everyone is constantly looking for fresh component and sub-assembly designs—designs that will help streamline both vehicles and the manufacturing and production systems which build them.

Simultaneous engineering is the latest revelation among the CAD, CAM and CAE tools now being used to optimize the performance of designs in the concept stage. Simultaneous engineering allows engineers to design for assembly and manufacturing while they design the product.

Perfected by the Japanese, simultaneous engineering was brought to America by Edward Deming. "Basically, simultaneous engineering helps designers to design their product to fit into the manufacturing process," says Mark Craig, president of Applied Computer Solutions in Detroit.

How does simultaneous engineering work? The computer simulates, through a process called Monte Carlo Analysis (a random number generator), the assembly process of a three-dimensional modeled component. It positions the designed part in its place on the assembly line, and then attaches the various fixtures to the new part.

The computer also attaches sub-assemblies to assemblies, and analyzes the variation between the computer

Above picture: Geometry entry in VSA program.

model form and what the real-world model should look like. And the analysis is all geared around the effect the newly-designed part has on the assembly process.

"The final result," says Craig, "is that the computer and simultaneous engineering software predicts how much the design of that part is out of spec with the rest of the assembly line."

If the model does predict an interference or problem with the new part, it goes back to the original design or manufacturing process—where the engineer was working—to correct the problem. The model points out how much variation there is for a given condition and where the variation is coming from.

The remarkable thing about simultaneous engineering, stresses Craig, is that **one** person is responsible for coordinating the entire manufacturing and design process.

"Eighty percent of the work done with simultaneous engineering is initiated in the design stage," says Craig, "so designers want to get it right from the start." Simultaneous engineering brings together manufacturing and design personnel early in the process—ideally at the design stage.

The designer in charge of the assembly or part can, with simultaneous engineering, get constant feedback from the manufacturing guys. And this can save him from doing countless iterations on the same design—which could take months.

Predicting production variations in the production of a complex assembly and determining final assembly dimensions is traditionally done through a process called limit stacking. "This involves simply adding up, in a linear mode, all of the maximum and minimum tolerances for each component," explains Craig. "These figures then tell you the maximum and minimum tolerances for the final assembly."

However, limit stacking does not account for the critical effects of the assembly sequence on the final assembly variation.

"Limit stacking is too conservative and doesn't comprehend the manufacturing process—such as locating fixtures," says Craig. "It simply takes the 'worse case' scenario and shows that a variation exists. But it doesn't tell you where the variation is.

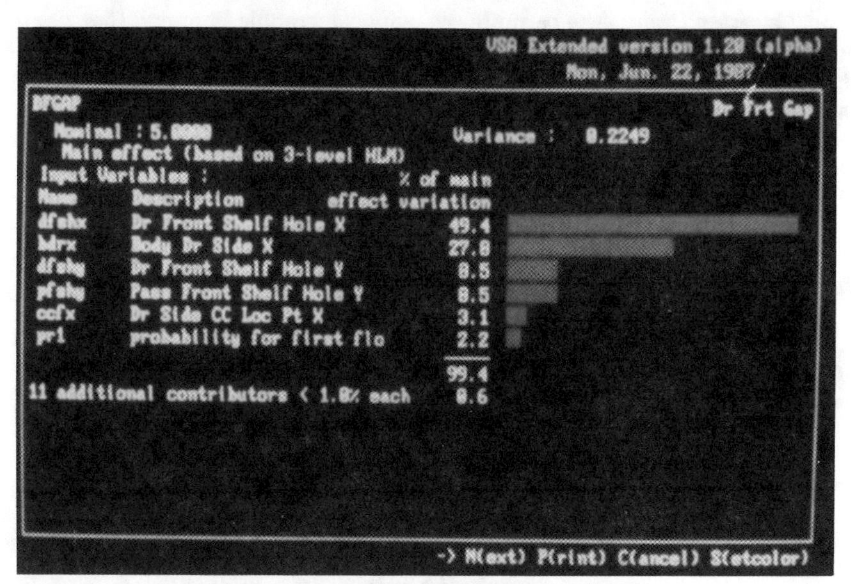

VSA program pinpoints cause of manufacturing variation.

Running Gear

"But VSA (variation simulation analysis—statistical program used in simultaneous engineering) is much more accurate—its model is based on probability and on build/distribution curves," explains Craig. "Because once the original model is built, 10 to 15 iterations can be done per day."

And by changing a limit stack model, you change the whole thing, not just one parameter. With VSA, you only have to change one parameter—the parameter that is shown on the model to be out of sync with the rest of the model.

The limitations of limit stacking makes the statistical technique of Monte Carlo simulation, when combined with 3-D modeling, a viable, cost-effective alternative for design and manufacturing engineers.

The hardware requirements for simultaneous engineering are no different than any other high-powered computer system. In fact, the requirements may be less than those the CAD/CAM system currently installed in your office requires. A typical program, says Craig, can be run on a host of hardware systems, including the IBM PC AT, Apollo or Sun workstation, or DEC VAX minicomputer.

"Whether the program is being run on a PC or a minicomputer usually doesn't matter," says Craig, "because the software is essentially the same." He says that the major difference between the smaller PC and more powerful minicomputer is the time it takes to run the program. "On the minicomputer, a routine could be run in a couple of hours, while on the PC it may have to be run overnight," he explains.

Obviously, the big benefits of simultaneous engineering is that it mandates communication between manufacturing and design engineers early in the process. And it forces them to work together as a team.

And it also saves both time and money. There's no "fixing what's broken;" the rework is done during the design stage—before the component ever hits the production mode.

The most dramatic example of just how much money simultaneous engineering can save crops up in the retooling budget. If the part, designed using conventional methods, fails during production, massive retooling might be needed to ready the redesigned part for production. Simultaneous engineering, by virtue of the fact that it has simulated the production assembly, anticipates, recognizes and "fixes" part and tooling problems before a single piece of metal is cast to start the actual production process.

As Good as the Ingredients

Yet even the most heralded of computer applications has met up with its share of drawbacks, and simultaneous engineering is no exception.

Oftentimes, explains Craig, the process can spit out inaccurate initial data. "You are working with a good engineering guess, especially with a new design or material," he comments. And with the influx of new composite materials the engineer is basically guessing on the component's properties. So, there are lots of hypotheses built into the initial design and manufacturing model.

The program is also time-consuming. It takes several weeks to several months to generate correct combinations of design and manufacturing. "But once the optimal design is in hand," stresses Craig, "the permutations and iterations on that design are as just as simple as changing a few variables."

If the future for simultaneous engineering is as promising as Ford, Budd (see sidebar) and Mark Craig say it is, then soon we should be seeing more and more software companies joining the bandwagon – simultaneously.

Into Statistics

"I think the next step is for the simultaneous engineering programs to supply a graphic front-end that ties to a CAD system," reasons Craig. The direct interface to CAD would allow engineers to take the process one step further – to design and modify directly with the CAD system. In essence, explains Craig, the ultimate system evolution is heading in the direction of a full CAD/CAM, design and manufacturing system – all in one.

Another hot area that could see a lot of benefit from simultaneous engineering is SQC – Statistical Quality Control.

Since simultaneous engineering and VSA can predict problems before they occur, it can also point out where to analyze the SQC data, and even read that data directly into the simulation model. And that, says Craig, would be the final tie between production, design, and manufacturing. AI

Below: Simultaneous engineering's ultimate evolution may be headed for a full CAD/CAM system.

A Study In Simulation

The Ford Motor Company and The Budd Company have teamed up to put simultaneous engineering into practice – with results both companies have termed excellent.

The two companies used simultaneous engineering and Applied Computer Solution's Variation Simulation Analysis (VSA) program to simulate the assembly of a door for Ford's Aerostar minivan. The original design effort was spearheaded at Ford – all design information originated on its in-house CAD system, and was then transferred to the Budd CAD system.

There, a three-dimensional VSA model "created" the entire Budd (Philadelphia stamping plant) assembly process – all the way down to the location of the control pins and stops.

After the door assembly was developed, the computer ran a simulation of a large production run by randomly selecting a point from each statistical distribution in the model for every simulated assembly.

The result of this "dry run" was both informative and productive. The two companies found that the hinge reinforcement for the door, as designed, was a major contributor to assembly variation. The VSA program pinpointed for Ford and Budd that a major source of assembly variation was coming from the door's newly-designed hinge reinforcement. Using this information as a starting point, Ford and Budd used the VSA software as an "electronic assembly line," to re-design the hinge, as well as the entire assembly process.

The final result was that new features were added to the vehicle, resulting in component addition and redesign. And this all happened before a single tool was stamped or a single line fault occurred – they had all been studied on the computer's "assembly line." – *CC*

John R. Coleman
Editor

Reprinted from *Assembly Engineering*, July, 1988
By permission of the Publisher copyright 1988.

Design for Assembly Users Speak Out

The moment of truth for many a design arrives when it's evaluated against assembly requirements. Concern about manufacturability belongs in the early stages of the design process, because the earlier a concept is refined and strengthened, the less costly it is to produce.

U.S. corporations are rapidly arriving at that same conclusion and are rushing to train their engineers in the methodology. General Electric, for example, has sent over 5000 engineers through design for assembly (DFA) programs and Xerox has trained over 1500.

The U.S. leader in DFA research is Boothroyd Dewhurst Inc., a company formed by two professors now residing at the University of Rhode Island. *Assembly Engineering* invited several DFA users, along with Dr. Geoffrey Boothroyd, to discuss their experiences. Here's what they told us.

Is design finally being recognized as the primary driver of total product cost?

Morrill: Yes, but not by all levels of management. It takes constant selling to make them realize what effect it has.

Branan: The importance of design and its role in containing product cost are now being realized. The computer-based tools and methods that optimize designs and help predict

WILLIAM A. MORRILL
Manager, Design for Manufacturability
Digital Equipment Corp.
Maynard, MA

Designers must realize that if we, as a company, as a country, want to survive, we need a strong manufacturing base.

helps ensure that designs are producible and cost effective.

Huthwaite: My firm is in the business of helping companies change how they develop products. During the past 12 months I have noticed a definite interest in DFA. Unfortunately, it's not being implemented properly in many firms. But, at least they are recognizing that if you don't get it right at the design stage you never will. You can't CAD it in, CIM it in, or SPC it in; you must design it in.

Sprague: I don't believe U.S. management really understands how to focus and completely manage design. They recognize the issue, but persist in emphasizing schedules and product performance, not overall cost. To do so requires a fundamental change in management culture.

Huthwaite: I agree, the challenge is cultural change, not technical change. By the way, we don't call it design for assembly or design for manufacturability, we call it design for competitiveness. That's what managers are interested in—the business of making money, competing.

Poletto: I sense a definite move toward understanding that design is a primary driver of product cost. We found that the difference between what is spent in the product design phase and what that leverages against down the road is significant.

Our focus is on quality/cost/delivery targets. Quality is the first goal, then cost and delivery, which is a tremendous turn-around from as recently as 10 years ago when delivery overshadowed the other two.

One more thing—all these acronyms confuse the real issue. Well defined, standard terminology is required so we all understand what we say.

Sprague: There is definitely a problem with words. Our corporate management formally stated that DFM is the strategy we are following. Just last year this definition was developed to minimize the confusion about what it means: "Design for manufacturability is a methodology that early in the design process addresses all those things that can impact production and customer satisfaction." Customer satisfaction says it's more than merely meeting assembly cycle time targets and reducing labor. It's the whole ball game.

Branan: The Motorola program is called Total Customer Satisfaction. Management must believe that de-

BILL BRANAN
Manager, Manufacturing Technology
Motorola Inc.
Ft. Lauderdale, FL

Most design engineers graduate with no training in statistics or basic design for experiments, which are fundamental skills necessary to even define what manufacturability is.

their effect on product manufacturability are mainly responsible.

Boothroyd: There are many ways of describing the methodology of evaluating product design alternatives in terms of manufacturability: simultaneous engineering, concurrent engineering, design for production, design for manufacture (DFM), and design for assembly are just a few. It doesn't matter what it's called, the objective is recognition that manufacturing costs are set at the design stage.

Hawiszczak: TI recently issued a design policy statement, which is a reaffirmation of its commitment to design for production as a key focus of the design process. Our management has essentially endorsed specific methodologies, including standard procedures for producibility and design to cost.

Design to cost takes into account that a customer is only willing to pay a certain amount for a product, thus it must be designed within a certain cost range. Cost targets are defined for each subassembly or major assembly. Cost targets then are similar to performance and reliability targets because they become parameters for judging the design effort. Moreover, it

signing for manufacturability—designing for assembly—is really an important way for achieving corporate objectives. In other words, DFM is a means to better satisfy the customer.

How was design for assembly introduced in your company?

Sprague: Top management saw its potential early on, as did engineers at the very bottom. After being sold to these groups, DFA squeezed through to the middle managers. You must get top-management endorsement because they control the purse strings.

Poletto: At Xerox we grew the DFA process from middle management, both up and down simultaneously. It's been successful for the last seven years. Still, I agree that if you don't get upper management's blessing, it won't continue.

Hawiszczak: When we started out, we tried going from the ground up, which is not very successful in the long term. When we eventually got design policy statements issued by upper management, we began to concentrate more on design for production at TI. It's made life significantly easier. Now the battle is being fought in the midlevels, where the

WILLIAM R. SPRAGUE
Principal Engineer
NCR Corp.
Cambridge, OH

Through teamwork and using DFA tools we developed a new terminal that cut part count and assembly time by 80%, and the number of suppliers by 70%.

crunch for satisfying design schedules, product performance, and design for production all meet.

Sprague: We've taken a pilot project approach. NCR has roughly 15 different manufacturing plants. The first complete DFM project was assigned to ours. We brought the right people together at the start. A DFA toolkit allowed quantification of what they were doing before there was anything on the CAD tube. They had basic definitions, knew how many parts there were going to be, and had an estimate of assembly time. Solids modeling relieved the design workload by providing visualization and electronic data transfer, and reducing documentation time. Through teamwork and using DFA tools we developed a new terminal that cut part count and assembly time by 80%, and the number of suppliers by 70%.

It's curious that we had the tools for a year, with no activity to speak of. It wasn't until the team was identified, their project solidified, and goals were set that anything happened.

What provided the motivation for adopting DFA?

Poletto: Economic pressure initially drove everyone to look for high-tech solutions to sliding productivity. We all felt there was some secret automation pill to cure our problems. For example, robots were just one, sometimes bitter, pill that was swallowed too fast.

Now there's a more conservative approach to investigating solutions. This has led to simultaneous engineering—teamwork for better design and manufacturing. It is what should have been done in the first place.

Branan: Competitive pressures have certainly accelerated the rate of change. The new tools to predict the effects of design on product manufacturability are much more prevalent, easier to use, and less expensive. And designers are accepting them more readily.

Hawiszczak: We've placed great emphasis on ensuring that we correctly use producibility tools such as DFA, because once a design is released, it's very difficult to change.

We also had to move toward CAE, especially for product producibility. One reason was a shortage of enough true manufacturing experts. There are few people in any company who really understand what goes on in manufacturing.

Because of this, we defined the rules, creating databases and other computer-based aids to extend the effectiveness of these experts. Our designers now have the benefit of accessing manufacturing knowledge right at the design workstation.

Of course, we had to deal with the issue of tightening design schedules. With the new producibility tools we realized as much as an eight-fold improvement in design cycle time compared to manual methods.

Have there been any pitfalls?

Hawiszczak: We've found that unless you understand and properly structure the design process, you won't be successful with the new automated tools, whether they're DFA, CAD, whatever. It's useless to automate chaos.

Morrill: I've seen strong support for the new DFA tools in upper management. Unfortunately, too many times they are installed without providing proper training. Then the tools are ineffective because the people who really need to use them, don't.

Huthwaite: A manufacturing executive I know says you may be able to get 20% from the software, and another 20% from the hardware. But, you must get 60% from the peopleware. The lesson is that you better make sure your people know what they are doing.

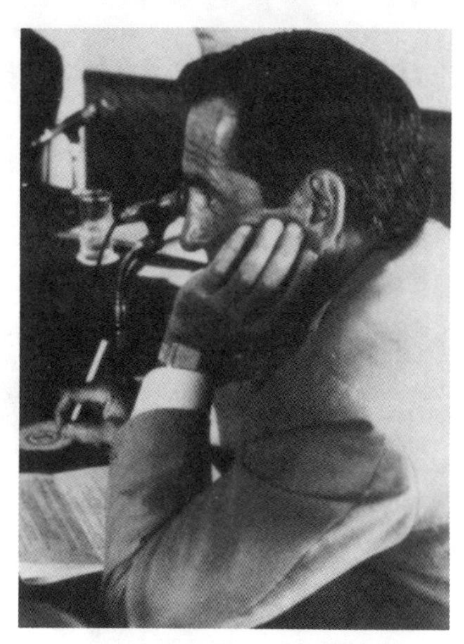

BART HUTHWAITE
President
Troy Engineering
Rochester, MI

What really is needed is a system that allows making trade-offs and converts design features to dollars and cents.

Poletto: The tools are easy to buy; it just means writing a check. But sometimes it takes up to six months to bring people up to speed. And by the time they are, the technology has changed. We are constantly retraining people, which is discouraging.

What about training?

Morrill: Design engineers typically don't understand manufacturing. They haven't worked in the factory, especially the younger engineers. That must change. Designers must realize that if we, as a company, as a country, want to survive, we need a strong manufacturing base.

Branan: Most design engineers graduate with no training in statistics or basic design for experiments, which are fundamental skills necessary to even define what manufacturability is.

Hawiszczak: New graduates come to us with not much of a design for production background, but they do come with an open mind. Our newest engineers are the easiest to train in design for production techniques.

DESIGN FOR FLEXIBLE ASSEMBLY

In 1984, Hewlett Packard's Vancouver Div., was making 3000 printers per month. Labor content was a small part of overall product cost and volume was low, so automation didn't make sense. All that changed with the introduction of the DeskJet product line.

"Because the markets that we were interested in were characterized by high volume and quality," reflects Brad Freeman, manufacturing engineering section manager, "we needed to rethink our manufacturing strategy for the new Deskjet printer. During feasibility studies, laboratory design and development of the Deskjet, manufacturing engineering was involved with the product designers to incorporate manufacturability into the design.

"We wanted a product that was designed to allow automatic assembly. Surface mount and conventional through-hole processes are well defined and highly automated—the DeskJet has 245 components of which 175 are surface mounted, 47 conventional through-hole automation and 23 that are hand inserted. Mechanical assembly, however, isn't as well defined. Therefore we adapted techniques for evaluating parts assembly such as the Boothroyd and Dewhurst system, IBM Design for Manufacturability, and the GE/Hitachi rating system—all can provide feedback to design engineers."

Freeman points out that the approach analyzes the relative assembly difficulty as it relates to product design. The analysis produces parameters that characterize "assemblability," assembly time, number of parts, number of difficult to automate assemblies, and so forth. The analysis method consists of defining motions and operations necessary to assemble each part, then assigning penalty points for anything other than a simple downward motion. This motion isn't assessed penalty points because it is assumed to be the fastest, easiest assembly operation for a human or machine to perform.

H-P modified several of the assumptions and penalty points to reflect some of their own philosophy. For example, original assumptions heavily penalized the use of screws, but H-P has successfully used automatic screw fastening technology for some time, thus they decreased the penalty. They did, however, increase it for using cables and E-rings because they are difficult for operators to handle.

Here are the design for manufacturability guidelines that H-P's manufacturing engineers stressed with the designers:

- Develop a modular subassembly design. It should be possible to build subassemblies independently of the remainder of the instrument.
- Use the largest component as an assembly base. Ideally, it will have a low center of gravity to help maintain orientation and stability and allow for precise, secure positioning on the pallet used to transport it from one operation to the next. Ideally, the base will act as the pallet.
- Minimize assembly levels.
- Design for one-dimensional assembly (build from the bottom up).
- Provide self-locating features in the components.
- Avoid designs calling for two-handed assembly.
- Choose efficient joining methods. Screws are okay; however, standardize on one size and length and design for downward insertion.
- Taper all parts to be inserted and chamfer the holes into which the parts are inserted.
- Facilitate parts handling. Avoid the need for orientation. Make parts symmetrical or very asymmetrical. Avoid tangling and nesting problems in part design.
- Eliminate wiring harnesses wherever possible.
- Avoid electrical and mechanical adjustments.
- Minimize the number of parts as well as the different parts used in the product. Combining many parts and functions is a basic tenet of any design for manufacturability strategy. H-P learned that the increase of functions per part past a certain point doesn't always result in lower absolute

Also, giving them first hand experience with DFA tools makes them better producibility engineers. By their second design they've gained knowledge they didn't have on their first. And on their third, they're even better.

Poletto: New graduates do not have the required experience. For us they must first work six months in manufacturing, three months in field service, and six months in engineering before reporting to their final department manager. This gives them a broad understanding of how the corporation works.

Branan: The single largest training effort Motorola has done to date is a DFM course required of all designers, engineers, and technical people. It covers the basics of SPC, how to estimate losses caused by poor quality and mismatched product and process, how to eliminate nonvalue added operations, and so forth. For many people, this is a first exposure to any of these subjects, even though they have designed or made products for years.

Morrill: We are setting up a design institute to facilitate DFA education. We also are taking engineers to boot camps so they can experience manufacturing first hand. They actually must put together the products they design, which is a real eye-opener.

Hawiszczak: We've developed a Design for Manufacture training series, which will have modules on subjects such as DFA, metal fabrication, casting, geometric dimensioning and tolerancing, etc. In effect, it tries to compensate for what engineers didn't learn in college about design for production.

Are design rules a reasonable substitute for DFA?

Boothroyd: The rule-based approach, i.e., a series of design rules that the engineer follows, seems to be favored in Europe. A typical rule, such as always design parts that are as symmetrical as possible, won't necessarily save anything in manufacturing costs. You simply must quantify these costs at the design stage.

Morrill: I agree. With the rules approach, points are assigned for following each one. Say, you come up with a total of 75. What does it mean? What's needed is a quantitative measure that can predict what happens when a variable changes and how it effects manufacturability. If there are design alternatives they must be quantifiable.

Branan: If you write down a bunch of rules and don't provide a quantitative method for predicting what the results of those rules will be, the designer will say, "I followed them to the best that the situation permitted."

Huthwaite: What really is needed is a system that allows making trade-offs and converts design features to dollars and cents. You can't identify trade-offs using design rules.

Boothroyd: One more thing about rule-based methods. At first we looked at DFM as an entire subject; however, at the time industry was more interested in DFA. We felt that the basic rules would be these: design a product so it is as easy to assemble as possible, then design the components so they are as easy

cost because of manufacturing and assembly constraints.

- Design so parts cannot be assembled wrong. If this isn't possible, then subsequent parts should be designed so they can't be put together if a previous part is misassembled.

"As it turns out," Freeman remarks, "we were fairly successful. The product is almost self fixturing; only three fixtures are used and no adjustments are required. Part count is low, and almost all are assembled in a top down fashion and are self locating. Many of the parts cannot be assembled wrong or subsequent parts cannot be attached if the previous part is misassembled.

"As is often the case, staying on schedule became one of our major objectives, so second iteration designs for improved manufacturability weren't feasible. Thus, there are operations such as pin insertion, motor and tubing installation that require the printer to be picked up and turned over, but the process design helps minimize the negative impact."

During development of the DeskJet, H-P learned that total top down assembly can have drawbacks when considering areas such as rework and service. It also was learned that designing snap details isn't straightforward—if it can snap in, it can snap out easily under shock loading. Consistently locking in all the snaps is also a problem.

Further, parts tangling was overlooked and should be considered for all components, not just springs. Elements such as belts, cables, flex circuits are difficult to assemble and even harder to design out.

At the center of the DeskJet assembly line is a Bosch modular transfer system. The conveyor system uses double belts in which workpiece carriers are transported from station to station on two continuously moving belts. The workpiece carriers aren't fixed to the belts, but merely rest on them. At each station the carrier is halted by stop gates while the belt continues to slide underneath it. In the case of automatic operations where precise location is necessary, the carrier is lifted slightly off the belts and accurately located with the use of lift modules and tooling pins. After a particular working step is completed, the stop gate opens and the part is released for transportation to the the next station's buffer. This system allows for working steps and processes to begin as manual operations, and then be automated after everything is well understood. The assembly system permits the flexibility of changing production capacity as necessary by easily adding or subtracting stations. H-P has changed line cycle time several times (from 60 sec. to 3 min.) as product forecasts have changed.

Examples of automation that were added to the Deskjet assembly line by virtue of the design for manufacturability focus and the Bosch modular transfer system include screw fastening automation, final test and auto-connect, laser engraving and serial number management, and on-line parts present and quality checks. Future applications such as automatic gear placement also are planned.

to make as possible. These sound like two good rules, but they are in conflict. Simplifying a product means fewer parts, but those parts may be very complicated and cost more.

What you really want is to design for simple product structure and then exploit the capabilities of the manufacturing processes. For instance, using injection molding to make parts that don't use separate fasteners.

Branan: Our experience so far suggests that designing something so it's easy to assemble does not violate any DFM tenet. In fact, in every case we've looked at, the material costs improved, as did every other variable.

What effect has design for assembly had on company communications?

Poletto: At Xerox, we use the Product Delivery Process, which is a DFM methodology that defines the design process and the tools to determine product manufacturability. We've found that it forces early involvement by team members, including our supplier base. We bring vendors in before we even put lines on a screen. They may even help us with the design.

Branan: I agree with bringing in suppliers early. It's an important part of the design process. DFA is only one area of manufacturing. What we're really trying to get to is a more predictive mode in general.

Hawiszczak: It's important not to confuse design reviews or drawing sign off with producibility or design for production analysis. The design reviews and drawing sign-offs should simply be a verification of all the work done as part of the development effort, from the day it was a glimmer in someone's eye to the day the drawing documentation is released.

Branan: I agree. When you do a drawing sign-off you're only looking at a small piece of the puzzle. That's really the wrong time to start raising questions that should have been dealt with at a more macro view.

Sprague: With the DFA methodology, we found people in engineering and manufacturing now know the person responsible for the specific area that they're interested in. DFA teams have gone a long way toward changing the organization, culturally as well as structurally.

Huthwaite: DFA is perceived as an early estimating tool by most companies. But when implemented, it becomes a learning tool and, more importantly, a concensus builder.

Boothroyd: DFA can easily turn people off, however, and inhibit communications if you're not careful. There is a great temptation for a company to say, "Here's one of our products. Analyze it and give us a presentation." Without a doubt, that's the fastest way to turn off designers. They must not feel that their designs are being criticized.

American tradition touts individualism and isolation, especially in the workplace. Can a true DFA team be achieved in tomorrow's factories?

Poletto: We all like baseball teams, yet everyone looks at each batter's average. When a time of crisis comes about, however, Americans work well as a team—but only when we all agree on the goal.

ANTHONY POLETTO
Automation Institute Manager
Xerox Corp.
Webster, NY

A few years ago we improved a design and actually cancelled a robotics application.

Branan: I can point to many projects that we've done in the last two years that wouldn't have been nearly as successful had we not been able to integrate all the different disciplines—manufacturing, marketing, industrial design and purchasing—into one team that was well focused on getting a specific product out the door.

Sprague: It's funny, but designers not involved in DFM teams quickly get frustrated because they are not able to be as successful as they want to be. They are simply tired of seeing design schedules compressed so they can't produce the expected results. Such designers will buck the system by doing DFM whether management wants it or not.

Huthwaite: I've dealt with hundreds of different firms and teams, and no two are alike. There typically is a core team; sometimes it's as small as four, or as large as a dozen. The players include all those people who will be responsible for manufacturing quality and costing, and all the various things that would normally have come later in the process.

Boothroyd: One DFA implementer told me that he likes to have a wild card on the team; someone who has nothing at all to do with the project, but can criticize the others' actions.

Branan: Computers are playing an increasingly important role in DFA, but a computer terminal is no place for a team. We want to make the designer a more broad-based expert very early so he can come up with the right concept. Then perhaps the DFM team can take over and do the fine tuning.

Where does robotic assembly fit in?

Boothroyd: During the first DFA development efforts 10 years ago, I was working on design for manual assembly and design for high-speed automatic assembly. At that time, though, industry only wanted to know about design for robot assembly. They were buying robots and didn't know what to do with them.

When we got around to doing that work, interest had waned. Everyone realized automation was not a panacea; it's the designing of the product in the first place that's the answer. Unless you do that, no amount of automation will be fully successful.

Branan: Automatic and robotic assembly systems are becoming more flexible and cost effective. They do more for the same dollar, which makes them economically applicable to a wider product range.

Boothroyd: Doesn't it make automation more difficult to justify when you've done a good job of designing the product?

ROBERT S. HAWISZCZAK
Manager, Producibility Engineering
Texas Instruments
Dallas, TX

Now the battle is being fought in the midlevels, where the crunch for satisfying design schedules, product performance, and design for production all meet.

GEOFFREY BOOTHROYD
Boothroyd Dewhurst Inc.
Wakefield, RI

It doesn't matter what it's called, the objective is recognition that manufacturing costs are set at the design stage.

Sprague: Not necessarily. You don't have as much meat to go after, but it's easy to apply something to what's there. You can better define the application, and the automation tool you're going to use. And you have much better results than the old "throw a robot out on the floor and make it do something" approach.

Poletto: A few years ago we improved a design and actually cancelled a robotics application.

Branan: What you've mentioned has been the knee-jerk response of most everybody. The reaction is "We've gotten it simple now so we don't really need robots." I believe what you're going to see is an opportunity to use the unique characteristics of these automatic systems to do operations that people can't, or won't.

Boothroyd: When I initially looked at product design for robot assembly, I asked user companies to submit a candidate product, one that wasn't currently being assembled by a robot.

The robots were used as any other piece of automation. Their reprogrammability simply reduced the develop-

CONCEPTUAL MODELING FOR AUTOMATIC ASSEMBLY

Critikon, a Johnson & Johnson Co., Tampa, FL, is improving both its concept-to-drawing cycle and automatic assembly process by using advanced modeling tools during initial design. The firm makes intravenous catheters, noninvasive automatic blood pressure monitors, and fluid delivery systems. Many of the parts are made by outside suppliers.

Richard Bloom, engineering services manager, notes: "We wanted a tool that would help visualize and create designs that are well thought-out from functional, appearance, and assembly standpoints. It also had to complement our existing system."

Pro/Engineer, from Parametric Technology Corp., Waltham, MA, met those requirements. It is intended to be used to automate design and interface with the CAD/CAM applications already used by an organization. What makes it suited to the problems of designing assemblies is that it is a parametric, feature-based solid-modeling system. The modeler captures functional relationships among assemblies, parts, features, and parameters. Interactive changes to these relationships propagate throughout an entire assembly. Users then can iteratively test and improve assembly models to optimize a final design.

"A major function is to support the design of assemblies, and then to consider parts as details of those assemblies," says Dr. Samuel Geisberg, Pro/Engineer developer. "The system allows designing the functional relationships of an assembly, then carrying those relationships on to a detailed design of individual parts."

Since receiving Pro/Engineer last January, Critikon has used it on two design projects. One is a new catheter that must be automatically assembled. Each of eight parts must be capable of easy orientation. "If you have a box full of parts," explains Bloom, "and you dump them into a vibratory feed hopper, they must be oriented so a mechanical arm can pick them up and place them in a nest for assembly."

The ability to automatically calculate a part's center of gravity with Pro/Engineer is used to evaluate handling alternatives. For example, one option is to have a part's center of gravity near one edge to force it through a gap in the hopper. In attempting to use that feature, Critikon engineers determined that the center of gravity was too near one part's middle, so they looked for another way of allowing it to orient.

The system's parametric capabilities allowed evaluation of many alternatives in less time than one could have been evaluated on a traditional system. In that way, the engineers were able to optimize the design early in the design-to-manufacturing cycle.

The visualization capabilities afforded by solid modeling also have helped design for assembly in two respects. First, by generating a shaded model from an initial design, engineering is able to communicate a strong concept to other people in the organization. That way, assembly experts in manufacturing can get a good idea of what features will be required for handling, before having parts in their hands. Second, the command to explode an assembly makes it easier for engineers to see potential problems before too much is invested in a design.

Critikon engineers plan to do at least some portion of virtually every new product design using the solid-modeling system. Bloom expects that they will do a significant amount of fixture design, as well. They generally complete the engineering portion of the design in Pro/Engineer. Only when they are confident of their direction, and the concept is relatively final, do they go over to drafting.

"One of the most important issues in choosing a solid modeler was that we had to be able to transfer files between it and our wireframe drafting package," says Bloom.

The company's wireframe CAD system is CAD3D (Grayteck, Chicago, IL) running on seven Apollo DN 3000 workstations. While using this software for over two years, the company had built up a sizeable database of drawings in the CAD3D format. To preserve this investment, it searched for a modeling package that could pass an IGES (Initial Graphics Exchange Specification) drawing file to CAD3D.

The data exchange between Pro/Engineer and CAD3D is accomplished through IGES version 2.0 and 3.0, which transfer drawings as flat images, as well as the entire 3D wireframe model.

ment time needed by the factory and more easily accommodated design revisions. The robots eventually were dedicated to one product, which is how the Japanese use them.

Morrill: The difference between the Japanese and us is that we tend to buy the higher price spread—robots that walk, talk, whistle, see, and feel—when they are not always required. The reason U.S. companies buy them is because they didn't design the product properly.

The Japanese are better at product design, so they can use a lower cost robot that they don't mind dedicating for simple tasks over the life of a product.

Poletto: More appendages are being added to robots to compensate for a lack of good design. For example, vision systems, even though they are inexpensive, require that you look to see if your shoe laces are tied. If you tied them to begin with you shouldn't have to look.

What about manual assembly?

Boothroyd: It's time we recognize that the human is the most versatile assembly worker, and always will be. An important issue is how you can help people perform assembly work.

I have seen examples where the experiences gained from developing automatic assembly machinery were applied to developing efficient manual workstations—workstations with significant assistance for the operator. If you combine that with proper product design, you often have a real winner.

One company came to me saying that its main competitors, both of them German, were installing automatic assembly equipment. The chief engineer asked if they should be doing the same thing. I told him to look at the product design and consider how efficiently it could be done manually. The effort resulted in a simple fixture and change in design.

Sprague: You're right in that most products can be redesigned to favor manual assembly. This approach can, in fact, tremendously reduce product cost while improving quality. The problem comes when you start really pushing the limits of manual assembly quality; take standard deviation of placement, for example. Still, maybe for the next five years simplified design for manual assembly will be a smart way to go for most products.

Presented at CASA/SME AUTOFACT '89 Conference, October 1989

Gaining Competitive Advantage by Using Simultaneous Engineering to Integrate Your Engineering, Design, and Manufacturing Resources

By John W. Foreman
RWD Technologies, Inc.

INTRODUCTION

The concept of Simultaneous Engineering is not entirely a new one. For many years, some companies have been taking advantage of the benefits of increased communication very early in the process of designing a new product and getting that product to market. This communication should be between all of the various functional groups within a company's organization involved in the new product and its related manufacturing process. The underlying message of the concept of Simultaneous Engineering is that it is extremely beneficial to design the processes that are required to produce a product at the same time that the product itself is being designed. Recent developments in both hardware and software are adding to the collection of tools available to increase the capabilities of many companies to utilize "paperless" Simultaneous Engineering systems. In these systems, all engineering work from design through production and inspection is carried electronically via interconnected workstations. In order to take full advantage of the available tools, it may require some changes in a company's organization and the way that people do their work.

Simultaneous Engineering is a blending of all different areas and disciplines within a company to improve the interface between design/engineering and manufacturing. This paper/tutorial reviews tools which can be used to resolve natural conflicts that arise between these organizations while designs are still on paper and not at the prototype or finished product stage.

The basic underlying philosophy of Simultaneous Engineering is to facilitate the concurrent design of a new product as well as the manufacturing process, tooling and gaging, and quality system for the production of the product. A key aspect of the Simultaneous Engineering concept involves the integration of a company's resources that perform these design activities very early in the design process to do the job right the first time.

BENEFITS

There are several benefits that a company can derive from using the Simultaneous Engineering approach:

- The process which will be used to produce the product can be optimized from the beginning, rather than being unduly restricted by product design requirements that do not take the process into account.
- The product can be designed to allow for the use of the proper levels and types of automated and integrated manufacturing and inspection techniques that are being developed.
- It is possible to consider the investment, marketing, and business plans early in the product design process; thus the design is not only manufacturable, it is cost-effective.
- The working environment promotes inter-disciplinary creativity. Typically, people of different types of expertise are located in close proximity.
- Managers have more complete information and a better control of inter-disciplinary activities. The manager's empire has expanded to include all activities related to a given product, rather than a particular function. This eliminates many of the political barriers that used to exist.

TOOLS OF SIMULTANEOUS ENGINEERING

To effectively implement Simultaneous Engineering, one has to address the tools that make this new corporate culture work. These tools fall into one of for major categories:

- Total Quality Control
- Computer Integrated Manufacturing
- Just-In-Time Productivity Improvement
- Human Systems

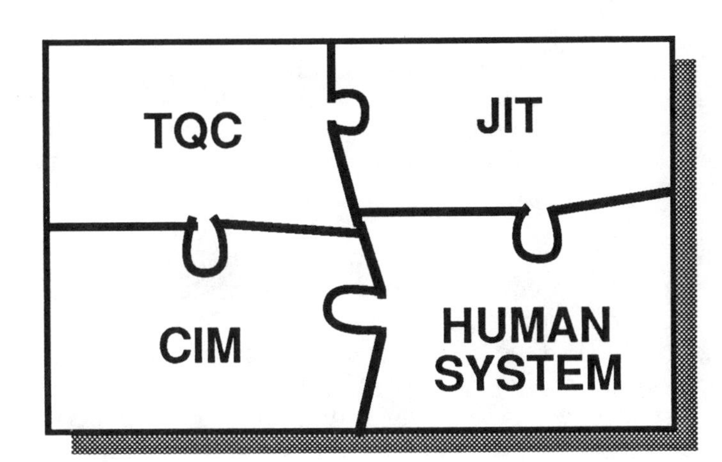

The effective implementation of some key tools in each of these areas provides a company with a competitive advantage by reducing engineering and manufacturing costs, shortening design and development time, reducing rework, and significantly reducing the number of engineering changes.

TOTAL QUALITY CONTROL

DESIGN STANDARDIZATION

There are two ways in which a design team can approach a design problem. The first and most common approach is to design the new part or product from scratch. The alternate approach is to begin with a set of existing parts and try to create the new product from these existing parts, modifying them if necessary, but creating new ones only as a last resort.

Design standardization enables engineers throughout a company to share design information more effectively. With design standardization, engineers can frequently find and reuse existing designs rather than redesigning the same part. Even when the existing designs are not exactly right, design standardization allows the engineer/designer to locate similar designs that can be used as a starting point. This can dramatically reduce the design effort.

The four components of a design standardization effort include a classification and coding system, a base of existing designs, a database linking these existing designs to the classification and coding system, and a user interface that allows the engineer/designer to ask questions to find similar designs.

TAGUCHI METHODS

Products have characteristics that describe their performance relative to customer requirements or expectations. The quality of a product is measured in terms of these characteristics. Quality is related to the loss to society caused by a product during its life cycle. A truly high quality product will have a minimal loss to society as it goes through this life cycle. The loss a customer sustains can take many forms, but it is generally a loss of product function or properties. Other losses are time pollution, noise, etc. If a product does not perform as expected, the customer senses some loss. After a product is shipped, a decision point is reached; it is the point at which the producer can do nothing more to the product. Before shipment the producer can use expensive or inexpensive materials, use an expensive or inexpensive process, etc.; but once shipped, the commitment is made for a certain product expense during the remainder of its life.

The Taguchi loss function recognizes the customer's desire to have products that are more consistent, part to part, and a producer's desire to make a low-cost product. The loss to society is composed of the costs incurred in the production process as well as the costs encountered during use by the customer (e.g. repair, lost business). To minimize the loss to society is the strategy that will encourage uniform products and reduce costs at the point of production and at the point of consumption.

The purpose of product or process development is to improve the performance characteristics of the product or process relative to customer needs and expectations. The purpose of experimentation should be to reduce and control variation of a product or process; subsequently, decisions must be made concerning which parameters affect the performance of a product or process. The loss function quantifies the need to understand which design factors influence the average and variation of a performance characteristic of a product or process. By using a Simultaneous Engineering approach to properly adjust the average and reduce variation, the product or process losses are minimized.

Since variation is a large part of the discussion relative to quality, analysis of variance (ANOVA) is the statistical method used to interpret experimental data and make the necessary decisions. ANOVA is not a complicated method and has a lot of mathematical basis associated with it. ANOVA is a statistically based decision tool for detecting any differences in average performance of groups of items tested. The decision, rather than using pure judgment, takes variation into account.

Engineers and designers are most often faced with one of two product or process development situations. One development situation is to find a parameter that will improve some performance characteristic to an acceptable or optimum value. A second situation is to find a less expensive, alternative design, material, or method which will provide equivalent performance. Depending on which situation the experimenter is facing, different strategies may be used. The first situation of needing to improve performance is the most typical situation for Simultaneous Engineering applications.

When searching for improved or equivalent designs, the Simultaneous Engineering team typically runs some test, observes some performance of the product or process, and makes a decision to use the new design or to reject the new design. It is the quality of this decision that can be improved upon when proper test strategies are utilized, in other words to avoid the mistake of using an inferior design or not using an acceptable design.

Not being aware of efficient, proper test strategies, experimenters resort to one of several approaches. The most common test plan is to evaluate the effect of one parameter on product performance. A typical progression of this approach, when the first parameter chosen doesn't work, is to evaluate the effect of several parameters on product performance one at a time. The most urgent and desperate of situations finds the experimenter usually evaluating the effects of several parameters on performance all at the same time. It is essential that a Simultaneous Engineering team does not use this approach.

The major steps that a design should use in designing, conducting, and analyzing an experiment are:

- Selection of factors and/or interactions to be evaluated
- Selection of number of levels for the factors
- Selection of the appropriate orthogonal array
- Assignment of factors and/or interactions to columns
- Conduct tests
- Analyze results
- Confirmation experiment

Probably the largest contribution to quality methodology that Taguchi has made is in the area of parameter and tolerance design. The designed experiments discussed above has existed since the 1930s. However, previous experimental approaches looked upon all factors as causes of variation. If these causes could be well controlled or eliminated, then product or process variation could be reduced, and therefore quality could be improved. But if a product is sensitive to ambient temperature variations, how can anyone control or eliminate temperature in a customer's environment? The answer is obvious; ambient temperature variations can neither be controlled nor eliminated without a large expense. Ambient temperature in a factory might be controlled very well to eliminate the temperature effect on a machine's performance, but the world's atmosphere cannot be controlled, so all kinds of devices are exposed to large temperature variations. Therefore a different approach is required if product quality is to be improved. This approach was entitled parameter design by Taguchi.

Parameter design is used to improve quality without controlling or eliminating causes of variation. Controlling or eliminating causes of variation may be expensive compared to a parameter design approach. Taguchi views the design of a product or process as a three-phase program:

- System design
- Parameter design
- Tolerance design

System design is the phase when new concepts, ideas, methods, etc., are generated to provide new or improved products to customers. One way to remain competitive in the world economy is to be a leader in utilizing technology. However, the technological advantage disappears quickly because it may be copied. If a competitor can fabricate the same new idea in a more uniform manner, then the technological advantage is more than lost. The parameter design phase is crucial to improving the uniformity of a product and can be done at no cost or even at a savings. This means certain parameters of a product or process design are set to make the performance less sensitive to causes of variation. The tolerance design phase improves quality at a minimal cost. Quality is improved by tightening tolerances on product or process parameters to reduce the performance variation. This is done only after parameter design.

Typically, when a problem is detected in product development, an engineer may jump directly to tolerance design; when tolerances are tightened, variation will be reduced and quality improved. However, tightening tolerances may be expensive and completely unnecessary if parameter design were used first. One serious mistake that a designer can make is to use expensive materials, components, or processes for a product when lower-cost items may be used if a parameter design approach is applied.

QUALITY FUNCTION DEPLOYMENT

Quality Function Deployment is a means of translating customer requirements into the appropriate technical requirements for each of the following stages of product development and production:

- Marketing Strategies, Sales
- Planning
- Product Design/ Engineering
- Prototype Evaluation
- Production Process Development
- Production

There are some distinct benefits of using the Quality Function Deployment methodology:

- Product objectives based on customer requirements are not misinterpreted at subsequent stages
- Particular marketing strategy or "sales points" do not become lost or blurred during the translation process from marketing through planning and execution
- Important production control points are not overlooked - everything necessary to achieve the desired outcome is understood and in place
- Tremendous efficiency because misinterpretation of program objectives, marketing strategy, and critical control points, and need for change is minimized

The Quality Function Deployment system provides a continuous translation and implementation flow from customer requirements plant operating instructions, i.e., a common purpose, priorities, and focus of attention or , in the words of W. Edwards Deming - " a clear operational definition".

PROCESS CONTROL PLANNING

A process control plan is a document describing those actions required at each phase of the process to assure all process output will be in a state of statistical control and in conformance with the customer requirements. A process control plan is the result of a series of structured activities conducted by a team of individuals associated with a particular product or process. After gathering preliminary data on the process, the team is able to logically construct a plan for achieving statistical control of the process.

The underlying basis of the process control planning process is the assembling of all pertinent information about the process in a single document which allows the analysis and improvement of the process. There are several prerequisites to constructing an effective process control plan:

- Management involvement
- Ability to identify resources
- Formation of department/ component teams
- Assembly of useful documents
- Developing a question log

The first activity the team conducts is the development of a process flow diagram. A structured format should be used to emphasize the impact of sources of variation on the process. The flow diagram helps to focus attention on the total process, rather than individual machines. It also ensures everyone is talking about the same process when designing the control plan.

After the process flow diagram is complete, the team lists all sources of variation potentially affecting each operation. The completion of this step is typical of most brainstorming sessions and is equivalent to developing a fishbone diagram for each operation. Next, the team should look at the outgoing dimensions/characteristics affected or changed at each step in the process.

Once the process flow diagram is complete, a control plan form is initiated for each operation. Every outgoing dimension/characteristic and its associated description is recorded. If gages are available for measuring process results, they are recorded on the control plan. After the gage inventory is complete, the team studies the measurement system variability. Gage repeatability and reproducibility is measured as the percentage of tolerance accounted for by the variation in the measurement taken. This technique is successful for identifying inappropriate measurement systems or those requiring modification or repair.

The next step of the team is the construction of a characteristic matrix which displays relationships between product requirements and operations. The greater the number of relationships, the more important control of a characteristic becomes. The matrix helps to evaluate the importance of characteristics while showing important upstream relationships.

COMPUTER INTEGRATED MANUFACTURING

SIMULATION AND ANALYSIS

This relatively new technique permits on-screen "testing" of a system before significant time, capital, and expense are incurred. The tool provides a means of determining whether or not the system will work and/or what is required to make it work efficiently. By using one of the commercially available simulation packages, the process and quality engineers can model their proposed systems to determine the effects of alternative product designs. Information about processing times and reject rates can be entered and used to evaluate these different alternatives and determine the optimum product and process design concurrently.

The latest applications of simulation include increasingly important design roles, not only in modeling and determining the correct design, but in generating the software controlling both material handling devices and other automated equipment in the system.

ELECTRONIC DATA INTERCHANGE

One of the biggest obstacles to effective Simultaneous Engineering programs has been the lack of a common graphics exchange standard. Much of the data generated on equipment supplied by one equipment vendor could not be entered directly into another vendor's equipment. Therefore, several vendors and federal agencies have developed a standardized format known as the Initial Graphics Exchange Specification (IGES) to at least allow dissimilar systems to "talk" to each other. In this approach, the transmitting system translates data into a second language - a so-called IGES neutral file - that can be sent to different systems; translators at the receiving end reformat the data into the appropriate form.

The initial geometric data-transfer capabilities of IGES have been extended to such engineering functions as finite-element modeling data. As this capability expands further, it will make it easier for engineers, designers, and manufacturing personnel to transmit data regardless of the individual pieces of equipment they are using in their own areas.

NETWORKS AND DATA COMMUNICATIONS

The new generation of workstations is providing direct links between areas such as design, analysis, manufacturing, and testing. Improved communications features allow data to be transferred rapidly; rates of 10 megabits per second are typical for the new 32-bit workstations connected in a local area network, while older communications systems were limited to the kilobit-per-second range. Therefore local area networks can quickly transmit huge data files to interconnected workstations, making the same information available almost instantaneously to a large number of users. This allows the rapid transmission of design information between product and process engineers, thus facilitating the Simultaneous Engineering approach.

GROUP TECHNOLOGY

Group Technology is an approach developed to improve the effectiveness of producing similar parts in small batches. A classification and coding scheme is used to identify parts that are manufactured using similar procedures and machine tools. These machine tools can be clustered together and all of the parts scheduled together. A Group Technology classification and coding scheme also allows easy retrieval of similar process plans, an essential step for variant computer-aided process planning.

A Group Technology classification and coding scheme can also be the basis for a design retrieval system. A Group Technology retrieval system can provide manufacturing with many benefits similar to design standardization. If designs can be reused as is, manufacturing documents such as process plans, shop and assembly drawings, NC programs, and machine instructions can also be reused rather than created from scratch. Similarly, whenever an existing design is modified, the corresponding manufacturing documents can also be modified. This saving also applies to any special tooling or fixtures needed for a design.

The benefits of a Group Technology scheme fall into two categories - one-time initial benefits and repeated benefits that occur whenever the parts are produced. The initial benefits of Group Technology are very similar to those of design standardization:

- Manufacturing engineering for new products is greatly reduced by reusing existing manufacturing documentation and tooling as is
- Development costs are lower
- Time to market for a new product is shortened
- Greater product variety with the same resources

The second type of benefit occurs repeatedly whenever the parts are produced. This savings comes from efficiently scheduling longer production runs by group rather than by individual part. Additional benefits occur if the company goes further and reorganizes its production facilities into work cells based on the Group Technology classification, but this is a more complex, long-term step.

VALUE ENGINEERING

The application of value engineering principles by design engineers will result in the development of goods and services which perform the required functions at the lowest possible cost. Because value engineering is function-oriented, it often increases the value of the product while lowering the cost of producing it. Maximum value is obtained when essential function is achieved for minimum cost.

Value engineering is defined as the systematic application of recognized techniques that:

- Identify the function of the product or service
- Establish a value for that function or service
- Provide the service at the lowest possible cost without diminishing performance or quality

A typical value engineering project involves the following five phases:

- Information
- Creative
- Evaluation
- Investigation
- Reporting

Value engineering is an excellent tool for integrating a company's design, engineering, and manufacturing resources. Value engineering techniques are tools to be applied by a multi-discipline team to each phase of a new product develop project. Some of these techniques are:

- Get all the facts
- Define the Function
- Work on Specifics
- Analyze Costs
- Use Your Own Judgment
- Create, then Refine
- Use Standards
- Use Specialty Products - Materials, Processes, Vendors
- Use Company and Industry Experts
- Think Creatively
- Put a Dollar Sign on Tolerances
- Overcome Roadblocks
- Use Effective Employee Relations
- Determine Engineering Value

In the final analysis, value engineering is an attitude - a desire on the part of the individual to want to design, to want to manufacture, and to want to purchase with value in mind.

SOLID MODELING

Many companies in the aerospace defense and automobile industries are using solid modeling in the design process. A solid model provides a complete geometric representation of a part or assembly. It includes more information about the part than either a two-dimensional drawing or a three-dimensional wireframe model. An engineer can design either simple piece parts or complex assemblies with a solid model. A solid model can be generated quickly by engineers without highly specialized computer training and presented to other members of a design team - designers, marketing people, manufacturing personnel - who can better understand the basic concepts included in the design.

Once the model is created, it can be used directly or indirectly in many other steps in the design and manufacturing process. Using it directly, the designer can determine whether the moving parts in a mechanism interfere with each other. Solid models of existing parts can be used to create the model for new products or assemblies. Complex assemblies can be automatically exploded to provide clearer assembly and maintenance drawings. When it is used directly, translation programs can take the data in a solid model and automatically generate much of the data needed by other applications.

The basic engineering drawing, including orthogonal views can also be generated from a solid model. All the information for dimensioning the draw are contained in the solid model, and some translators include automatic dimensioning capability. There are programs available today that take a solid model and create the meshes for a finite element analysis or generate the Numerical Control program to fabricate the part.

JUST-IN-TIME PRODUCTIVITY IMPROVEMENT

DESIGN FOR ASSEMBLY/DESIGN FOR MANUFACTURABILITY

Design for Assembly/Design for Manufacturability (DFA/DFM) is a process which optimizes the relationship between the materials, the technologies discussed, the manufacturing processes, and the costs in the design stage. It is a process to which technical and non-technical personnel can relate and contribute. DFA/DFM applies to virtually all fabricated components and mechanical assemblies including electrical and electronic components. The greatest opportunities for DFA/DFM exist in product designs with the focus on improved cost and quality performance and team building.

There are several important benefits to DFA/DFM beyond simple cost savings. These include:

- Fewer parts = Fewer failures + Better quality
- Minimization or elimination of production problems while still in the design phase
- Reduced maintenance costs
- Improved product function
- Improved design and production flexibility
- Increased customer satisfaction and value
- Improved company image

DFA/DFM as Simultaneous Engineering tools, break down the barriers of manufacturing stagnation. They force different engineering and manufacturing functions to examine each other's needs and, as a team, solve the overall problems of quality, function, production, and profit.

SYNCHRONOUS MANUFACTURING

Synchronous Manufacturing is a manufacturing management technique which focuses on the systematic acceleration of material flow through a production operation. It is also an analytical technique that provides a framework for the systematic application of quality and productivity programs. Synchronous Manufacturing will help an organization become more competitive by providing the following:

- A set of global measurements to evaluate the management of material flow
- Rules for understanding and classifying complex production operations and their resource interactions
- A systematic procedure for analyzing and managing material flow
- An environment and the initiative for the process of continuous improvement

In recent years, the focus on cost reduction and labor efficiency has contributed to a loss of competitive position in U.S. manufacturing companies by the following:

- New product introductions have been delayed
- Quality problems have been buried in high inventories thus raising the cost of managing quality problems
- Deliveries have been unreliable and the responsiveness to the market poor
- Expenses have run high due to overtime, inventory costs, and freight charges

The Synchronous Manufacturing approach to business strategies allows for new measurements of a business's success:

- Throughput - the money generated through sales, not through production
- Inventory - the money invested in purchasing materials intended to be sold (this figure contains only raw material and has no labor value added
- Operating Expense - the money spent in converting Inventory into Throughput

The use of Synchronous Manufacturing techniques in a Simultaneous Engineering environment directs the thought process and actions of all staff and operating groups to a common goal. It provides for the identification of critical resources and constraints.

Synchronous Manufacturing provides the following:

- Links between Manufacturing and Marketing strategies
- Material Management and Control System
- Improved Man/Machine effectiveness
- The ability to identify and prioritize process improvements such as automation, SPC, CIM, setup reduction, preventive maintenance, etc.
- Improved employee involvement

CONTINUOUS IMPROVEMENT PROGRAMS

Continuous Improvement is an on-going, systematic effort to improve day-to-day operations to remain competitive and sustain profitability. Continuous Improvement is process-oriented and people-oriented; all levels of employees constantly analyze the workplace to identify ways to improve operations. As teams or as individuals, these employees identify problems and causes of waste and develop solutions to improve the process for producing products or services.

Continuous Improvement is based on the following major principles:

- The customer must be satisfied.
- Everything can be improved.
- Every problem identified in the process is an opportunity to improve.
- An ongoing effort is needed in which everyone is allowed to help achieve the primary business goals of improved:
  - Quality
  - Cost
  - Delivery
- People want to be involved and do their job well.
- The person performing the job is the most knowledgeable about the job.
- A systematic approach to evaluating processes produces better results than an unsystematic approach.

Continuous Improvement has several advantages over traditional management approaches that rely primarily on innovation to change operations. Continuous Improvement:

- Enables everyone - designers, engineers, and manufacturing personnel - to contribute suggestions and ideas to improve a company's operations
- Continues to produce improvements in the process and the product or service provided
- Benefits all employees of the company's operations - management, production, sales, and support - thereby increasing customer satisfaction

HUMAN SYSTEMS

PROGRAM/PROJECT MANAGEMENT

Consider the role that project management plays in your organization. As stated earlier, you will probably find that well run projects are essential to the survival and growth of your organization. Whether or not a formal project management system is in place, technical environments are generally defined by a collection of "projects."

Good organizations view their "projects" as the building blocks in the design, development, and execution of their organizational strategies. Customer goals are satisfied by well constructed corporate goals. Corporate goals are translated into project goals. Properly managed projects produce products/deliverables which meet these goals, thereby satisfying the customer, which leads to business success.

Some companies (such as architect-engineering firms) are so completely immersed in a project environment that their corporate organizations become project organizations. The dependency here on good project management is obvious in this case, and reaction times are rapid. In many companies, however, the immediate impact of project management is far less obvious. In these cases, the potential for long term (and possibly irreversible) damage due to poor project management is strong.

The point is that all of the players have a lot at stake in achieving excellence in project management. When all goes well - the project team wins, which means the project manager wins, which means his corporate management wins, which means the customer wins, which means - more projects and the chance to do it all again.

During the performance of any technical project, tensions naturally occur between the quality, cost, and schedule requirements of the task. For example, a perfectionist may want to make a satisfactory technical report "a little bit better" and, unhindered, this individual might pour additional time and money into "polishing" the report while exceeding the project cost estimate and ensuring late delivery of the end product. Similarly, a "results oriented" manager might push a shoddy report out the door too soon, all for the sake of on-time delivery. Both situations clearly fall short of the objectives of delivering quality technical work, on time and within the cost estimate.

Experience has shown that the most effective way to resolve conflicting requirements is to concentrate technical, budget, and schedule responsibility in the Project Manager, thereby providing the freedom and organizational flexibility to ensure high quality performance. This approach is also highly desirable from the functional manager's point of view, because he can quickly obtain from one well-informed source the most up-to-date information available on the project.

The project management environment also provides the flexibility to assemble personnel with the required technical expertise to meet specific project requirements. When staffing projects, managers should be encouraged to look at the entire Division or Group's resources in assembling the best project team. By assembling a project team from all available resources and heading it with a Project Manager who has the necessary responsibility and authority, it is possible to attain the greatest flexibility and performance controls.

A project represents a definable scope of work. Describing the work desired before the project starts is a fundamental requirement for project management to exist. The essential need is for the project manager, the project team, and the sponsor (customer) to understand and agree to the definition of scope for the project.

Projects produce an end product. This may be a structure, item, device, or other form of hardware; or it could be a study, a plan, set of specifications, drawings, software, a report, or presentation of results. It is usually best to define the product of a project as a set of deliverables.

Resources are needed to perform the work of a project. Labor is always required. This may just be the time and effort required of the project manager and a few individuals, or it could involve large numbers of people within different parts of the organization and from subcontractors. Resources also include materials, equipment, and other items which will require expenditures to complete the project.

A very important attribute of a project is that it has a definite beginning and a definite end. This implies that the work to be performed can be scheduled and budgeted, and that resources can be assigned over specific periods of time.

COMMUNICATIONS

Proper communications are vital to the success of a Simultaneous Engineering project. Communications are the process by which information is exchanged. Communication is deceptively simple; it is the act of sending ideas and feelings in such a way that the receiver can recreate those ideas and feelings the way the sender expected them to be interpreted. If individuals cannot communicate effectively, they find it difficult to relate well to other people or to influence them from a leadership role. Effective team members are effective communicators.

There are some characteristics of the communication process that are important to understand in order to be able to effectively communicate. Communication is:

- Purposeful
- Transactional
- Symbolic
- Complex
- Personal
- Irreversible

The average person spends over 50% of his or her waking hours listening. Studies have shown that in American business, there is only 30% listening efficiency. Imagine if 2/3 of the instructions of the project manager to the project team were not retained - some pretty important things could fall through the cracks. Likewise, consider the impact of the team member failing to retain 2/3 of the important issues being communicated to him! Project team members need to improve their listening skills to become more effective.

Effective listening can do quite a few things for a member of a Simultaneous Engineering team:

- It can make him feel better about himself
- It can make him a better manager
- It can encourage project team and team member development
- It promotes cooperation
- It helps the project manager be more concerned and tuned in to people
- It assures the project manager of receiving more complete and accurate information

With a little knowledge and practice, a team member can as much as double his listening ability. The power of effective listening should not be underestimated. Great benefits can be obtained from some very basic and practical techniques.

We are normally so absorbed in what we are saying, or plan to say after the other person stops talking, that we fail to hear what is being said. We sometimes hear a few facts and assume we know the whole story. We tend to jump ahead of the speaker and assume we know what he or she is about to say. Many times we get turned off by something we don't like about the speaker, and we fail to get the message.

The communications process is more than simply conveying a message; it is also a source of control. Proper communications let the project team members in on the act because the team members need to know and understand. Communication must convey both information and motivation. The problem, then, is to how to communicate.
Nonverbal behavior communicates more than people often realize. Project team members send the "real message" to others through nonverbal means. Understanding the generally associated meanings of certain gestures can help to communicate more effectively.

Communication is an art. It is also one of the most formidable devices a project manager has to use in the process of completing a project successfully. Communication is implemented through both oral and other forms like "body language." However, to communicate effectively in a project environment, the team member must keep in mind the objective of the message, the timing, and most of all how to get the undivided attention of the person with whom he is communicating.

PROBLEM SOLVING

Data exists that shows that 80% of available manpower works on solving problems and 20% of available manpower works on making decisions or plans. Data also shows that most problems are solved more than once. Problem solving has been engrained in the American way of doing business. Perhaps this stems from our enthusiasm to "get things done", or to throw money and technology at a problem without really understanding it, or just to make a quick profit. The results of our problem solving orientation is that we spend 80% of our talent in working on issues related to the past. Which leaves us behind the curve in dealing with present or future issues. To make things worse, because of misunderstanding a problem, we usually solve it more than once. This means that we use three times the manpower to work on past oriented tasks as we do to work on present or future oriented tasks.

Obviously, there will always be problems that need to be identified and solved. However, when the need occurs the required manpower resources should not be 80% but rather 20% or less. This can happen, but it will take time and a powerful interim commitment. Two things are needed:

- The ability to identify the problems for what they really are
- To solve the problems only once

The savings from exemplary problem solving now, will enable a more productive and creative pool of manpower for the present and future oriented activities.

Problem solving is often ineffective in the manufacturing and engineering environment. What might cause this may be the lack of descriptive information regarding the problem. "It's not working!' does not give a thorough description. Likewise, a description such as "The defect is on the left side of the casting" is not a description of the problem. Sometimes problems are described by their symptoms:

- Manpower is down due to absenteeism
- Perishable tool cost is skyrocketing
- Quality rejections are in an upward trend
- Choke points in the line are holding up production speeds

A clear, thorough description of the problem is necessary. A problem must be adequately described and be narrow enough in scope for the team to handle effectively. To describe a problem, we can use the 5W, 3H questioning method:

- <u>What is the problem</u> is simply "What happened?"
- <u>Who is complaining</u> identifies what customer has identified a symptom or problem.
- <u>Where was/is the symptom or problem occurring</u>; where in the factory is it happening, and/or where on the product is it happening.
- <u>When did it happen</u> requires an answer relating to the what time, what shift, what season, etc.
- <u>How, how much, how many</u> are all used to assess the scope of the potential problem, little or large.

Use all available indications to tell what, where, when, why, who, how many, and how much. Include all symptoms in either technical or nontechnical terms. Use your own words, customers' words, engineers' words, etc.

There are several tasks to get from the problem as you know it to an operationally defined problem. The first step is to state the problem as you understand it. Next, assume you have several problems combined to look like one. Separate each concern and prioritize the way in which you will work on them. The third step is to collect information which requires some decision regarding what information is needed; is there a way to measure it; what can be done with the information after it is collected. Next is to describe what was expected from the product or process; what was gotten; the differences; and try to illustrate the differences.

Step five is for clarifying and recording. Be more specific and describe the defect and/or problem again. Answer the W&H questions. Write it down. For the analyze and compare step, continually compare "what is happening" to "what could or should be happening" but is not. For example, using W&H's - where on the part could the defect have been observed but was not.

The last step is to restate the problem. To do this, it is appropriate to use terms that are operationally defined. These terms and definitions come from the collected information, the described differences, the clarifying details, and should include a flow diagram. This last step becomes the problem documentation for the problem solving activities and for later reference as historical records.

CONCLUSION

There are many benefits that can be derived from implementing a Simultaneous Engineering to designing and producing products. Simultaneous Engineering is a blending of all different areas and disciplines within a company to improve the interface between design/engineering and manufacturing. There are a number of tools available which can be used to resolve natural conflicts that arise between these organizations while designs are still on paper and not at the prototype or finished product stage.

The basic underlying philosophy of Simultaneous Engineering is to facilitate the concurrent design of a new product as well as the manufacturing process, tooling and gaging, and quality system for the production of the product. A key aspect of the Simultaneous Engineering concept involves the integration of a company's resources that perform these design activities very early in the design process to do the job right the first time.

BIBLIOGRAPHY

Bray, Olin H., Computer Integrated Manufacturing, The Data Management Strategy, Digital Press, 1988

Duncan, Acheson J., Quality Control and Industrial Statistics, Richard D. Irwin Inc., 1965

Foreman, J., Taking Advantage of Advances in Simultaneous Engineering to Improve Your Product/Process Designs, Specifications, and Acceptance Criteria, September 1987, Engineering Society of Detroit Test Measurement & Inspection for Quality Control Symposium, Detroit, Michigan

Ross, Philip J. Ross, Taguchi Techniques for Quality Engineering, McGraw-Hill, 1988

The Spectrum Management Group Inc., Regaining Competitiveness Through Synchronous Manufacturing, 1987 -1988, New Haven, Connecticut

Reprinted from *Production Engineering*, June 1987

Partners for Productivity

By J. David Griffin and Charles F. Myers

Teamwork. The people who produce parts need to find common ground with the people who design parts. All too frequently design engineering and production engineering are treated as a linear process. Design engineers do their thing and pass it on to the production engineers who do their thing. This concept of linear flow is wrong; the functions need to be concurrent to be most effective.

But, in many companies the design engineering group and production engineering group have an adversarial relationship. This is unfortunate because these two groups working together hold the potential to make significant contributions to corporate productivity and profitability.

Consider a few examples where teamwork isn't used:

● A manufacturer of printed circuit boards uses a variety of integrated circuits (ICs) on boards. One particular IC, a dynamic RAM chip, is made by three different manufacturers, each with a unique part number for it. On the engineering drawings, one manufacturer's part number is shown on one drawing, another part number on a second drawing, and so on. In the production stockroom there are three boxes, one each for ICs from each vendor. Production tends to run out of the particular IC called for on a drawing. This causes production to stop until additional ICs are flown in.

A solution of course would be for engineering to assign one in-house part number for this IC. The in-house part number would appear on all prints and all inventory would be consolidated into one box. A list would be maintained of what vendors' part numbers corresponded to the in-house number. Purchasing could then reference this list when buying parts.

● Design engineering develops a new printed circuit board using surface mount components. Production, however, does not have the equipment to build this type of circuit board and cannot get the approval to buy the equipment necessary to do so. The design engineer is thus forced to redesign the board using different components. This could have been avoided if the design engineer knew what equipment manufacturing had.

● Build instructions for a company's color workstations call out options for these products low in the product structure. As a small company, it built few workstations; each was built from the ground up with the options included. As the company grew, volume increased until manufacturing had to set up an assembly line to turn out the volume required. Employees found, however, that there was no "standard" product that could be built on the assembly line. The way the options were called out on the build instructions required employees to continue to build one at a time. This resulted in high overhead and eventually made it difficult for the product to compete. A solution would be to restructure the product and redo all documentation and all build instructions. But, a little information from manufacturing at the beginning of product design could have avoided this altogether.

As you can see from these examples, it can be vitally important that design engineering understand the basics of production. However, the design engineer may not get this understanding from college. In many colleges the design engineer is not exposed to how a production operation works. Similarly, the production engineer may not be that familiar with the concepts of design engineering. It is therefore up to the company to train both groups to what the other's functions are all about. By doing so a company should be able to realize these benefits:

- Reduction in quantity of engineering change orders.
- Improvement in producibility of a product.
- Reduction in cost to produce a product.
- Reduction in time to get a new product on-line.
- Improvement in reliability of a product.
- Improvement in quality of life in a company.

These benefits can be achieved with minimal cost and effort. All that is needed is understanding, and to achieve that understanding, all that is needed is training.

How to achieve this training? There are are a number of steps which can be taken:

- Design an internal training course to teach engineering and manufacturing to consider questions such as:
    - What is a part number?
    - Who assigns new part numbers? How are they assigned?
    - How are alternate part numbers assigned?
    - What is a product structure?
    - What is the difference between make-to-stock and make-to-order?
    - When should a modular bill be used?
    - How should options be structured?

- Let the design engineer spend time on the production floor to become familiar with the equipment available. Let the production engineer spend time in the design engineering group to become familiar with the design process.

- Set up meetings between design engineering, production engineering, and marketing to review product structure before part numbers are assigned. These meetings should concentrate on:
    - How is the product to be sold?
    - How is the product to be forecasted?
    - How is the product to be manufactured?
    - How is the product to be tested?
    - Will the product have options?
    - Will the options be shipped separately or installed in-house?
    - How will the options be forecasted, stocked, and tested?
    - How will the product be packaged and shipped?

After discussing these items, an appropriate product structure can be developed which meets the needs of all three groups.

- Document all of the technology which is available within production. This would include types of equipment that are in use and the capability of each piece of equipment.

- Hold regular meetings between production, engineering, and quality to discuss new technology coming on the market. Consider when and if each new type of technology should be implemented.

- Let the design engineer work in the test department for a while, to become familiar with the test equipment available and the test procedures which are used. Let the quality people work in design engineering for a while to learn how products are designed and tested. In many cases, there are things which can be done before the product is released to production which will improve the testability and reliability of the product. One such thing might be using Monte-Carlo computer simulation to test printed circuit boards.

- Document all chemicals used in the production process. The way a printed circuit board is cleaned, for example, may affect how close traces can be layed out on the board. It could also cause a shorting effect on the board which could cause reliability problems.

- Take tours of other factories in the area. Visit low volume job shops. Visit high volume repetitive shops. Knowing the difference could make a difference in:
    - How the plant is laid out.
    - How the product should be structured.
    - Whether options are appropriate.

- Attend a lecture on MRP-II. Even companies that never intend to put computers in their operations will benefit from the discussions of the relationships between the various organizations within a production operation.

- Discuss the possibility of setting up a total quality control system (TQC) within the company. A TQC program deals with more than just product quality; it deals with perfecting the communication between all organizations within the company.

- Learn the difference between push manufacturing and pull manufacturing. This knowledge could significantly affect product structure and test procedures as well as process.

- Encourage top management to learn the relationships between engineering, marketing, and production. Not everyone in a company has the same goals all the time. It is up to top management to ensure that individual departments' goals are in line with overall corporate goals.

J. David Griffin is a Member of the Technical Staff of the Missle Systems Div., Rockwell International, Duluth, GA. Charles F. Myers is a Manufacturing Consultant with the Corporate Education Dept., ASK Computer Systems, Los Altos, CA.

Presented at SME Simultaneous Engineering Conference, June 1987

Early Designs for Manufacturing Quality

By John P. Hinckley, Jr.
Chrysler Motors

TODAY I'M GOING TO TALK ABOUT WHAT THE LIBERTY APPROACH MEANS IN TERMS OF MANUFACTURING QUALITY IMPROVEMENT THROUGH SIMULTANEOUS FIRST-TIME PRODUCT AND PROCESS DESIGN AND USE OF UPSTREAM QUALITY TOOLS SUCH AS QUALITY FUNCTION DEPLOYMENT, FAILURE MODE AND EFFECT ANALYSIS, DESIGN FOR ASSEMBLY, MODULAR DESIGN AND ASSEMBLY, ASSEMBLY VARIATION ANALYSIS, PROCESS SIMULATION, FEASIBILITY ASSESSMENT, AND EARLY SUPPLIER INVOLVEMENT.

BEFORE I GET INTO WHERE WE'RE GOING, LET'S TALK ABOUT WHERE WE'VE BEEN AND WHAT WE'RE DOING RIGHT NOW.

THE RECENT HISTORY OF CHRYSLER HAS BEEN WELL-DOCUMENTED, AND I'M SURE EVERYONE IN THIS ROOM IS AWARE OF OUR RAGS-TO-RICHES STORY. FROM THE DAYS OF THE LOAN GUARANTEES AND BILLION-DOLLAR LOSSES TO OUR RETURN TO PROFITABILITY, THE STORY OF CHRYSLER HAS BEEN ONE OF EQUAL SHARING OF SACRIFICES BY MANAGEMENT, EMPLOYEES, AND SUPPLIERS.

DURING THOSE GRIM DAYS WE BROUGHT OUT THE K-CARS WHICH HAVE PROVIDED THE BASIS FOR ALL OUR SUBSEQUENT FRONT-WHEEL-DRIVE PRODUCTS, AND AS TIMES IMPROVED WE BROUGHT BACK THE CONVERTIBLE. WE ARE THE ONLY DOMESTIC MANUFACTURER THAT BUILDS OUR OWN CONVERTIBLES IN-HOUSE, AND WE'VE JUST LAUNCHED OUR NEW LEBARON CONVERTIBLE, DESIGNED FROM THE BEGINNING AS A CONVERTIBLE RATHER THAN A CONVERSION.

IN 1984 WE INTRODUCED OUR PLYMOUTH VOYAGER AND DODGE CARAVAN MINI-VANS WHICH CREATED A WHOLE NEW MARKET SEGMENT, AND WE'VE JUST ADDED ANOTHER PLANT'S WORTH OF CAPACITY TO SATISFY CONSUMER DEMAND FOR THESE UNIQUE VEHICLES, NOW AVAILABLE WITH A V-6 ENGINE AND TWO WHEELBASES.

LATE IN 1985 WE INTRODUCED THE CHRYSLER LEBARON GTS AND DODGE LANCER SPORT SEDANS, BUILT AT OUR STATE-OF-THE-ART STERLING HEIGHTS ASSEMBLY PLANT, AND WE RECENTLY EXPANDED AND RE-TOOLED THE PLANT TO ADD THEIR SMALLER COUSINS, THE PLYMOUTH SUNDANCE AND DODGE SHADOW.

IN MID-1986 WE CONVERTED AND RE-TOOLED OUR WARREN TRUCK FACILITY TO CREATE OUR NEW HIGH-TECH DODGE CITY TRUCK COMPLEX TO INTRODUCE OUR

NEW DAKOTA MID-SIZE PICKUP LINE,

AND WE ARE RE-ENTERING THE LUXURY SPECIALTY MARKET WITH OUR ALL-NEW LEBARON COUPE AND CONVERTIBLE.

WE HAVE ALSO REDESIGNED THE DODGE DAYTONA FRONT-WHEEL-DRIVE SPORTS CAR,

AND THIS FALL WILL SEE THE INTRODUCTION OF OUR ALL-NEW FRONT-WHEEL-DRIVE LUXURY SEDANS BUILT AT OUR BELVIDERE, ILLINOIS ASSEMBLY PLANT,

AND OUR EXCITING NEW TURBO CONVERTIBLE BY MASERATI, WHICH WILL BE BUILT IN ITALY.

THE PRIDE _IS_ BACK - WE'VE ALL BEEN HARD AT WORK LAYING THE FOUNDATIONS FOR A TREMENDOUS NEW PRODUCT LINE, AND WE'RE MAKING MAJOR STEPS FORWARD IN QUALITY.

BUT NO MATTER HOW NEW AND EXCITING THESE PRODUCTS ARE, THE AUTOMOBILE INDUSTRY IS ENTERING AN ERA OF COMPETITION LIKE NOTHING WE'VE SEEN BEFORE, WHERE _QUALITY_ WILL BE THE PRICE OF ADMISSION.

WE'RE WELL AWARE OF THE IMPACT OF THE FOREIGN AND TRANSPLANT INVASION AND THE RAPIDLY RISING LEVEL OF QUALITY AMONG ALL AUTOMOTIVE PRODUCERS. TEN YEARS AGO THE AMERICAN MARKET WAS COVERED BY THE BIG THREE AND SOME FOREIGN PRODUCERS OF QUESTIONABLE QUALITY. SINCE THEN,

THE PICTURE HAS CHANGED DRASTICALLY. THERE ARE NOW TEN AUTO MANUFACTURERS IN THE UNITED STATES, AND THERE ARE 30 BRANDS OF VEHICLES AVAILABLE TO THE BUYING PUBLIC. THE COMPETITION IS FIERCE IN SPITE OF THE CHANGING VALUE OF THE YEN, AND THE NAME OF THE GAME IS QUALITY.

ACCORDING TO RECOGNIZED INDEPENDENT RESEARCH, THE TWO TOP REASONS FOR BUYING A CAR LAST YEAR WERE "DURABILITY AND RELIABILITY" AND "A WELL-MADE CAR". IN 1978, THESE TWO REASONS RANKED "NOWHERE" AND "TENTH" ON THE LIST, WAY BEHIND "STYLING" AND "PREVIOUS EXPERIENCE WITH MAKE". THESE LAST TWO REASONS ARE NOW EITHER WAY DOWN THE LIST OR HAVE JUST DISAPPEARED, INDICATING A MAJOR SHIFT. THAT SHIFT CLEARLY SHOWS THAT TODAY'S CUSTOMERS WANT QUALITY AND VALUE FOR THEIR MONEY, AND THEY'LL BUY FROM ANYBODY WHO MEETS THEIR NEEDS. THAT MEANS A WELL-DESIGNED CAR THAT'S PUT TOGETHER RIGHT AND STANDS UP UNDER USE SO IT DOESN'T FALL APART BY THE TIME IT'S PAID FOR.

THIS CHART INDICATES OUR PROGRESS IN REDUCING WARRANTY REPAIR CONDITIONS PER 100 VEHICLES BY 50% SINCE 1979 THROUGH IMPROVED QUALITY PERFORMANCE, SAVING OVER $600 MILLION IN WARRANTY EXPENSE DURING THAT PERIOD BY DOING MORE THINGS RIGHT THE FIRST TIME. THIS KIND OF QUALITY PERFORMANCE ALSO DRIVES IMPROVEMENTS IN OUR PRODUCTIVITY, AS IT SQUEEZES WASTE OUT OF THE WHOLE PRODUCTION SYSTEM AND THE RESOURCES ONCE ABSORBED IN PURE WASTE (INSPECTING, REJECTING, FIXING, AND DOING THINGS MORE THAN ONCE) CAN BE APPLIED TO ADDING VALUE TO THE PRODUCT AND IMPROVING THE BUSINESS.

I'M PROUD TO SAY THAT CHRYSLER HAS IMPROVED ITS QUALITY PERFORMANCE 50% OVER THE LAST SEVEN YEARS, BUT THE DOWN SIDE OF THE STORY IS THAT THE "BEST IN CLASS" STATUS WE'RE ALL TRYING TO ACHIEVE IS RARELY FOUND IN VEHICLES PRODUCED BY THE "BIG THREE". THIS POINT OF EXCELLENCE MOST OFTEN COMES FROM ACROSS THE SEA, BOTH FROM JAPAN AND EUROPE. WHILE CHRYSLER AND THE OTHER U.S. PRODUCERS HAVE ACHIEVED SIGNIFICANT GAINS IN QUALITY, WE ALL KNOW THAT WE HAVE TO MAKE EVEN GREATER IMPROVEMENTS IN THE NEXT FIVE YEARS - JUST TO STAY IN THE GAME! "TO BE THE BEST" ISN'T JUST A SLOGAN - IT'S IMPERATIVE FOR OUR SURVIVAL IN THE MARKETPLACE.

WHAT IS QUALITY? QUALITY IS MORE THAN SHINY PAINT, GOOD DOOR FITS, MAJOR IMPROVEMENTS IN RELIABILITY, AND HIGH SCORES ON THE ROGERS REPORT. QUALITY IS ALSO CUSTOMER-PERCEIVED EXCELLENCE OF FUNCTION AND FEATURES. THINGS LIKE RIDE AND HANDLING, PERFORMANCE, DRIVEABILITY, NOISE-VIBRATION AND HARSHNESS, DOOR CLOSING EFFORT AND SOUND, HEATING AND AIR-CONDITIONING SYSTEM PERFORMANCE, TACTILE FEEL OF SWITCHES AND CONTROLS AND ERGONOMICS ARE ALL PART OF THE CUSTOMER'S PERCEPTION OF QUALITY. QUALITY IS NOT AN ILLUSION; IT IS THE TRUE SUBSTANCE OF SOUNDLY ENGINEERED PRODUCTS WHOSE DESIGNS AND MANUFACTURING PROCESSES HAVE BEEN THOROUGHLY DEVELOPED AND OPTIMIZED IN PARALLEL BEFORE THEY ARE LAUNCHED. THE WORLD'S BEST CARS AND TRUCKS ARE DEVELOPED, NOT JUST INTRODUCED.

I MENTIONED A FEW MINUTES AGO THAT QUALITY WILL BE THE PRICE OF ADMISSION IN THE COMPETITIVE MARKETPLACE; THE OTHER HALF OF THE TICKET TO SURVIVAL WILL BE COMPETITIVE PRICE, WHICH DRIVES MAJOR COST

REDUCTIONS IN ORDER TO REMAIN PROFITABLE. AS MOST OF US HAVE DISCOVERED, QUALITY IMPROVEMENT AND COST REDUCTION GO HAND-IN-HAND, WITH PROPER PLANNING AND EXECUTION. WE'RE LOOKING TO TAKE $2,500 OUT OF OUR TOTAL UNIT COSTS TO MAINTAIN OUR POSITION AS THE LOWEST-COST PRODUCER. THIS MEANS REDUCED VARIABLE COSTS, THOSE OF MATERIALS AND LABOR, ELIMINATION OF WASTE IN ALL FORMS, AND REDUCED INVESTMENT THROUGH LEANER, MEANER, SIMPLIFIED MANUFACTURING SYSTEMS AND FACILITIES, MADE POSSIBLE BY PROCESS-DRIVEN DESIGNS.

COST REDUCTION ALSO MEANS LOOKING AT NEW MATERIAL AND PROCESS TECHNOLOGIES AND HOW WE CAN APPLY THEM IN AN OPTIMAL MANNER SO THAT EACH ITEM MAKES SENSE FOR ITS SPECIFIC VEHICLE APPLICATION AND IN TERMS OF MEETING LONG-TERM STRATEGIC BUSINESS OBJECTIVES. COST REDUCTION MUST BE MORE THAN "SINGLE-CAR" FOCUSED - IT MUST ALSO CONSIDER THE DOWNSTREAM COSTS INVOLVED IN CHANGING THAT VEHICLE FOR THE NEXT MODEL YEAR, AND THE LEAD TIME REQUIRED TO EXECUTE THOSE PRODUCT AND FACILITY CHANGES.

LIBERTY OBJECTIVES FOR QUALITY IMPROVEMENT AND COST REDUCTION ARE QUITE SPECIFIC, AND TRANSLATE INTO MATERIAL COST IMPROVEMENTS, REDUCTIONS IN MANPOWER THROUGH THROUGH IMPROVED MANUFACTURING AND ASSEMBLY PROCESSES, MAJOR INVENTORY COST REDUCTIONS THROUGH THE COMBINATION OF IN-LINE SEQUENCING, JUST-IN-TIME DELIVERY, DEFECT-FREE SUPPLIER PERFORMANCE AND IMPROVED MATERIAL HANDLING TECHNOLOGY, AND PLANT SIZE REDUCTIONS PERMITTED BY THESE CHANGES. PROCESS-DRIVEN DESIGN, SIMULTANEOUS ENGINEERING, AND USE OF A WHOLE NEW "TOOLKIT" OF METHODS TO IMPROVE THE PRODUCT DESIGN AND ITS RELATED MANUFACTURING

PROCESSES "UP-FRONT" IN THE PRODUCT DEVELOPMENT CYCLE ARE THE KEYS TO ACHIEVING OUR QUALITY OBJECTIVES FOR THE 1990'S.

IN YEARS PAST, THE ROLE OF THE QUALITY DEPARTMENT WAS THAT OF A POLICEMAN, LOOKING AROUND THE PLANT FOR VIOLATIONS. TICKETS WERE WRITTEN AND FINES WERE ASSESSED, BUT LITTLE WAS DONE TO FIND AND ELIMINATE ROOT CAUSES OF PROBLEMS OR TO CHANGE THE BEHAVIOR OF PROCESSES TO PREVENT THE RECURRENCE OF CHRONIC PROBLEMS. THIS "FIRE-FIGHTING" MENTALITY LEFT NO WINNERS; THE QUALITY OF THE PRODUCT SUFFERED - NOT ALL DEFECTS WERE FOUND, AND THOSE THAT WERE WEREN'T NECESSARILY REPAIRED TO REQUIRED QUALITY LEVELS. WORSE THAN THIS, NOTHING WAS PUT IN PLACE TO PREVENT THE NEXT VEHICLE FROM HAVING THE SAME DEFECT. THE CONCEPT OF "IRREVERSIBLE CORRECTIVE ACTION" HADN'T SURFACED YET, AS ONLY THE SYMPTOMS OF THE DISEASE WERE ADDRESSED AS THEY POPPED UP.

UNDERSTANDABLY, NO MAJOR CORPORATION IS NOW OPERATING IN THIS MODE. SUCH AN APPROACH WOULD INSURE DISASTER IN TODAY'S MARKETPLACE. WE'VE ALL COME A LONG WAY - THE IMPROVEMENTS IN QUALITY IN U.S.-PRODUCED VEHICLES HAVE BEEN DRAMATIC IN RECENT YEARS, DRIVEN BY MARKET REQUIREMENTS. THE CUSTOMER IS DEMANDING HIGHER QUALITY, AND THE COMPETITION HAS PROVIDED WHAT THE CUSTOMER WANTS - WE'VE HAD TO IMPROVE OR PERISH.

IN ORDER TO ACHIEVE OUR CURRENT LEVEL OF QUALITY WE'VE HAD TO MOVE THE IMPROVEMENT PROCESS UPSTREAM. THE "FIX AND REPAIR" MENTALITY HAD TO BE ELIMINATED. WE HAD TO GET OUR PRODUCTION PROCESSES IN

CONTROL AND DIG TO FIND THE ROOT CAUSES FOR POOR QUALITY. ONCE THE CAUSES WERE IDENTIFIED, WE BEGAN TO DEVELOP NEW PROCESSES AND DESIGNS TO MATCH THEM TO ELIMINATE THE CAUSES OF FAILURES. WHEN PROPERLY IMPLEMENTED, THIS APPROACH LEADS TO MAJOR IMPROVEMENTS IN FIRST-TIME-THROUGH CAPABILITY OF OUR PROCESSES, WITH CORRESPONDING REDUCTIONS IN BOTH FACILITIES AND MANPOWER REQUIRED FOR REPAIR; OUR NEW PROCESS-DRIVEN ASSEMBLY PLANT LAYOUTS NO LONGER ALLOCATE SPACE FOR THE TRADITIONAL IN-LINE INSPECT AND REPAIR STATIONS. IF WE DO IT RIGHT THE FIRST TIME, EVERY TIME, WE GET IMPROVED QUALITY, REDUCED COST, AND SATISFIED CUSTOMERS.

IN THE PRODUCTION AREA WE'VE IMPLEMENTED STATISTICAL PROCESS CONTROL TO DETERMINE THE LEVEL OF VARIABILITY IN OUR PROCESSES AND WORKED TO REDUCE THOSE VARIATIONS. WE TRACK OUR PERFORMANCE THROUGH S.P.C., AND WHEN WE ASSURE OURSELVES OF ACCEPTABLE REPEATABILITY, WE CAN THEN USE "X-BAR" AND "R" CHARTS TO MONITOR THE SYSTEM. IN TRACKING THESE VARIATIONS, WE CAN DETERMINE WHEN SYSTEM ADJUSTMENTS HAVE TO BE MADE BEFORE PRODUCING UNACCEPTABLE PARTS AND ASSEMBLIES, AND THIS INFORMATION IS BEING USED TO REFINE FUTURE PROCESSES, PRODUCT DESIGNS, AND SPECIFICATIONS FOR TOOLING AND MACHINES TO FURTHER REDUCE VARIATION.

WE HAVE ADDED ON-LINE FACTORY INFORMATION AND PERFORMANCE FEEDBACK SYSTEMS THAT MONITOR THE FLOW OF THE PRODUCT THROUGH THE MANUFACTURING PROCESS; WE MONITOR THE RESULTS OF EACH PROCESS AND THE PERFORMANCE OF THE EQUIPMENT AND RECORD EACH UNACCEPTABLE VARIATION. THIS INFORMATION IS RECORDED AGAINST THE RECORD OF THAT VEHICLE FOR

CORRECTION BEFORE IT CAN BE OK'D FOR SHIPMENT, AND THE EQUIPMENT INFORMATION IS USED TO DEVELOP MORE EFFECTIVE PREVENTIVE MAINTENANCE AND MACHINE DESIGN PRACTICES. WITH OVER 200 ROBOTIC WELDERS IN A MODERN BODY SHOP MAKING OVER 3000 WELDS ON EACH VEHICLE, IT'S IMPERATIVE THAT A MALFUNCTION BE REPORTED AND CORRECTED IMMEDIATELY.

WE HAVE INSTITUTED JUST-IN-TIME DELIVERY OF MATERIAL TO IN-LINE SEQUENCE ON MANY COMPONENTS SHIPPED TO OUR PLANTS; OVER 70% OF THE DAILY DOLLAR VALUE OF MATERIAL USED AT OUR STERLING HEIGHTS AND WINDSOR ASSEMBLY PLANTS IS CURRENTLY RECEIVED JUST-IN-TIME, SOME WITH AS LITTLE AS FOUR HOURS' NOTICE TO THE SUPPLIER, AND THIS DISCIPLINE IS EXPANDING RAPIDLY AT OUR OTHER FACILITIES. THE IN-LINE SEQUENCE APPROACH, WHERE FINISHED VEHICLES ARE DRIVEN OFF THE FINAL LINE IN THE SAME ORDER THAT THEIR MAJOR STAMPINGS WERE FIRST WELDED IN THE BODY SHOP, DRIVES PREDICTABILITY THROUGH THE ENTIRE SUPPLY SYSTEM, AND IS SUPPORTED BY A STABLE PRODUCTION SCHEDULE. ASIDE FROM THE OBVIOUS REDUCTION IN INVENTORIES AND THE ASSOCIATED CARRYING COST AND FLOORSPACE SAVINGS, THIS ESSENTIAL DISCIPLINE DRIVES DEFECT-FREE SUPPLIER QUALITY PERFORMANCE. WE NO LONGER HAVE A 5-DAY IN-PLANT SUPPLY OF PARTS TO DRAW FROM IF DEFECTIVE MATERIAL IS DISCOVERED; THERE ARE NO "SPARES" - IF MATERIAL ISN'T DEFECT-FREE, THE LINE GOES DOWN.

THESE MEASURES HAVE HELPED US MAKE MAJOR QUALITY GAINS OVER THE LAST FEW YEARS, BUT THEY AREN'T ENOUGH TO GET US WHERE WE HAVE TO BE IN ORDER TO REMAIN COMPETITIVE IN THE FUTURE. WE HAVE TO CHANGE THE WAY WE APPROACH THE ENTIRE PRODUCT DEVELOPMENT PROCESS AND GET

EVERYONE INVOLVED UP-FRONT TO INSURE THAT THE END RESULT MEETS THE QUALITY OBJECTIVES REQUIRED TO KEEP US COMPETITIVE.

HISTORICALLY, COMMON PRACTICE IN OUR INDUSTRY HAS CALLED FOR A SEQUENTIAL PRODUCT DEVELOPMENT PROCESS WHERE SENIOR MANAGEMENT OUTLINED THE CONCEPT OF A NEW PRODUCT. THEN THE PRODUCT PLANNERS AND STYLISTS DID THEIR JOB AND HANDED THAT PACKAGE "OVER THE WALL" TO ENGINEERING. THE PRODUCT ENGINEERS PUT DETAILS TO THE THEME, ESTABLISHED SPECIFICATIONS, BUILT AND TESTED PROTOTYPES, CREATED THE RELEASE DRAWINGS, AND HANDED THAT PACKAGE OVER ANOTHER WALL TO MANUFACTURING AND PURCHASING TO BUY THE PARTS, DEVELOP THE PROCESSES, DESIGN AND BUILD THE TOOLS, BUILD OR REVISE THE FACILITIES, AND LAUNCH PRODUCTION OF THE NEW PRODUCT. THIS SEQUENTIAL METHOD OF OPERATION -- THE "HAND-OFF CYCLE" -- IS A THING OF THE PAST. THIS TRADITIONAL APPROACH MEANT LAST-MINUTE CHANGES, HASTY COMPROMISES, HIGH COSTS, AND LONG LEAD TIMES. IT WAS GENERALLY INEFFICIENT, AND, MOST OF ALL, RESULTED IN POOR QUALITY.

SIMULTANEOUS ENGINEERING IS OUR NEW WAY OF DOING BUSINESS WHERE MANUFACTURING, PRODUCT ENGINEERING, PROCUREMENT AND SUPPLY, PRODUCT PLANNING, FINANCE, AND THE OTHER DISCIPLINES NECESSARY TO DEVELOP AND LAUNCH A NEW PROGRAM ALL WORK TOGETHER, IN PARALLEL, RIGHT FROM THE CONCEPT STAGE WITH A COMMON FOCUS TO BRING TEAMWORK, COORDINATION, AND TIMING DISCIPLINE TO THE PROGRAM.

LIBERTY WAS FORMED TO MEET THIS CHALLENGE, AND IS WORKING WITH A NEW DESIGN APPROACH THAT PUTS MANUFACTURING UP-FRONT IN THE FIRST

PHASES OF THE DESIGN PROCESS -- "PROCESS-DRIVEN DESIGN". PROCESS-DRIVEN DESIGN GOES A STEP BEYOND SIMULTANEOUS MANUFACTURING AND PRODUCT ENGINEERING, BY DEFINING THE REQUIRED END RESULTS FIRST, AND THEN DEFINING THE MANUFACTURING PROCESS AS THE FIRST STEP IN NEW MODEL DEVELOPMENT. MANUFACTURING LEADS WITH DEFINITION OF THE "BEST" PROCESSES FOR BUILDING THE PRODUCT WHICH WILL ACHIEVE OUR GOALS FOR QUALITY, RELIABILITY, PRODUCTIVITY, AND COST. THESE MANUFACTURING PROCESSES THEN GENERATE THE PRODUCT DESIGN GUIDELINES, OR THOSE FEATURES REQUIRED IN THE DESIGN TO ALLOW USE OF THE "BEST" PROCESSES. THEN THE STYLISTS AND PRODUCT ENGINEERS DETERMINE THE BEST WAY TO DESIGN AND ENGINEER THE PRODUCT AND ITS COMPONENTS TO TAKE ADVANTAGE OF THOSE PROCESSES, AND THE VEHICLE IS DEVELOPED IN A SYNERGISTIC "TEAM" APPROACH WITH MANUFACTURING, USING THE MANUFACTURING PROCESS-DRIVEN DESIGN GUIDELINES.

NOT SURPRISINGLY, THIS APPROACH MESHES PERFECTLY WITH THE CHRYSLER QUALITY POLICY OF "TO BE THE BEST" BY DEFINING REQUIREMENTS FIRST, PREVENTING DEFECTS INSTEAD OF DETECTING THEM, APPROACHING OUR TASK WITH A "DEFECT-FREE" MENTALITY, AND MEASURING THE RESULTS IN TERMS OF THE COSTS OF NON-CONFORMANCE TO THE REQUIREMENTS.

ONCE WE'VE DEFINED OUR REQUIREMENTS AND UNDERTAKEN DEFECT PREVENTION, WE CAN THEN GO AFTER "DEFECT-FREE" PERFORMANCE AS AN ATTITUDE. FOR MANY OF US, THIS CONCEPT BOGGLES THE MIND - 100% QUALITY? MAYBE 98 OR 99%, BUT EVERYTHING RIGHT? ISN'T 99% GOOD ENOUGH? THINK ABOUT IT FOR A MINUTE - THERE ARE ABOUT 3000 ASSEMBLY PLANT-LEVEL PARTS IN A CAR; AT A 99% QUALITY LEVEL, THAT WOULD MEAN

30 POSSIBLE IN-PROCESS DEFECTS IN EVERY CAR. JUST LOOKING AT ONE SMALL CATEGORY - LIGHT BULBS - A 99% QUALITY LEVEL WOULD MEAN ONE FAILED BULB ON EVERY OTHER CAR. NO, 99% ISN'T GOOD ENOUGH - WE HAVE TO APPROACH THE ENTIRE SYSTEM WITH A "DEFECT-FREE" ATTITUDE.

EARLIER IN THE PRESENTATION I MENTIONED USE OF A "TOOLKIT" OF NEW METHODS TO IMPROVE THE DESIGN OF THE PRODUCT AND ITS RELATED MANUFACTURING PROCESSES "UP-FRONT" IN THE PRODUCT DEVELOPMENT CYCLE AS BEING KEY TO ACHIEVING OUR QUALITY OBJECTIVES FOR THE 1990'S; LET'S TAKE A LOOK AT SOME OF THESE TOOLS AND HOW WE'RE USING THEM AT LIBERTY TO DRIVE QUALITY UPSTREAM.

OUR MOST IMPORTANT RESOURCE IS OUR PEOPLE, AND THEY NEED "PEOPLE SYSTEMS" THAT PROVIDE THE ENVIRONMENT, TRAINING, AND SKILLS TO UNLEASH EACH PERSON'S INNOVATIVE AND CREATIVE CAPABILITIES AND TO ENCOURAGE THEM TO "REACH" AND TAKE RISKS WITHOUT FEAR OF FAILURE. RECOGNIZING THAT LIBERTY IS THE FOUNDATION FOR OUR NEXT GENERATION OF PRODUCTS, WE HAVE CREATED A UNIQUE ORGANIZATION TO ADDRESS THIS CHALLENGE. LIBERTY IS A "TEAM" OF HIGHLY-SKILLED PRODUCT ENGINEERS, MANUFACTURING ENGINEERS, PROCUREMENT AND SUPPLY EXPERTS, PRODUCT PLANNERS, FINANCE PEOPLE, AND OTHER AUTOMOTIVE PROFESSIONALS. THIS TEAM IS JOINED IN A FOCUSED, CONCENTRATED EFFORT, UNDER ONE ROOF, TO DEVELOP THESE NEW PRODUCTS, MANUFACTURING PROCESSES, AND FACTORY-OF-THE-FUTURE CONCEPTS IN A CONCURRENT FASHION. HAVING THESE DIVERSE SKILLS AND VAST REAL-WORLD EXPERIENCE TOGETHER IN ONE LOCATION ALLOWS US TO DEVELOP APPLIED TECHNOLOGY THAT WORKS, NOT JUST "NEW TECHNOLOGY", AND ALLOWS US TO "CLOSE THE LOOP" ON PRIOR PROBLEMS

SO THEY CAN BE DESIGNED AND PROCESSED OUT OF OUR FUTURE PRODUCTS -- THE "PREVENTION" APPROACH THAT'S SO IMPORTANT FOR QUALITY.

NEW PRODUCTS WON'T DO US MUCH GOOD IF THEY DON'T MEET CUSTOMER NEEDS AND EXPECTATIONS. WE ARE USING A NEW TECHNIQUE KNOWN AS "QUALITY FUNCTION DEPLOYMENT" WHICH USES THE "VOICE OF THE CUSTOMER" FOR ESTABLISHING PRODUCT AND PROCESS PRIORITIES. THIS TECHNIQUE IDENTIFIES CUSTOMER REQUIREMENTS IN TERMS OF SPECIFIC QUALITY CHARACTERISTICS WHICH ARE THEN CORRELATED TO PRODUCT AND PROCESS FEATURES TO ESTABLISH WHAT THE IMPORTANT PARAMETERS REALLY ARE THAT SHOULD DRIVE THE DESIGN. FOR INSTANCE, IT ISN'T SUFFICIENT TO SPECIFY AN AIR-CONDITIONING COOLING CAPACITY OR COOL-DOWN TIME IF THAT PERFORMANCE IS ONLY ACHIEVABLE WITH AN OBJECTIONABLE LEVEL OF BLOWER NOISE. THE RESULTS OF USING THE "VOICE OF THE CUSTOMER" CAN INCLUDE BOTH IMPROVED CUSTOMER SATISFACTION AND COST REDUCTION BY DOING THE JOB ONCE WITH THE RIGHT CRITERIA KNOWN UP-FRONT.

F.M.E.A., OR FAILURE MODE AND EFFECT ANALYSIS, IS ANOTHER TOOL TO INSURE THE JOB GETS DONE RIGHT THE FIRST TIME; THIS TECHNIQUE, PREVIOUSLY USED PRIMARILY FOR PRODUCT DESIGN VALIDATION, IS ALSO APPLICABLE TO OUR MANUFACTURING PROCESSES. DONE ON A "TEAM" BASIS, A GIVEN DESIGN OR PROCESS IS EXAMINED TO DETERMINE ALL POSSIBLE MODES OF FAILURE, INCLUDING CAUSES, FREQUENCY, DEGREE OF CRITICALITY, AND POSSIBILITY OF DETECTION. FOR INSTANCE, WE CAN EXAMINE A PROPOSED ROBOTIC OPERATION AND, AS A TEAM, DETERMINE ALL OF THE POSSIBLE MODES OF FAILURE OR BREAKDOWN. WE CAN THEN RANK THE SEVERITY OF EACH FAILURE MODE, DETERMINE WAYS TO ELIMINATE THE CAUSE, AND RE-EVALUATE

THE SEVERITY AFTER MAKING IMPROVEMENTS, ALL ON PAPER. IF THE CHANGES ACHIEVE THE REQUIRED IMPROVEMENTS, WE CAN INCORPORATE THE CORRECTIONS, AND ENSURE THAT A CONTROL PLAN IS IN PLACE TO VERIFY THE RESULTS. THIS PROCESS ALLOWS US TO HIGHLIGHT PROBLEMS AND MAKE CORRECTIONS UP-FRONT IN THE DEVELOPMENT CYCLE, PREVENTS COSTLY PRODUCTION DESIGN AND TOOLING CHANGES, AND IMPROVES QUALITY AND RELIABILITY.

"DESIGN FOR ASSEMBLY" IS ANOTHER IMPORTANT TOOL IN OUR QUALITY KIT. THIS TECHNIQUE RANGES FROM ANALYSIS OF DETAIL COMPONENT DESIGNS TO REDUCE NUMBER OF PARTS, FASTENERS, AND ASSEMBLY DIRECTIONS TO DESIGNING MAJOR COMPONENTS AND ASSEMBLIES AROUND SPECIFIC PROCESS AND AUTOMATION REQUIREMENTS. AS AN EXAMPLE, WE JUST HAD A "DESIGN FOR ASSEMBLY" TEAM EXAMINE AN INSTRUMENT CLUSTER PROPOSED FOR A FUTURE MODEL. AFTER ONLY TWO DAYS OF ANALYSIS, THE DESIGN WAS REVISED FROM A TOTAL OF 60 DISCRETE PARTS TO 25, A POTENTIAL 60% REDUCTION IN THE NUMBER OF PARTS TO BE DESIGNED, TOOLED, AND ASSEMBLED.

IN OUR INDUSTRY TODAY, THERE ARE A NUMBER OF RECENT EXAMPLES OF "AUTOMATION FOR AUTOMATION'S SAKE". AT LIBERTY, WE'RE BEING BOTH INNOVATIVE AND PRACTICAL; WE'RE NOT JUST "DESIGNING FOR AUTOMATION" -- WE'RE "DESIGNING FOR ASSEMBLY". WE'RE DEVELOPING DESIGNS AND PROCESSES THAT CAN BE AUTOMATED IF THAT MAKES SENSE; THIS APPROACH ENSURES SIMPLER ASSEMBLY WITH IMPROVED QUALITY AND RELIABILITY WHETHER AUTOMATION IS EMPLOYED OR NOT, AND MAKES THE QUESTION OF WHETHER OR NOT TO AUTOMATE SIMPLY A BUSINESS DECISION.

ANOTHER FACET OF "DESIGN FOR ASSEMBLY" IS MODULAR DESIGN AND ASSEMBLY TECHNIQUES, WHICH INTEGRATE MANY DISCRETE DETAIL PARTS INTO A SUBASSEMBLY "MODULE" WHICH CAN BE BUILT-UP OFF-LINE EITHER IN THE PLANT OR BY A SUPPLIER AND DELIVERED TO THE MAIN LINE JUST-IN-TIME. THIS APPROACH ALLOWS INTEGRATION OF DETAIL PARTS BY DESIGN TO REDUCE TOTAL PART COUNT, REDUCES COMPLEXITY AND THE NUMBER OF PARTS THAT MUST BE SCHEDULED AND HANDLED BY THE PLANT, AND REDUCES THE NUMBER OF INDIVIDUAL ASSEMBLY OPERATIONS PERFORMED ON THE MAIN LINE. THE MODULAR APPROACH ALSO OFFERS OPPORTUNITIES FOR QUALITY IMPROVEMENT THROUGH CERTIFICATION OF SYSTEM FUNCTION PRIOR TO FINAL ASSEMBLY. A TYPICAL ASSEMBLY PLANT TODAY HANDLES ABOUT 3,500 PART NUMBERS -- WE HOPE TO REDUCE THIS "END ITEM" PART COUNT TO 1,500 THROUGH A COMBINATION OF MODULAR DESIGN AND ASSEMBLY AND COMPLEXITY REDUCTION. WE HAVE IDENTIFIED 28 POTENTIAL MODULES FOR OUR FUTURE PRODUCTS, INCLUDING SUCH APPLICATIONS AS THE INSTRUMENT PANEL, COMPLETE DOOR ASSEMBLIES, AND MAJOR CHASSIS SYSTEMS.

NEW COMPUTER-AIDED ENGINEERING TECHNIQUES ALSO PLAY A KEY ROLE IN OUR "FACTORY-OF-THE-FUTURE" EFFORT. ONE OF THE NEW TECHNOLOGIES WE'RE EMPLOYING IS ASSEMBLY VARIATION ANALYSIS. THIS METHOD EMPLOYS UP-FRONT COMPUTER-BASED SIMULATIONS OF ASSEMBLY PROCESS VARIATIONS AND PART DIMENSIONAL VARIATIONS TO DETERMINE IF DESIGN INTENT CAN BE MET. IT IDENTIFIES VARIATIONS AND HIGHLIGHTS THEIR CAUSES IN RANK ORDER, AND SHOWS THE RESULTS OF CORRECTIONS BEFORE TOOLING IS BUILT. THIS TECHNIQUE HELPS GUIDE THE DESIGN OF THE PRODUCT AND ITS PROCESSES TOWARD A QUALITY END RESULT BY HELPING DECIDE THE MOST COST-EFFECTIVE CHANGES IN EITHER THE PRODUCT OR THE PROCESS TO REDUCE

TOTAL VARIABILITY, AND HELPS VERIFY OUR CAPABILITY TO ACHIEVE DESIGN INTENT.

FACTORY FLOOR PROCESS SIMULATION IS ANOTHER EMERGING COMPUTER-BASED TECHNOLOGY FOR OPTIMIZING THE DESIGN OF INDIVIDUAL PROCESSES AND PRODUCTION SYSTEMS. IT IDENTIFIES POTENTIAL DYNAMIC PROCESS BOTTLENECKS AND HELPS GUIDE US IN DETERMINING THE PROPER SIZE AND LOCATION OF ACCUMULATORS TO MAXIMIZE TOTAL SYSTEM UP-TIME CAPABILITY WITH THE LEAST INVESTMENT AND FLOORSPACE. POTENTIAL APPLICATIONS RANGE FROM INDIVIDUAL WORK CELLS TO ENTIRE DEPARTMENTAL SYSTEMS. MAINTAINING A SMOOTH, CONTINUOUS FLOW THROUGH OUR PROCESSES IS ESSENTIAL TO ACHIEVING OUR GOALS FOR FIRST-TIME-THROUGH CAPABILITY.

TODAY, AS A NEW VEHICLE OR SYSTEM PROGRESSES THROUGH THE DESIGN STAGES AT CHRYSLER EACH ENGINEER PROVIDES PERIODIC PROJECTIONS OF THE SUCCESSFUL DEFECT-FREE LAUNCH OF HIS SYSTEM. THIS IS BASED HIS OWN ENGINEERING EVALUATION AND THOSE OF OTHER AFFECTED AREAS SUCH AS MANUFACTURING FEASIBILITY, VENDOR QUALITY ASSURANCE, TRANSPORTATION, AND SERVICE. THIS ASSESSMENT, THE PRODUCT CONFIDENCE INDEX, IS USED BY PROGRAM MANAGEMENT TO EVALUATE THE PROGRESS AND STATUS OF A GIVEN NEW PROGRAM AND TO HIGHLIGHT POTENTIAL PROBLEM AREAS. AT LIBERTY, WE'RE TAKING THIS APPROACH ONE STEP FURTHER UPSTREAM. WE'RE WORKING ON DESIGNS AND PROCESSES IN SOME CASES LONG BEFORE THEY ARE TARGETED FOR SPECIFIC PRODUCT APPLICATIONS; AS THESE PRODUCT AND PROCESS CONCEPTS EVOLVE, WE'RE DOING PERIODIC FEASIBILITY ANALYSIS -- HOW CONFIDENT ARE WE THAT THE CONCEPT WILL BECOME A VIABLE PRODUCTION

PROCESS? OUR MULTI-DISCIPLINARY "TEAM" ORGANIZATION AT LIBERTY SIMPLIFIES THIS CONSENSUS APPROACH AND ALLOWS THE CONCEPT LEADER TO COMMUNICATE EASILY WITH THE REST OF HIS "TEAM". THE RESULT IS OUR LIBERTY FEASIBILITY RATING. WHEN THAT CONSENSUS RATING IS ACCEPTABLY HIGH, THE DESIGN AND PROCESS ARE READY FOR INCORPORATION IN THE NEXT VEHICLE PROGRAM WITHOUT THE RISKS OF "INVENTING ON THE FLY" AND HAVING TO MAKE MAJOR PROCESS CHANGES ON THE PLANT FLOOR.

OUR SUPPLIERS ALSO HAVE A MAJOR ROLE IN OUR QUALITY IMPROVEMENT EFFORTS, AND THEY MUST BE INCLUDED IN OUR PROCESS-DRIVEN SIMULTANEOUS ENGINEERING PROCESS AS WELL. WE PURCHASE 70% OF OUR PARTS FROM OUTSIDE SUPPLIERS, SO WE NEED TO WORK WITH THEM AS PARTNERS, NOT JUST VENDORS, TO STAY AHEAD OF THE COMPETITION BY DEVELOPING NEW CONCEPTS AND SYSTEMS AND PUTTING THEM "ON THE SHELF" TO HELP NEW PRODUCTS AND FEATURES FLOW RAPIDLY INTO OUR PROGRAMS WITH HIGH QUALITY, LOW COST, AND KNOWN TIMING.

IN THE PAST, MOST AUTOMOTIVE KNOWLEDGE WAS WITHIN THE WALLS OF THE OEM'S -- NOW THIS EXPERTISE IS EVERYWHERE, AND MANY WORLD-CLASS SUPPLIERS CAN DO THE JOB AS WELL OR BETTER THAN WE CAN, WITH CLOSER AND EARLIER INTERACTION AND NEW LONG-TERM RELATIONSHIPS. THESE INTEGRATED SYSTEM SUPPLIERS, WITH R & D ORGANIZATIONS CAPABLE OF DEVELOPING ORIGINAL CONCEPTS AND THE ABILITY TO CARRY THEM THROUGH TO HIGH-QUALITY, LOW-COST PRODUCTION CAN BE A VALUABLE ASSET TO US FOR INVENTIVE, CREATIVE APPROACHES TO MACHINERY, TOOLS, COMPONENT DESIGN AND PRODUCTION.

ALL OF THIS LEADS US BACK TO QUALITY AS THE PRICE OF ADMISSION

TO THE FUTURE MARKETPLACE -- THE TRUE SUBSTANCE OF SOUNDLY-ENGINEERED PRODUCTS WHOSE DESIGNS AND MANUFACTURING PROCESSES HAVE BEEN THOROUGHLY DEVELOPED AND OPTIMIZED IN PARALLEL BEFORE THEY ARE LAUNCHED, LEADING TO COST REDUCTIONS IN MANPOWER, MACHINERY, MATERIAL, AND METHODS -- THE 4-M'S OF MANUFACTURING. THE LIBERTY APPROACH, USING THE UP-STREAM QUALITY TOOLS WE'VE DISCUSSED THIS MORNING, WILL RESULT IN PRODUCTS WHICH ARE EASIER TO ASSEMBLE WHETHER AUTOMATION IS EMPLOYED OR NOT; THAT ARE LESS SENSITIVE TO BUILD VARIATIONS; THAT INTEGRATE COMPONENTS TO REDUCE PART COUNT, NUMBER OF SUPPLIERS, DIRECT LABOR, FLOORSPACE AND MATERIAL HANDLING REQUIREMENTS; AND WHICH RESULT IN PROCESS REPEATABILITY AND RELIABILITY THAT TRANSLATE INTO DEPENDABLE, QUALITY PRODUCTS FOR THE CUSTOMER.

FIVE YEARS FROM NOW WE MAY WELL BE WORKING WITH PRODUCTS AND PROCESSES WE DON'T EVEN KNOW ABOUT TODAY AS THE LIBERTY PROCESS CONTINUES TO FEED ON NEW CONCEPTS AND IDEAS AND GENERATES NEW MANUFACTURING, ASSEMBLY, AND PRODUCT TECHNOLOGIES THROUGH THE PROCESS-DRIVEN DESIGN APPROACH.

IN SUMMARY, WE AT LIBERTY ARE DEDICATED TO EXPLORING THE CONCEPTS AND DEVELOPING THE PRODUCTS, PROCESSES, AND PEOPLE INVOLVEMENT THAT WILL LEAD CHRYSLER MOTORS INTO THE TWENTY-FIRST CENTURY WITH WORLD-CLASS PRODUCTS -- PRODUCED IN NEW, MODERN, FLEXIBLE FACILITIES -- THAT ARE "THE BEST" IN QUALITY, COST, AND TECHNOLOGY.

THESE ARE THE BENEFITS OF "PROCESS-DRIVEN DESIGN" -- THE FOUNDATION OF LIBERTY.

Reprinted from *Plastics World*, Newton, Mass., December 1988

Meet two architects of design-integrated manufacturing

Quantifying manufacturing costs early on can help product designers achieve major savings. Here's how it's done with new software for personal computers

BY CARL KIRKLAND, SENIOR EDITOR

It saves Ford Motor Co. $700 per vehicle and reduces Ford's manufacturing costs by 30%. It was used in the design of IBM Corp.'s award-winning ProPrinter, where assembly time was reduced from 30 minutes to three minutes. And by using it, NCR Corp. will save $1.1 million in total lifetime manufacturing labor costs for its new model 2760 point-of-sale terminal, which is assembled in 75% of the time of a previous model with 85% fewer parts.

What is "it?" Nothing less than quantitative methods for evaluating the cost and manufacturability of a design during the conceptual stage of product development. The methods are called "design for manufacture and assembly," and they are sweeping through U.S. industry and turning it on its ear. In general, users have found that these methods can improve product cost and reliability, reduce inventory and paperwork, and improve productivity and competitiveness.

When you meet them for the first time it's hard to believe two unassuming, expatriated British professors, Geoffrey Boothroyd and Peter Dewhurst, could be responsible for igniting a revolution in American design and manufacturing. But if you ask any one of the more than 160 U.S. firms presently applying their methods about the effectiveness of Boothroyd and Dewhurst's techniques, you'll hear glowing testimonials. For example, Ford engineers call it a leap ahead in technology. And, in some cases, Ford requires vendors to submit analyses of parts based on Boothroyd and Dewhurst's methods before accepting bids on a project.

Although the methods were originally developed for improving the competitiveness through more efficient assembly procedures, Boothroyd and Dewhurst have recently turned their attention to injection molding. They have developed a comprehensive software package that, when combined with their earlier designed for assembly (DFA) system, analyzes injection molding product conceptual designs for fully accounted costs, component consolidation, and overall ease of manufacture and assembly. And that could prove to be cause for rejoicing among plastics designers and manufacturers alike for years to come.

In the beginning

In 1964, Boothroyd was invited to spend a year at Georgia Tech. Before he came to the U.S., however, he had already developed an interest in improving product assembly methods in his native England at the University of Salford, near Manchester, where he taught manufacturing engineering. In 1967 he moved to the University of Massachusetts at Amherst, became a permanent U.S. resident, and received funding from the National Science Foundation to continue research work he had started in Georgia on automatic assembly, feeding, and orienting of small parts.

Over the next few years, Boothroyd realized that manufacturing costs are usually fixed at a product's design stage. "It had long been recognized that there was a need for designers to take more account of manufacturing costs when they design products, because there are a lot of avoidable costs at that stage of development," he says. After manufacturing starts on a product, he adds, it is usually too late to do much about its design without incurring substantial cost penalties.

Subsequently, in 1977, he began to work on developing techniques for helping designers anticipate problems in manufacturing—work that would eventually lead to the design for manufacture and assembly methods in practice today. But it wasn't until the summer of 1980 that "things really started to take off," as he puts it.

Webb Chappell

Peter Dewhurst (right) and Geoffrey Boothroyd, experts in design for assembly and manufacture.

■ ■ ■ ■

'Design for manufacture and assembly is attempting to foresee, at the product design stage, what the manufacturing problems will be and to quantify them so that they can be corrected on the drawing board, rather than later, during manufacturing.'

—Geoffrey Boothroyd

That's when he was joined by Dewhurst.

Curiously enough, Boothroyd and Dewhurst both taught at Salford, but their paths never crossed. Dewhurst brought expertise in computer systems design to the team, as well as a background in manufacturing engineering. Says Boothroyd, "Back in 1980 we started working together on developing software that led to our design for assembly toolkit, and it soon became evident that we couldn't possibly develop commercial software in a university environment." That's because demand from industry for their methodology had increased.

"So we decided to set up a partnership," continues Boothroyd. "We moved to the University of Rhode Island (URI) just three and a half years ago. About the time we moved, we'd just taken some big steps—hiring a part-time programmer instead of doing it all ourselves. Now the business is funded so that we have three full-time programmers, a full-time graduate engineer, and an office staff."

Although both professors continue to teach at the Department of Industrial and Manufacturing Engineering at URI in Kingston, R.I., they also are the driving forces behind their company, Boothroyd Dewhurst, Inc. (BDI), in nearby Wakefield, R.I. In addition to developing software and handbooks on their methodologies, BDI conducts monthly one-day workshops, annual conferences and three-day short courses, as well as providing consulting services. BDI licenses its software to a division of a company, entitling the licensee to two years of technical support and enhancements at no extra charge.

Designing in the dark

What's new about design for manufacture and assembly? A good way to find out is to compare the new methods against "the way we were."

Historically, experienced design engineers have tried their best to develop products for cost effectiveness, simplicity, and efficiency. Usually, they have relied on traditional engineering techniques to do so. Such techniques often include qualitative engineering methods, which combine well-known value and producibility engineering.

Value engineering is concerned with product design—that is, the function, performance, and cost of a product. On the other hand, producibility engineering is more concerned with the manufacturing phase of product development—determining the best method of producing a product at a cost target.

Yet, qualitative engineering efforts are usually implemented on a part-by-part, or component-by-component basis *after* a product is designed as a unit. Sometimes these methods are even used as a post-manufacturing analysis. Unfortunately, manufacturing productivity is usually chipped in stone at the earliest stages of product design. For example, Ford Motor Co. estimates that although product design constitutes only 5% of total product cost, design's influence on product cost is 70%.

'What a designer needs to know is what the alternatives cost.'

Peter Dewhurst

So designers using traditional qualitative engineering methods may be shooting in the dark regarding the true production costs and manufacturability of their concepts at the critical, early phase of product development. In fact, designers may not become aware of their product's manufacturability, or "un-manufacturability," until much later, when the product is designed as a unit, or in the worst case, when it's in production. By that time changes are often too expensive to make.

According to Dewhurst: "We came to recognize through case studies in design for assembly that 'producibility' really misses the main gains that are possible at the design stage—the cost reductions—because producibility guides designers to look at details too soon for manufacturing. A producibility handbook will tell a designer to avoid cross features in injection molded parts that would require side pulls, for example, because of tooling costs."

"But," he continues, "what the designer really wants to know is that if he or she *wants* to put cross features in a part, what will it cost. If you follow the producibility guidelines you may avoid putting side projections on an injection molded part. Instead, the piece that would have been the side projection will be a separate part with added assembly processes and costs associated with it. So what a designer really needs to know is not what the preferable practice is, but what the alternatives will really cost."

Light in the darkness

Before the advent of personal computers and computer-aided design (CAD), gathering and manipulating all the existing information and design analysis required for developing accurate manufacturing cost estimates was difficult. Computers allow such factors as material costs, set-up costs per component, non-productive cost, and molding costs to be brought to bear on a concept in a variety of "what if" scenarios quickly, easily, and economically. Therefore, computer power has helped to make a newer design method called "quantitative engineering" a practical reality.

Unlike qualitative engineering, quantitative engineering allows a holistic analysis of design, materials, costing, and manufacturing processes at the conceptual stage of product development. Moreover, quantitative engineering involves a *product level* review of all the various design and manufacturing factors, rather than a component-by-component analysis. In the latter method, designers can run the risk of losing sight of the individual part's function within the total product system.

"The key problem was that designers lacked the tools to quantify manufacturing costs," Boothroyd says. "Once you put such tools in the hands of the designer it's amazing what can be done. When designers can identify where the major cost problems are in manufacturing their products, they can begin to cut costs quite significantly."

Boothroyd and Dewhurst's methods are the exact opposite of traditional producibility engineering, Dewhurst says. "Producibility would guide you to avoid all the complicated, difficult features. And the outcome of that would be a larger number of simpler parts with high assembly costs and, maybe, poor reliability. You know, if you follow producibility rules faithfully, then the best injection molded part is a cylindrical cup."

New design tools

Two powerful quantitative engineering tools have been developed by Boothroyd and Dewhurst. In addition to DFA, they have also developed design for manufacture (DFM) tools. The professors refer to the overall concept and software packages as design for manufacture and assembly (DFMA).

DFM identifies the appropriate materials and manufacturing processes for component parts being considered in a product's design. It also allows components to be designed based on the combination of various capabilities and limitations inherent in both the material and process selected for use. On the other hand, DFA identifies the most effective and cost-efficient assembly system for a new product's design. It carefully analyzes the structure and suitability of the design for a firm's chosen assembly method, whether it be manual, fixed automation, or programmable robotic assembly, for instance.

This fall, *PW* introduced you to Boothroyd and Dewhurst's newest personal-computer-based DFM software package (see September issue,

NCR saves megabucks with new design methods

Extraordinary reductions in overhead, product development time, and manufacturing costs were recently achieved by NCR Corp., Cambridge, Ohio. It used Boothroyd and Dewhurst's design for manufacturing and assembly (DFMA) analysis in the development of its new point-of-sale terminal, model 2760. Here's just a sample of benefits NCR achieved compared to a previous model:

- An overall manufacturing labor cost reduction of 44%, amounting to a projected $1.1 million in savings;
- 85% fewer parts;
- 75% quicker assembly time;
- a 100% reduction in assembly tools; and
- 65% fewer suppliers.

"Boothroyd and Dewhurst's software is nothing more than an organized approach to common sense methodology," says Bill Sprague, senior advanced manufacturing engineer at NCR. He spearheaded the application of DFA in the 2760 project. "It was neither difficult to learn nor to apply even though it was our first direct application of the software." The entire terminal snaps together without screws or fasteners. In fact, Sprague recently demonstrated its ease of assembly by manually assembling it himself, in two minutes, *blindfolded.*

Although Boothroyd and Dewhurst's software played an integral role, Sprague also credits early vendor involvement and solids modeling computer-aided design and manufacturing (CAD/CAM) technology as contributing to the new product's successes. Its injection molded polycarbonate parts were molded by Sajar Plastics, Inc., Middlefield, Ohio. Sajar also assembles key components and subcomponents, like the terminal's cabinet top, bezel, and base. Sajar ships assembled modules on a just-in-time basis.

Molds were manufactured by Diamond Tool & Die Corp., Dayton, Ohio. Both companies contributed major input into the design of the product, with an emphasis on its manufacturability, at the conceptual design phase. The entire project went directly from sketches and roughed-out models to hard-steel production tooling. There were no hard-copy drawings, and none exist to this day. "And, basically, there were no engineering changes once we began cutting steel," says Diamond Tool.

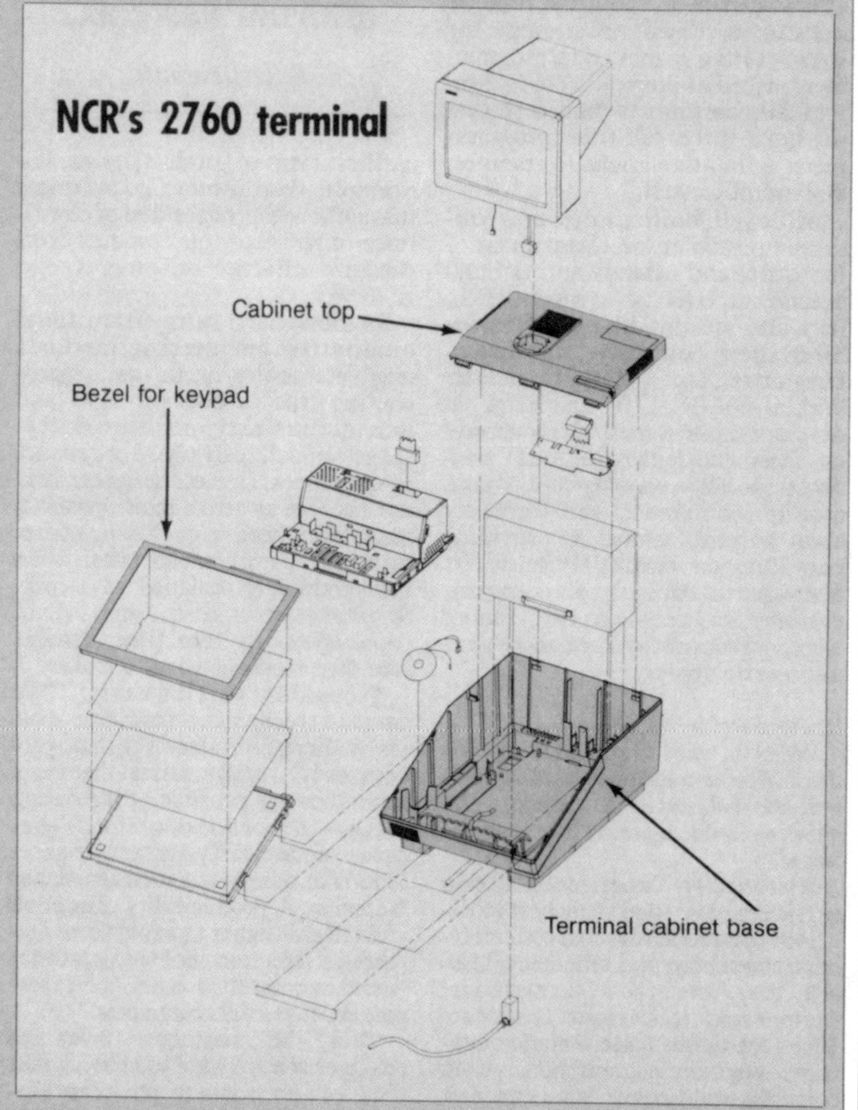

Major new plastics parts on NCR's 2760 terminal (shaded in blue) were designed for assembly. The terminal cabinet base, for instance, consists of nine separate parts that snap together. Completed assemblies are shipped by the molder to NCR on a just-in-time basis.

p. 85). Designed for the IBM AT and PS/2 type desktops, it was developed for cost estimating injection molding component designs during the conceptual stage of the design process. An overview of the quantitative engineering system helps to bring Boothroyd and Dewhurst's concepts, and the power and versatility of their software, into sharper focus.

When making design for assembly trade-off decisions, designers need accurate cost estimates of the various injection-molded components that go into a product. Proper application of DFA techniques usually results in component consolidations and more "elegant structures," as Dewhurst puts it. Therefore, individual component cost savings can be more substantial than savings in assembly costs. With minimal information on final design specs, BDI's new DFM software allows these comparative savings to be quickly and easily quantified.

Based on a product concept, the software automatically selects an appropriate injection molding machine from its machine data base. The machine data base, gathered from a variety of manufacturer's specification sheets and personal discussions with manufacturers and molders, contains clamp-force values, clamp strokes, power ratings, and hourly rates. Moreover, the software advises users if a machine is required with capacity—clamp force, clamp stroke, or shot size—exceeding those available in its onboard memory. Therefore, designers need no prior knowledge of molding machines or molding machine performance to run the program.

A materials data base is also included. It contains data on engineering thermoplastics and commodity materials commonly used in industry. And both the materials and machinery data bases can be added to or deleted from through easy-to-use menu driven procedures. Other features of the software include the following elements:

- A part geometry complexity evaluator;
- A graphics spreadsheet system for volume and area calculations;
- Algorithms for relating mold manufacturing costs to part size, geometric complexity, tolerances, and appearance factors;
- Procedures for determining filling, packing, cooling, clampstroke, and ejection times;
- An optimization routine to identify the optimum number of cavities for minimum cost, depending on machine rates, cycle time estimates, part size, and core and cavity manufacturing costs;
- "Associative" editing capabilities for instant cost recalculation across the board when changes are made in either material selection, part definition, or production volume; and
- An interactive costing data base used to establish mold making hourly rates, molding shop efficiency values, machine hourly rates, molding shop efficiency values, machine loading limits, and other operational variables.

Ford Motor Co. estimates that design's influence on product cost is 70%

Even though their current customers rank among the bluest of the blue-chip, large multinational product manufacturers, Boothroyd and Dewhurst claim their methods can be used by smaller firms. And that includes custom molders beefing up their design engineering departments for simultaneous engineering work. "One of the advantages of using these quantitative techniques is that it gives smaller manufacturing firms something concrete to talk about when they discuss a project with their customers—the bottom line," explains Boothroyd. "If you've got a handle on costs at the early stages of design, you can get your quotes together more accurately and have something to talk to customers about. So this material provides a framework for bringing the manufacturers and the designers closer together."

As a matter of fact, Boothroyd and Dewhurst relied heavily on the expertise of custom molders and mold makers in their area in developing the injection-molding DFM software.

Blue sky time

The professors have started looking at integrating DFMA software with computer-aided design (CAD) technology. What Boothroyd and Dewhurst envision as the "ultimate" is the capability of simultaneously designing a product on a CAD workstation and obtaining cost information for manufacturing the product from that same system. "Tools are becoming available in the CAD field that will facilitate that kind of link between CAD and DFMA," Dewhurst says.

But, integration with computer-aided engineering (CAE) software, like finite element, flow, and cooling analysis, is not a high priority item at the moment, Dewhurst says. "The reason is that those are either tools to make sure that the design satisfies the function, or they are tools to make sure that a design is efficiently producible. That's further downstream than what we're concerned with. Such techniques are important, because you've got to make sure that you're not producing rejects once you get a product going on the shop floor. But it's not where the major cost savings are."

Could DFMA be the key to successful implementation of the lights-out, remote control, factory of the future? "Well," Dewhurst says, "if you're going to have a lights-out factory, you'd better design your products so that they can be manufactured and assembled in that factory. It becomes even more important to ensure that the design is a suitable one. And if people have become somewhat disenchanted with robotics and computer-integrated manufacturing in this country, we would claim that one of the main reasons is that not enough attention was given to product design. Many of those systems failed to work satisfactorily because they were trying to accomplish tasks that were too difficult for them, and that should have been designed out of the product."

Dewhurst reports that plans are in the works to develop DFMA software for plastics processes other than injection molding—blow molding in particular. "Blow molding is no longer just for bottle manufacturers. There are all sorts of wonderful components starting to be made by blow molding. I saw one, for example; a frame for a backpack. The traditional backpack frame is an assembly nightmare. It's all aluminum struts, and tons of screws and washers. The new one is a single piece blow molding. It's probably much more comfortable, too."

"We're looking at a number of processes, and the approach is essentially the same in every case even though they present different challenges," says Boothroyd, adding with a smile. "But we're both engineers, and engineers can do anything."

Concurrent Product/Process Development (CP/PD)
A Concurrent Design Methodology: Making It Happen

By Rick Norman
Manager, QFD Software Marketing
International TechneGroup Incorporated

When Dr. W. Edwards Deming approached the American automobile industry after World War II with some fundamental ideas on quality management, he encountered a management philosophy directly opposed to many of his ideas. Rather than adopt Dr. Deming's Statistical Quality Control (SQC) techniques to support a strategy to produce high quality, long-lasting automobiles, it seemed that "design for obsolescence" was perhaps the best name for the underlying business practices of the time.

Dr. Deming found a different way of thinking and a more receptive audience in Japan. The automotive world is now well aware of the results of and the fundamental differences in the philosophies which led to today's market share situation.[1] Other manufacturing industries have learned equally painful lessons: consumer electronics, machine tools, and semiconductors to name a few. Time has proven that taking a very long-range view at the fundamental process of manufacturing is indeed a better way to do things. As American business managers and consultants began to assess the Japanese manufacturing methods, the message began to trickle out: "Quality control is a management decision," as Deming used to often say. The lessons are clear.

I think it can be safely said, now, that setting high quality goals for manufacturing, implementing the operational techniques to achieve that quality, and coupling them with a long-term business strategy, for many companies, is becoming the essential first step to remain competitive in today's global business environment.

THE CONVERGENCE OF IDEAS

Some of the techniques being employed to achieve those quality objectives are now becoming disciplines in many companies. Philip B. Crosby's battle cry is "do it right the first time,"[2] a way of thinking that he feels is fundamental to achieving high quality. Boothroyd Dewhurst, Inc. has developed a set of software tools and methods that are concerned with designing products with explicit considerations for assembly and manufacturability. Statistical Quality Control techniques have been in place for some time now, allowing quality management to monitor the manufacturing process more closely.

In the early 70s Structural Dynamics Research Corporation pioneered the concept of MCAE, using computer modeling techniques to better enable engineers to predict performance behavior of complex mechanical systems before committing to costly prototypes. In the 80s, International TechneGroup Incorporated is taking the MCAE concept a few steps further and is developing a model that includes a broader simulation along product, process and business lines.

---

1 Jeanes, William, "The Idea That Saved Detroit," Northwest Magazine, September 1987, pp.15-19.

2 Crosby, Phillip B., Quality Without Tears, p.59.

Also in the 80s, a methodology known as Quality Function Deployment (QFD) has become popular. The QFD process assures that customer needs drive the product design and the production process. This system has been used by Toyota since 1977.[3] It is now being widely used throughout the US manufacturing industry to help focus development team efforts on what is important to the customer.

All of these methods are focused at considering manufacturablity (taken in its broad sense) in the early stages of design. As we move into the 90s, these related techniques are beginning to merge into a unified whole. A broader range of computer simulation tools are now available to predict not only product performance, but process performance and product cost as well. "Next-generation" CAD/CAM tools are beginning to emerge that address designing a product for manufacturability, so it really can be done right the first time.

DESIGN FOR MANUFACTURABILITY: GETTING IT RIGHT THE FIRST TIME

What does this statement really mean? Taken literally it implies that if one designs a product "right" on the first attempt, then that product will be manufacturable. But what does manufacturable mean? Depending on one's knowledge and depth of understanding, that term can mean a lot of things. If one takes a very broad view of an entire manufacturing process, from initial product concept through product retirement in the field, the meaning of "manufacturable" becomes extremely fuzzy.

This view, which is the perspective necessary for manufacturing enterprise executives who are interested in being competitive in today's global economy, takes into account the entire product development process, the production process itself, the distribution strategy development, and the product support strategy. The costs to manufacture a product are embedded in each of these fundamental areas. When given this corporate view of the problem, getting it right the first time becomes a significant challenge.

CONCURRENT PRODUCT/PROCESS DEVELOPMENT (CP/PD)

Many companies today are beginning to employ this broader view of manufacturing in the attempt to get it right the first time. This complex process is known by many names: design for manufacturability was perhaps one of the first. Simultaneous Engineering, Concurrent Engineering, Concurrent Design Methodology, among others, are becoming more commonplace and normally have much broader implications than design for manufacturability.

In a nutshell, CP/PD is the process of simultaneously considering all stages of the product's life cycle during the initial design stage; product cost and performance are engineered to meet the desired objectives. Heavy emphasis is placed upon assuring that the product's target market needs are thoroughly analyzed so that the proper focus is achieved by the product team. Many product and process alternatives are considered early to assure that the most competitive alternative (i.e. cost effective) has been chosen for further development. CP/PD implies that

---

3 Sullivan, Lawrence P., "Quality Function Deployment," Quality Progress, June 1986, pp.39-50.

the organizational structure is in place to allow for more work up front, when conceptual product and process changes cost much less.

The CP/PD process has three major components:

- Methodology
- Tools
- People

The methodology is one of working in multi-disciplined teams rather than the traditional departmental approach. All product and process-related costs and performance are considered by the team as it develops different product alternatives for evaluation. The people on the team come from product engineering, manufacturing engineering, finance, and marketing disciplines. This is the staffing expertise and view required to evaluate the product and process alternatives thoroughly enough to gain a high degree of confidence of the results. Other disciplines can be added to the team structure, if necessary. It all depends on the size, scope and type of project.

The tools involved vary widely from company to company. (See page 7 for further discussion of these tools.) Automated tools are the enabling technology that allow CP/PD to happen.

HOW DO YOU DO IT RIGHT THE FIRST TIME?

For complex products it is doubtful that all tasks will be done correctly the first time. A very good way to get extremely close, however, is to implement a CP/PD methodology. The CP/PD approach has been successfully used in several cases.[4] Without going into the detail of reference 4, Scenario C in Figure 1 shows that at the time of initial production, traditionally we never meet our customers quality expectations. This is a costly scenario; engineering changes in the production phase cost 5 times more than in the build/test phase and 100 times more than in the simulation phase.

[4] Fox, Robert W., "Strategic and Financial Justification For CIM," Proceedings of the Sixth Annual Control Engineering Conference, May 19-21, 1987, Rosemont, IL.

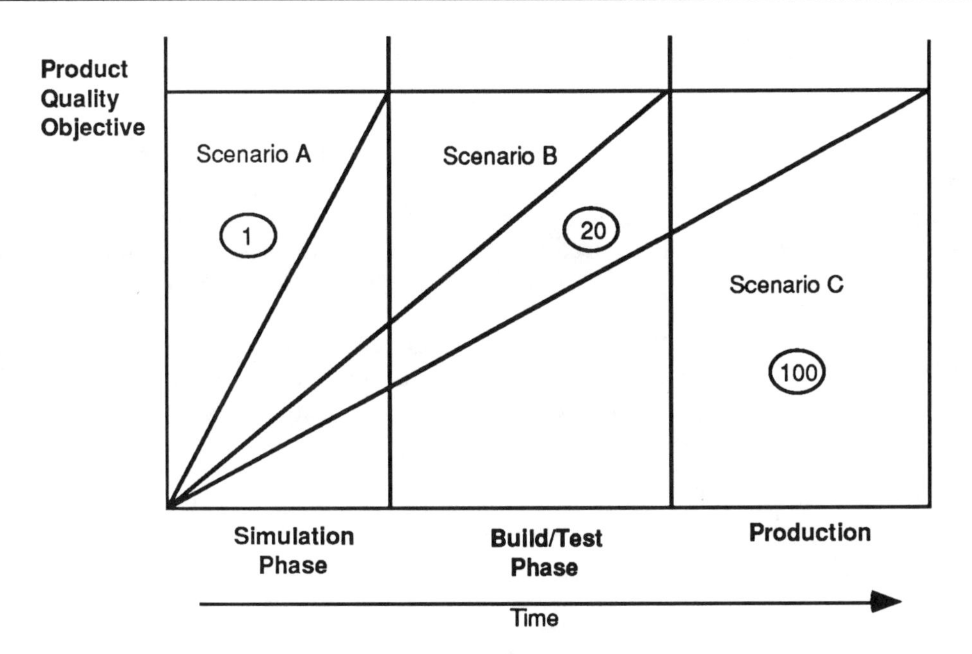

Figure 1
Quality Objectives Can Be Met Using CP/PD

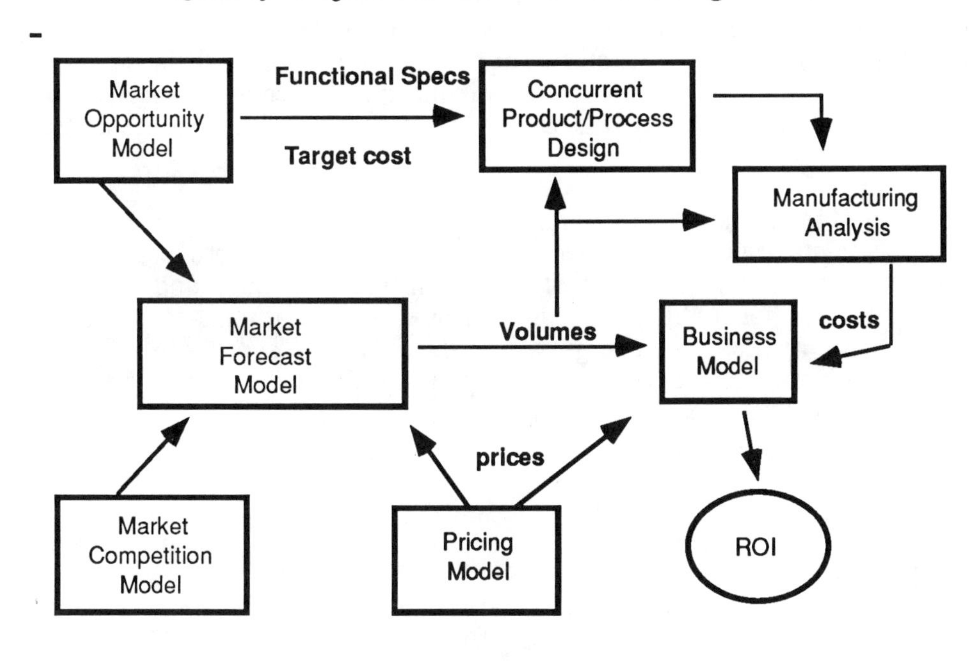

Figure 2
CP/PD is a Structured Methodology Driven by Market Models

Scenario B, representing a strategy of using CP/PD methods with todays tools, shows that quality objectives can be met before reaching the production stage. Scenario A, representing the ultimate goal of achieving quality standards as early as possible, is the result of using

CP/PD with the help of automated tools. Products developed in this way will more likely be done right the first time.

The CP/PD approach using the team concept allows several things to happen. First, the entire manufacturing development and production process is simulated (see Figure 2). Simulation here is taken in its most general sense. Much like early MCAE efforts to more precisely predict product performance, this model attempts to predict product and process performance with all their related costs.

Exercising the model by the team allows many product and process alternatives to be evaluated. This broad-scope "what if" type of process takes much of the guess work out of the required business decisions early in the development process. Second, communication is enhanced and the quality of decisions is higher because more complete information is available in a timely manner. Third, the structured approach, driven by market models in a multi-disciplined environment, provides the people with a much broader view of the problem and brings the focus of everyone closer to the desired end. QFD matrices are applied throughout this process to help focus the effort.

HOW DO YOU KNOW YOU HAVE DONE IT RIGHT?

A variety of indicators make it clear that the products developed using these methods are the right ones. Probably the most clear sign is cash. When corporate profits start to rise, it is probable that something is going right. Another indicator is market share. This is perhaps the best indicator that all facets of the entire model were simulated and executed correctly. Finally, there is corporate growth. A healthy growth pattern is a very good sign that the underlying principles of the team effort are on track.

Of course, none of these indicators are short-term. The results of the CP/PD process cannot be seen overnight. Implementation of CP/PD is a very slow and deliberate process that evolves with time and much hard work. Companies could use many quality and performance measurements to gage how their CP/PD process is progressing. But customer satisfaction, measured in terms of cash, market share and growth, is the only true "success test" for CP/PD. (Reference 4 describes in more detail a case study using the CP/PD process; ROI is expected to exceed 35 percent on an investment of $50 million)

HOW DO YOU MAKE IT HAPPEN?

There are no magic formulas to ensure success. CP/PD takes on the many faces and peculiarities of the many corporations using it. Its culture-changing nature assures this characteristic. However, based upon experience, a formula has emerged that works well in most situations. Several steps are involved:

1. Get Top Management Commitment and Involvement

Without question the most critical factor in the entire process is top management commitment. The CP/PD process, like CAD/CAM implementations yesterday, requires much cultural change. Without visible support from the top, these changes probably won't happen. Many unexpected barriers will be encountered along the way. The process is costly and difficult. It requires a much longer view to achieve the payback. Without top management support, all these nuances could be show stoppers.

The most important reason for upper management support, however, is to ensure that the process is part of a long-range business strategy. In 3000 BC, Sun Tsu wrote, "All men can see these tactics whereby I conquer, but what none can see is the strategy out of which victory is evolved." CP/PD implementation is a series of battles. To win the war, a sound strategy must be in place and supported for the long haul. Only top management can provide this support.

2. Establish Strategic Vision

This concept of vision goes beyond the establishment of a long-term strategy, although that is part of it. This vision implies the establishment of a more general awareness of the CP/PD activity on the part of all concerned. It is one of the most significant cultural changes required for CP/PD. At every level in the organization, contributors must gain an overall view of the development process so that problem identification and solution happen more naturally and immediately. This general improvement of personal awareness of a larger chunk of the overall process is one of the prime contributors to that elusive term "worker productivity."

3. Begin Awareness Projects

One of the most effective means of establishing this awareness is through the conduct of an Information Flow Study. Effective and sensible interviewing techniques not only collect information flow knowledge, but more importantly inform contributors of what is going on. Contributor "buy-in" is an important goal.

When conducting these studies it is a good idea to take a functional approach. Identify organizational functions so that contributors can speak in their own language. Much of the computer jargon of the day is incomprehensible to most people in the manufacturing process. Keep the focus of the interviews on a person's job function, and you would be surprised how much useful "design information" can be gained.

4. Establish Pilot Projects

Pilot projects are an essential first step for CP/PD. They should be carefully selected and sized to ensure that a good sampling of typical functions is included but not so large or risky that success would be seriously jeopardized. A pilot allows "real-world" resource commitment without overextending in a new situation. Many of the information system tools that must be used today are not really integrated, so a small pilot will assure more timely, clear, and complete communication among team members. Manufacturing automation alternatives should be fully evaluated in the pilot and used to the extent practical; don't assume full automation. The financial model may show that there is a manual or semi-automated solution more desirable.

Establish project goals clearly and insure understanding. A functional decomposition of sorts should occur from the strategic vision. Everyone should be able to see clearly how his or her own work contributes to the whole. Clear goals and a renewed culture of building quality into everything make the "teaming" concept work to its fullest potential.

The pilot serves as the training ground for people becoming familiar with the many parts of the CP/PD process.

5. Use Available Simulation Tools

One of the things that makes CP/PD possible is computer technology. As suggested in Figure 1, the next generation of simulation tools promises powerful capabilities. Today, an adequate set of tools exist to allow the methodology to succeed admirably. PC spreadsheets, project management software, 4th generation languages, CAE software, CAD/CAM systems, factory simulation languages, and other previously mentioned capabilities, all can contribute significantly to provide the tremendous amount of information necessary for the team to conduct the appropriate amount of "what-if's."

The important thing to remember is to use tools because they make the task(s) easier, not because the tool is available. Adequate knowledge of the tool's capability is essential for its productive application.

6. Select An Internal Champion

The concept of "champion" has emerged over the past few years, but the meaning of the word varies. Used here, a champion is the person who has the ultimate responsibility to see that things get done. Champions can and should exist at all levels in an organization. Champions are the internal catalysts for change. They are the "reformers" that Niccolo Machiavelli refers to in The Prince:

> "Thus it arises that on every opportunity for attacking the reformer, his opponents do so with the zeal of partisans, the others only defend him half-heartedly, so that between them he runs great danger."

Machiavelli has merely said very eloquently that there are three kinds of people in this process of initiating "... a new order of things." There are those that make things happen, those that watch things happen, and those that wonder what happened. Champions make things happen.

7. Use Experts

Outside consultants can play a very significant role in the CP/PD process. Used most effectively they can be catalysts for change. Sometimes it is very difficult to effect change from inside the corporate hierarchy. Organizational "battle lines" are often too entrenched for an internal "champion" to operate effectively. Consultants can provide an unbiased neutral ground to help break down these old organizational boundaries. This is essential if the CP/PD process is to function effectively. Consultants can also ease the heavy initial burden of awareness training, education program development and execution, and actual hands-on training programs for the many computer tools needed.

SUMMARY

In order for today's US manufacturing industry to maintain its world-wide competitiveness in the 1990s and beyond, corporations must develop a sound business strategy that has as its fundamental underpinnings the principles implied in "do it right the first time." The CP/PD process described here is a set of tools and methodologies that people can use to realize those principles. CP/PD takes simultaneous engineering and design for manufacturability concepts along with Quality Function Deployment methods to a more fundamental level of manufacturing management.

Implementation of the CP/PD is not a trivial undertaking. It requires high-level corporate commitment that presumes significant cultural change. Lead by an internal champion, the CP/PD process requires automated tools, a practical level of automation, the help of outside experts. All this effort must be implemented in the team framework to allow for the lack of adequate integration of available computer tools, to facilitate the cultural changes, and to manage the initial pilot projects.

Given a balanced effort along these guidelines, there is no reason why significant progress shouldn't be made toward "getting it right the first time." Indeed, the results of these efforts should be increased profitability, higher quality, and increased customer satisfaction (market share).

REFERENCES

Crosby, Phillip B., Quality Without Tears, p.59.

Fox, Robert W., "Strategic and Financial Justification For CIM," Proceedings of the Sixth Annual Control Engineering Conference, May 19-21, 1987, Rosemont Illinois

Jeanes, William, "The Idea That Saved Detroit," Northwest Magazine, September 1987, pp.15-19.

Sullivan, Lawrence P., "Quality Function Deployment", Quality Progress, June, 1986, pp.39-50.

Reprinted from *Assembly Automation*, May 1987

Bridging the gap between design and assembly

Management, design and new flexible assembly systems came into focus at the recent International Conference on Assembly Automation. Brian Rooks reports.

IT IS becoming increasingly evident that a major stumbling block to the implementation of assembly automation is more managerial than technological. This message came through on the opening day of the International Conference on Assembly Automation held in Copenhagen last month, the eighth in the series organised by IFS (Conferences) of Bedford, England. One of the critical managerial factors is the integration of the product design process with the manufacturing process. Bringing design and manufacture closer was one of the aims of the conference and a reason for taking the conference to Denmark under the chairmanship of Professor Myrup Andreasen.

That the advances achieved by technological experts in assembly automation have not been realised was jumped upon by the conference opener, Jan Bendix, managing director of the Danish company Sant and Bendix. Pointing a finger directly at the conference audience he said, 'Improvements promised by you professionals have had only limited success. At least until now you have not greatly influenced the trade figures of Europe'. The problem is that most solutions presented at such conferences are from 'star companies', to quote Bendix, whereas most European companies are in the small to medium category - Denmark has less than 100 large companies. These smaller companies do not have the resources of the 'stars' to implement high technology solutions. In Bendix' experience those who have production engineering teams with an understanding of, market conditions and customer needs, production methods, job design and working conditions are most likely to succeed. But, in Bendix' view, 'Success occurs through careful but tough cooperation, where only a few standard solutions will help, and where new principles and methods can be balanced with the capabilities of the people, the organisation and the equipment'. His parting advice to the conference was, 'Look at assembly automation as one important piece in the jigsaw puzzle called the future European production unit'.

Meanwhile on the other side of the Atlantic the USA also is looking to the future of manufacturing automation. The conference had a status report from Frank Riley on flexible assembly in the USA.* Being senior vice president of The Bodine Corporation, one of the country's longest standing suppliers of automatic assembly equipment, and vice president and president-elect of SME, the leading US professional society for manufacturing engineers, Riley is in a unique position to comment on the US scene.

His first comments were to the point, 'The year 1986 will be recorded as the year in which factory automation in the US had to face up to harsh competitive challenges. It was a year in which perceptions had to face reality and a year in which older movements came to an end and new directions started. In short, it was the year that user industry, equipment manufacturers and academia, as well as government truly recognised that the problem of implementing new technology was primarily managerial not technological'.

Riley traced some of the problem back to the early 1980s when several large US corporations already in new manufacturing technology found new captains at the helm. The new leaders of GE, IBM, Ford, Chrysler and GM were not content with a slow evolution of these technologies and saw them as an opportunity to not only improve productivity but also capture new

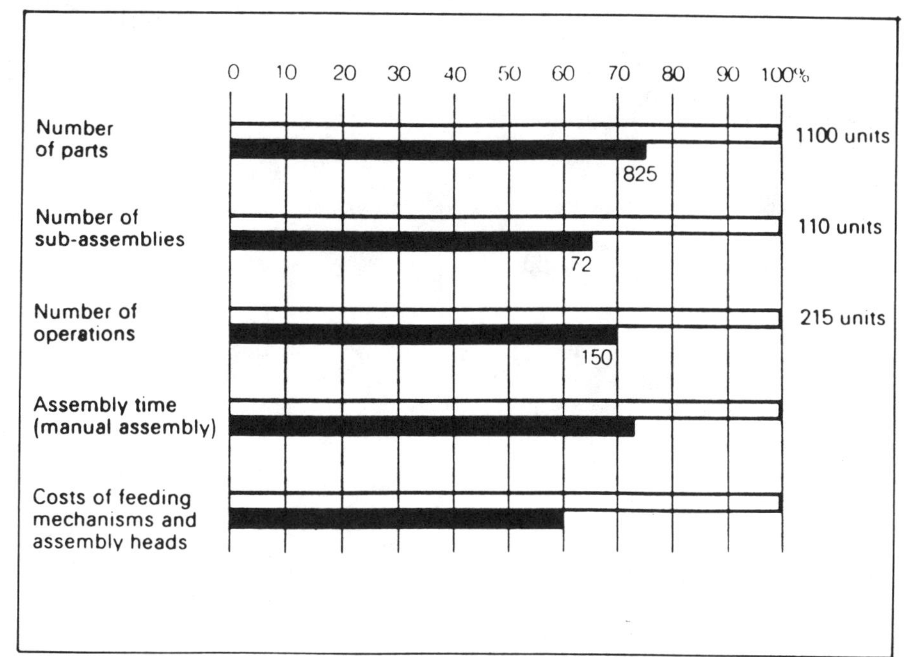

Using the Boothroyd system of DFA Philips have been able to influence product design for the better.

*The full text of the paper by Frank Riley as well as all the other papers presented at the 8th ICAA are now available in the Proceedings published by IFS (Publications) Ltd, 35-39 High Street, Kempston, Bedford MK42 7BT, England.

standing technology such as robots cannot achieve competitiveness alone. The economic realities that led to the decision to combine the 1988 Robot and Machine Vision shows will ultimately lead to a decision to combine them all under the banner of computer-integrated manufacture'.

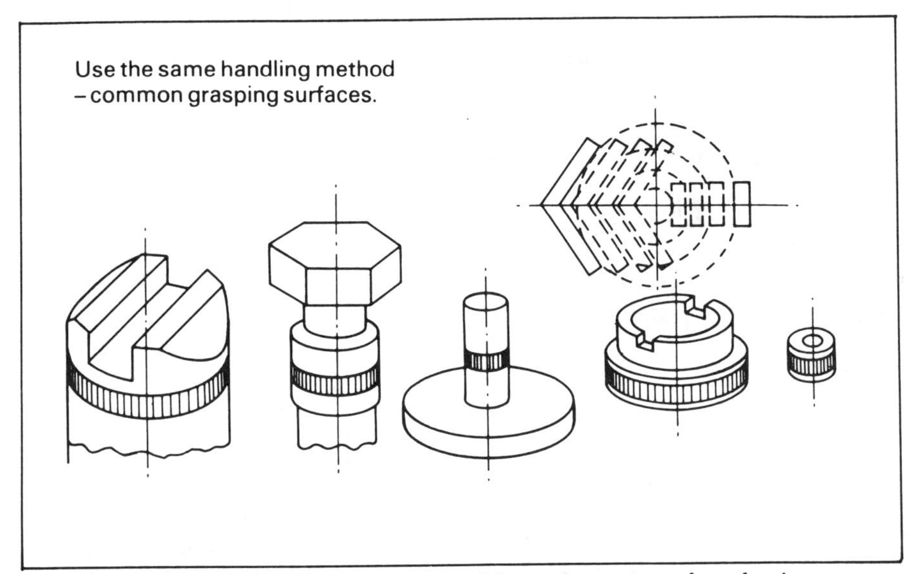

Making variants transparent to the assembly equipment, such as having a common gripping surface, was stressed by Myrup Andreasen.

markets. Such moves were sufficient to trigger a flood of venture capital into new technology companies. And, the press and market analysts hyped up the whole atmosphere to proclaim an era of re-industrialisation and competitiveness in which automation would replace costly and inefficient direct labour and improve profitability.

By 1985 enough had been done to measure the economic benefits of the new investments and they were found wanting. Costs were higher than anticipated, market projections were not being met and often installed equipment did not work. It started to emerge that the perception of direct labour was not correct. And, Riley pointed to the often overlooked facts that US labour is highly productive - higher than Japan - and direct labour costs are only 10 to 12% of total factory costs. The problem is the overhead which in the US contributes to almost half of factory costs. Management found that applying conventional return on investment methods to new technology did not work.

The impact of this was to dramatically slow the introduction of new technology. Riley commented, 'GE decided to phase out its factory automation efforts and close down its robotic and vision units, IBM reorganised its internal structure and made its products capable of communicating with each other and GM severely cut-back on its spending leaving the long range future of EDS unclear'.

But, all is not gloom, Riley seeing many encouraging signs. Some robot companies are doing well, notably Adept, Milacron and Cimflex as well as GMF which seems to be weathering the storm of last year's cancellations. More importantly there appears to be a new thrust that will impact on the American drive for computer-integrated flexibility. He went on, 'There is now a sensitivity that free-

New procedures

This new thinking has also spread to the US academic grant awarding body, the National Science Foundation (NSF). New procedures, with government funding equally matching that from industry, will force university research to be more in tune with industrial needs, similar to the European and Japanese patterns. And, now there is encouragement being given to removing the barriers between product design and process planning. A strong message from a recent NSF symposium spelt out that flexibility will only come about by joining the product design and process planning functions, which is where the 8th ICAA came in.

It was on the second day of the conference that discussions turned to

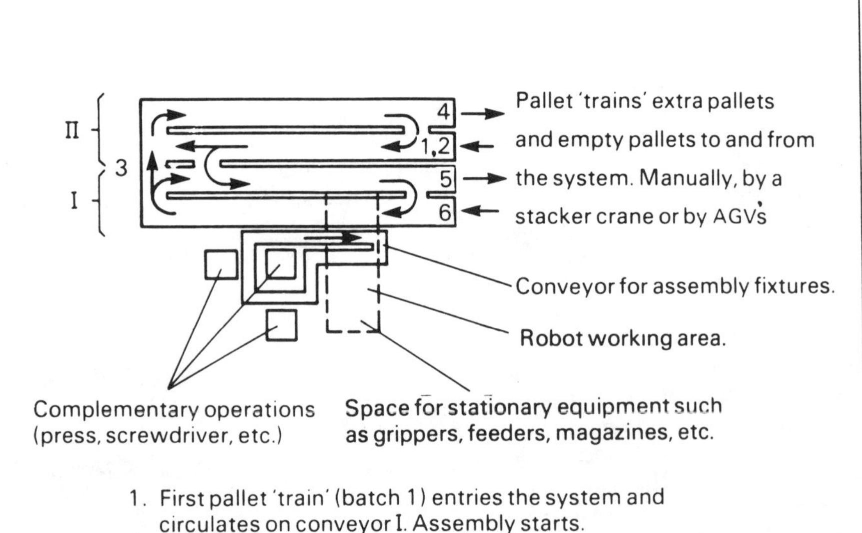

Layout of the IVF Mark II flexible assembly system.

'design for assembly'. Opening the session Thomas Lund of Dansk Teknologi recommended a simple approach, 'First eliminate and then automate'. Reducing the number of parts in a product eases the assembly process. But, there is one 'danger', after elimination there may not be any need for automation - could this lead to redundancy amongst the conference delegates?

The problem of design and manufacture being separate functions was stressed by several speakers. Professor Graves of the University of Massachusetts centred his paper on reducing costs by integrating design and manufacture. He discussed the approach taken and the design aids used by one large company, which resulted in reductions of the time and risk required to bring a new product to the market.

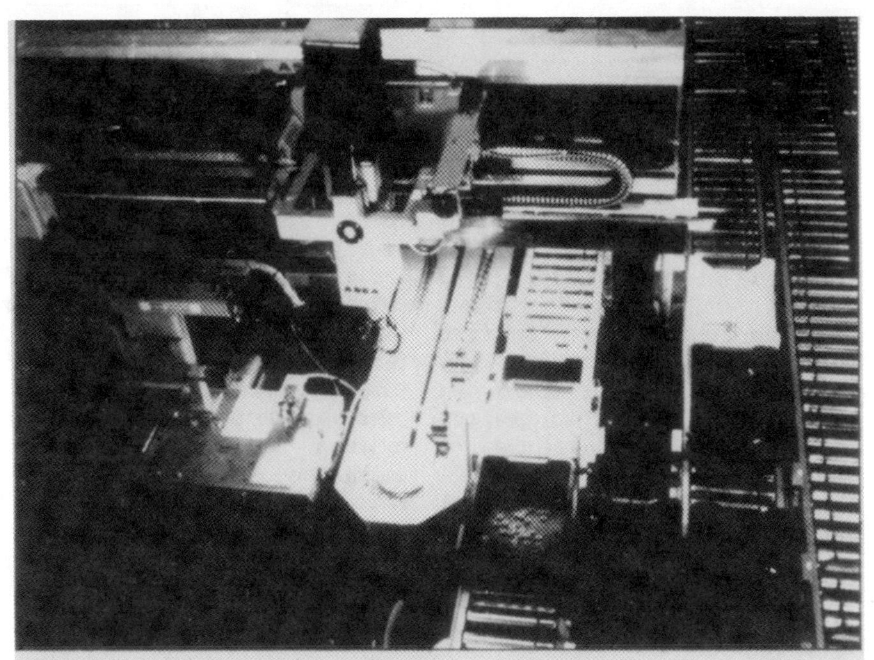

The IVF system uses two circulating conveyors to enable a robot to assemble a very large number of different part-types whilst minimising cycle time.

Improvements

Educate the designer seemed to be the plea of Jan Boorsma, of Philips Holland. He said, 'Designers only think of the function of a product and not its manufacture. The result is a complex product'. He uses the Boothroyd system of design for assembly (DFA) to cut through the complexity. Tests of the method have shown significant improvements of around 30% in several areas including reduced component count, shortening of manual assembly time and increased effectiveness of cooperation at the development and starting stages. With design lives of audio products falling well below one year Philips appreciates the value of the latter. But, Boorsma pointed out, there is resistance to the use of DFA; product redesign is often in conflict with the image of future products; DFA is seen as a criticism of a designer's professional expertise; and risky designs seem to emerge.

To overcome these 'roadblocks', he recommends producing a 'naked assembly diagram' of a product before it is detailed. Then, the analysis can be partly applied at this early stage. And, if this is done for all the expected members of the product family the various diagrams can help to minimise the parts diversity. At the same time the total number of parts can be reduced before going into fine detail. In a later stage the analysis can be completed which gives more precise guidance on part shape and ease of assembly. In Boorsma's experience this method not only removes the objections to DFA but also shortens the throughput time because it generates detailed product diversity planning.

Speaking in the DFA session, conference chairman Myrup Andreasen, himself a designer by training, also put the designer in the 'limelight'. He said, 'The designer is the problem', but continued, 'and he is also the solution to the assembly problem'. However, Andreasen's paper was directed to more than just design for assembly. He concerned himself with design for flexibility and distinquished between flexible production and the products for manufacture by flexible means. He defined the degree of variation which is built into the product as the variance. The ease with which variance can be handled in production is related to the flexibility of the manufacturing system. But, by manipulating the design of the product it is often possible to reduce the need for flexibility in the system. The aim is to 'fool' the assembly equipment into thinking it is only producing one product type and not a family of variants.

Discussion of flexible assembly equipment was divided into two sessions; electronics and mechanical products. Significant steps have been made in the former over the last two years. Electronic assembly is mainly concerned with PCB manufacture and standard solutions are becoming available. With a greater variety of parts and with no common base the assembly of mechanical products has not progressed as rapidly. So it was with some interest that the conference heard of new ideas for the assembly of these products.

Solution

One of the most exciting ideas came from the Swedish Institute for Production Research (IVF) in Stockholm. Professor Anders Arnstrom presented the IVF Mark II system as a solution based on the assembly of 'part batches'. This development follows experience with the Mark I version presented at the 14th ISIR in Gothenburg in 1984. The major step forward was to eliminate the coupling between the number of parts and the system capacity imposed by the feeding of those parts to the assembly robot. Also, the feeding equipment is not product specific and changeover from one batch to another within the system is automatic.

The new arrangement has two conveyor systems, one for the circulation of the product under assembly (an continuous link conveyor) and the other for transportation of robot grippers, fixtures and parts (a roller conveyor). The objective is to minimise the time for the robot - an ASEA IRb 1000 in the Mark II system - to pick up and assemble a part. Thus, the two conveyors run in parallel adjacent to the assembly station.

Tote boxes circulate on the roller conveyor. The first tote holds batch-specific equipment such as product pallet fixtures robot grippers. The second tote holds the first part to be assembled, the third one the second part and so on. The final tote is empty ready to receive the completed assemblies when the whole batch has been made up. The system is arranged so that totes for two batches can circulate enabling an automatic switch from the first to the second batch. Parts are stored in the totes either simply on flat plates unorientated but the right way up or in part-specific formed trays. A vision system is used to detect the position and orientation of the parts within the tote.

When a new batch arrives the first function of the robot is to pick out a fixture holding the robot's grippers for the parts of the batch and place it into a designated area. Then, the assembly fixtures are handled again by the robot and placed onto standard pallet carriers on the link conveyor. The fixture tote is replaced by the first part tote when assembly can commence. As each assembly pallet presents itself a part is added. Between each assembly operation the pallet circulates on the interlink conveyor allowing other assembly operations, such as pressing, screwing, welding and the like, to be performed.

Experience of the system has been gained by assembling two commercial products; an Atlas Copco air motor and a steering. These tests have demonstrated the principle that one robot can feed an almost unlimited number of different parts and that cycle times can be very short. The latter comes about by having short gripper exchange times - Arnstrom claims it is 7-10 times faster than most flexible assembly systems - and the short distance between the pick up point from a tote and the assembly point; only about 400mm.

The 8th ICAA was refreshing in the way that it brought out both the managerial problems and some technical solutions particularly in flexible assembly. It highlighted the need to bring design and manufacture closer together. Perhaps when the 9th conference in the series is held next year in London (15-17 March 1988) more designers will attend and help bridge the divide that unfortunately still exists. □

Reprinted from *Machine Design*, June 8, 1989

Designers Gain Insight into the Factory

When engineers work from the same database as manufacturing, both sides find it easier to talk to each other.

NANCY E. ROUSE
Senior Editor

Engineers traditionally have a hard time communicating their ideas to manufacturing. The invisible barricade that prevents clear communication is cultivated by both corporate culture and the lack of good technology. The information barrier is so palpable that engineers often refer to throwing designs "over the wall" into manufacturing.

Computer technology was supposed to tumble this separating wall. Yet, almost twenty years since computer design and manufacturing methods took root in industry, it still stands.

Engineers, however, have chipped away at it by using new modeling techniques, better computer hardware, computer networks, and expert systems. When they are used, such methods let engineers gain access to manufacturing information before they finalize designs. Thus, designers make early decisions that prevent later delays in production.

Solids or surfaces

Design usually communicates geometry to manufacturing in the form of surfaces. Within the year, however, many vendors will move to solid rather than surface models as the primary means to represent designs. Use of solid representations may make it easier to generate some toolpaths from the design model. But solids will not support machining of molds, tools and dies where the cavity does not exactly match part shape.

Even though present machining software still operates similarly on solid models as on surface models, solids eliminate the ambiguity that exists when surfaces are used. "With a solid you get one surface, rather than a series of surfaces. You don't have the same problem determining the completeness of a surface in solids," says Jim McDonald of Intergraph. Intergraph recently moved from a surface-based modeling system to a solids-based one.

When surfaces taken from most geometric models are used, the NC programmer must specify each section of the surface to be machined. This is done because such surfaces have no sense of being a single en-

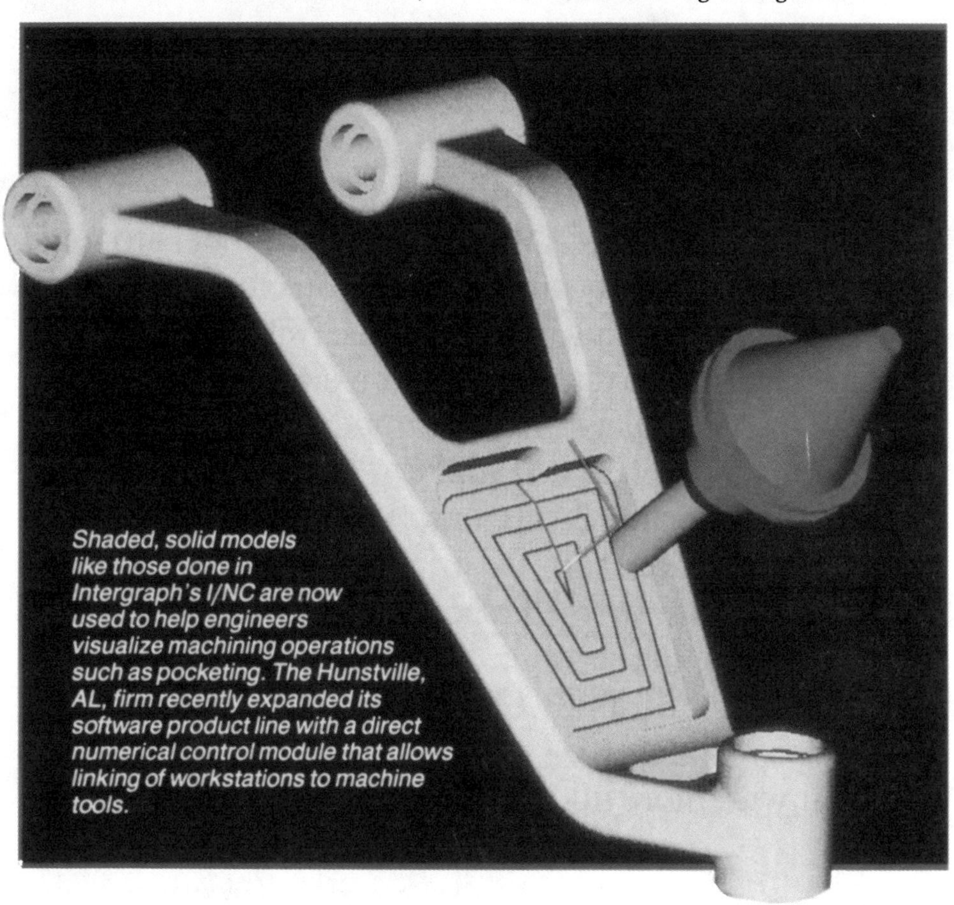
Shaded, solid models like those done in Intergraph's I/NC are now used to help engineers visualize machining operations such as pocketing. The Hunstville, AL, firm recently expanded its software product line with a direct numerical control module that allows linking of workstations to machine tools.

tity. The individual parts of the surface are not connected. NC programmers or engineers must blend the surfaces together before a proper toolpath can be machined.

A surface taken from a solid model, however, does not need to be blended because of the way solid models are formulated. The description of their geometry is more complete mathematically than surface model geometry. Thus, experts refer to solid models as more "robust" than surface models.

Experts see robust solids as opening the door to the use of different algorithms to produce machine toolpaths. Today, most toolpaths are produced using the point-to-point method. This means that the programmer must specify each surface to be machined so that the software can generate a toolpath.

A problem with this method is that it is hard to control the tool over a complex surface because the three components of the toolpath — part, drive, and check surfaces — must be differentiated manually by the programmer. The part surface is the section that is actually machined. The drive surface runs parallel or alongside the tool. Check surfaces are places that the tool may bump against as it moves along the part. Specifying surfaces is difficult, and it is easy for programmers to make mistakes.

"There's a lot of people interaction with computers today. By the time you've designed in 3D, and then gone into the NC program to make all the fixture, clamp and tool selections, it sometimes seems as if you could have done the programming easier by hand," says Bailey Squier of CAM-I, an industry group sponsoring an Expert Manufacturing Programming System (EMPS) project. EMPS will develop a system architecture to automatically generate data and machine instructions to produce and verify parts.

Part of the CAM-I strategy is to develop a rule-based system that uses the characteristics of solid models rather than surfaces. For instance, the rule-based processor

INDUSTRY LEADER INSTALLS SECOND GENERATION SOFTWARE

Ingersoll Milling Machine Co., Rockford, IL, often is cited for its leadership in CAD/CAM. The company, which produces custom-made machine tools, won a Society of Manufacturing Engineers Lead Award in 1982 based on its use of a Cadam database in both design and manufacturing.

But Ingersoll's Cadam installation, which implements traditional surface-based technology, does not optimize savings because design and manufacturing are still separate operations, even if they use the same database.

"We realized that our CIM installation, no matter how well done, still did not have CAD and CAM fully integrated. They were separate functions tied together very well to a common database, but the two functions operated independently," says Stephen K. Lewis, Ingersoll vice-president of manufacturing. "Our goal now is to eliminate dual effort. New systems must make engineering and manufacturing one function."

The company is moving to a full 3D solids database, which will provide more complete design data than surface-based systems. Thus, NC programming and other manufacturing tasks can be made more automatic and accurate.

For example, data in a solid model may be used to do 3D collision-checking of toolpaths at Ingersoll. Such checking, which cannot be done well on surface-based systems, verifies an NC tape before it is used in production. At Ingersoll, where programmers produce 25,000 NC tapes per year, such checking could prove a godsend since programmers presently do not have time to do dry runs of tapes before they are moved to production. Such interference checking will eliminate much factory-floor debugging.

Solids models at Ingersoll also will abolish much of the detail drawing production and checking that occurs in their present system. And they will provide software prototypes. "With every one of our machine tools being custom-designed, there is never time nor money for prototypes. Therefore, we have relatively high costs related to correcting fits and interferences in productions. These occur today because errors are not caught in 2D views," says Lewis.

Toolpaths are represented as solids in ATP G/NC. ATP's software eventually will generate complex toolpaths automatically, without much user intervention.

The solid modeling software chosen by Ingersoll is a features-based one from ATP, Campbell, CA. In the ATP system, each feature of a part design such as a drilled hole or a surface carries with it a list of attributes like tolerances, finishes, machining processes, and tools. This information is present even in design, so the engineer knows immediately whether something he has thought up can be made.

Ingersoll and ATP have extended the features concept to a Generative NC package that automatically produces an NC program and a process plan from attribute information attached to features. This G/NC package just was installed in production at Ingersoll.

"Using the new software, NC tape programming is done merely by using statements such as 'mill side 1, drill and bore side 2.' This is exciting, since today's CAD/CAM systems require users to exert much more effort to prepare NC tapes from CAD models," says Lewis.

ATP also has the commitment of Boeing Commercial Airplanes, Seattle, and Chrysler Motors Corp., Detroit, for G/NC projects. Installations at these two companies, however, are not as far along as the Ingersoll project.

would automatically define the part a series of smaller volumes.

In this method, called volume decomposition, the small pieces are delta and sub-delta volumes. Delta volumes carry information about the features of a part. The part must be broken into these volumes to make it easier for NC software to analyze part information and specify correct processes and toolpaths automatically.

But algorithms that use solids to do complex machining are still years away, and for the near future most complex machining will use surface technology. "We are not investing in solids as the only basis to prepare parts for machining," says Ken Kutz of Camax, a leading vendor of high-performance machining software. Part of the problem, says Kutz, is that present solid modeling systems cannot develop smooth, sculptured surfaces used in applications such as car bodies and aircraft wings.

Kutz says Camax is putting most of its effort into making it easier to develop toolpaths from sculptured surfaces. He agrees with critics of surfaces that present systems often are hard to use. "To use some systems today, you practically have to have a Ph.D. because you have to understand the mathematics of the software to set up parameters. We've added features that let users get to a toolpath more quickly," says Kutz.

Features that improve efficiency in Camax include roughing routines for large parts, automatic surface blending that allows creation of sculptured surfaces with fewer entities, and routines that can do 5-axis, simultaneous machining. Software that contains simultaneous machining routines can program the tool and part to move at the same time. Routines with 5-axis positioning rather than simultaneous machining allow only one or the other to move. Camax transfers surfaces from solids done in systems such as those from MatraDatavision, SDRC, and Parametric Technologies Corp.

Making systems smarter

Most CAD/CAM systems are used primarily to model part geometry, but this geometry must be mass-

AEROSPACE TECHNOLOGY FOR BETTER GOLFING

Irons made by Karsten Mfg., Phoenix, are so accurate that they soon may be banned from some professional golf tours. In 20 years, the Karsten Ping club has gained a reputation for making great golfers stupendous and average golfers great.

What gives Ping clubs, named for the sound they make when they hit a golf ball, so superior? Karsten uses computer analysis, design, and manufacturing methods founder Karsten Solheim first became familiar with as an aerospace engineer at General Electric. Finite-element analysis is done on club heads, for example, to determine how the club and ball deform in the split second when contact is made. Aerodynamic analysis is done on woods so engineers can figure out ways to minimize wind resistance.

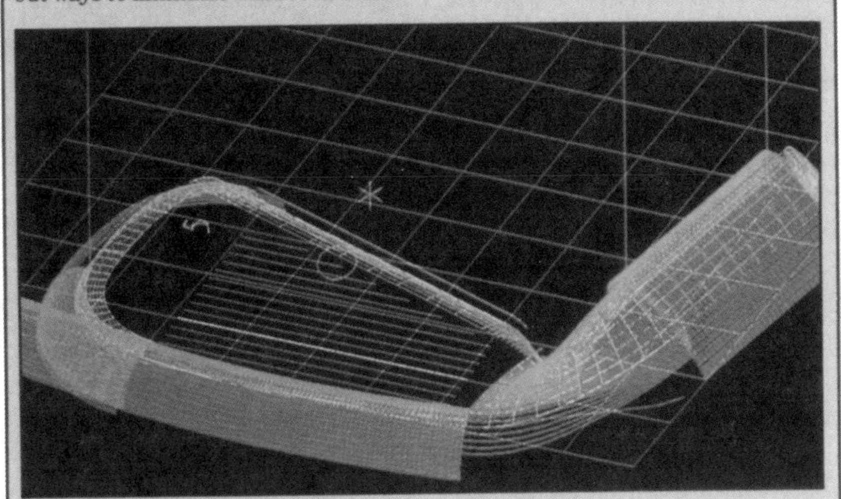

Software from Camax Systems, Minneapolis, produces NC instructions used to machine the complex surfaces of golf club heads.

NC programs are sent directly to machine tools at Karsten. Here, a prototype club head is machined.

Because any manufacturing defect can affect club performance, special attention is given to how club heads are processed. Karsten Mfg. owns its own foundry and heat-treating facilities. And head shape is controlled through use of software for surface design and complex NC programming.

Karsten recently began using Camax surface-based design and manufacturing software on Silicon Graphics workstations. The software lets programmers develop graphic toolpaths for multiaxis machining. This means that programmers and engineers can model tool and part movement in three dimensions. Engineers can then see the movement using the dynamic rotation features of the Silicon Graphics machines.

Unlike most computers, which repaint images each time a part is moved, SG machines contain special microprocessors that manipulate graphics so fast that the image looks as if it is moving in real time. Engineers control movement of images from a small console.

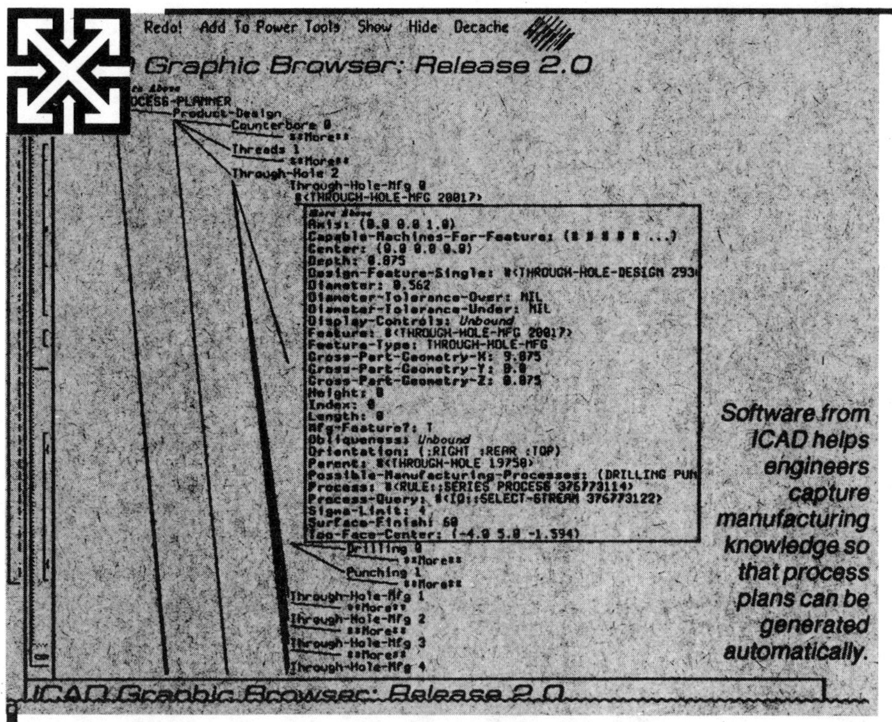

Software from ICAD helps engineers capture manufacturing knowledge so that process plans can be generated automatically.

ICAD Graphic Browser

aged a great deal before it can be used in manufacturing. Improved links to manufacturing data through use of database management and rule-based systems will allow engineers to take processes into account before a design is completed. In this way, some manufacturing planning is eliminated downstream.

For example, today a designer might specify a 0.379 in. hole in a design, based on analysis results. The spec represents the engineer's best attempt to optimize the design. What the designer does not know is that the odd-sized hole is more difficult and expensive for the factory to make because it is drilled and bored. A slightly smaller hole that would require only drilling might be a better choice. But in most cases, the designer does not have access to manufacturing process information. Software that allows engineers to tap into the manufacturing database would give them access to information about processes.

In the above case, for instance, software might flag the hole to let the engineer know it requires double processing. The engineer then might check a relational database to see what drills are used in the factory. After checking this information, the engineer might decide to change the hole size to 0.0375 in. to match the available tool size, possibly eliminating a manufacturing step.

Software that moves manufacturing information up into design is being introduced by many vendors. McDonnell Douglas, St. Louis, for example, is putting together an Engineering Information Management (EIM) system that it hopes will improve the ability of engineering to release designs with 100% assurance that they can be manufactured. EIM will include links to Manufacturing Resource Planning (MRP) systems and relational database managers used in manufacturing. Thus, engineers will have access to information about the resources available in the factory.

Rule-based systems that capture manufacturing knowledge early in design may soon be standard in design and manufacturing software. Manufacturing and Consulting Services (MCS), Irvine, CA, for example, spent several years developing an expert system generator, Metexsys, for its Anvil-5000 program. Using the system, the engineer will specify design rules and create a knowledge base for design of parts or assemblies. The system draws on the knowledge base to make judgements, or inferences, about designs.

If Metexsys infers that the design is poor, it disapproves it. The engineer redesigns and rechecks the part. This iterative process continues until a design is approved. An advantage of such systems is that they can include, in the knowledge base, manufacturing information that is checked along with the design.

Metexsys should be fully implemented in Anvil this year. Patrick J. Hanratty, president of MCS, be-

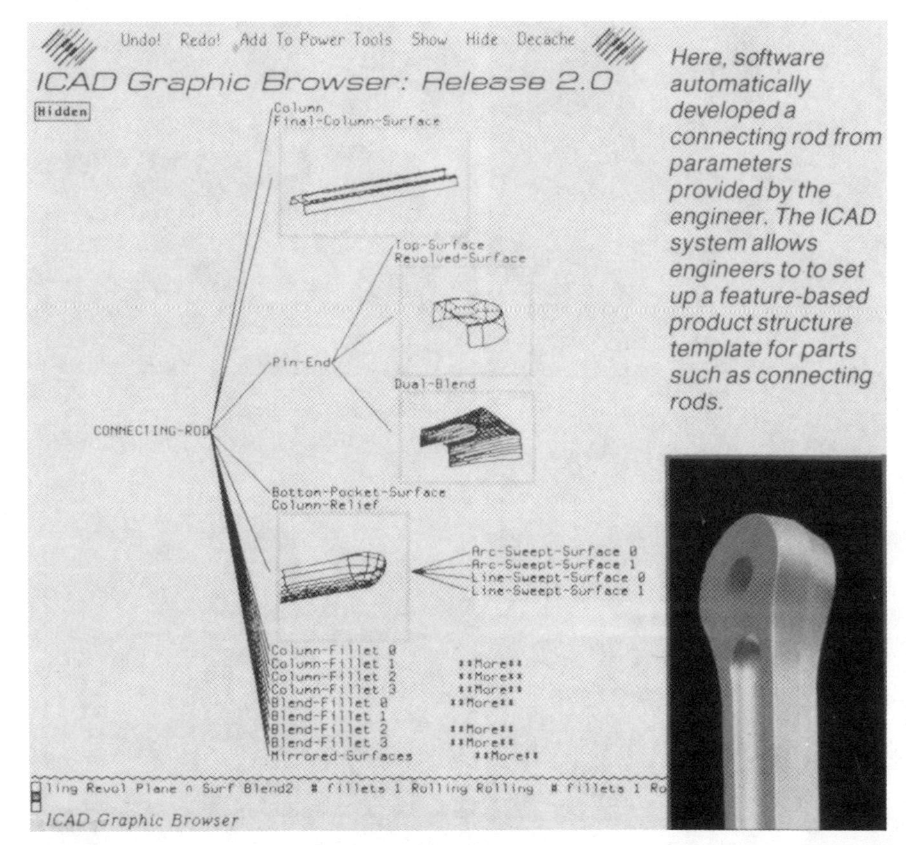

Here, software automatically developed a connecting rod from parameters provided by the engineer. The ICAD system allows engineers to to set up a feature-based product structure template for parts such as connecting rods.

ICAD Graphic Browser

lieves that within two years most vendor software will include ways to generate and access knowledge about part. "Knowledge will be built in so that engineers can control the entire process, " says Hanratty.

McDonnell Douglas is incorporating software from ICAD, Cambridge, MA, into its CAD/CAM system that will allow engineers to model any design or manufacturing process using a Lisp-based description language. In the past, ICAD systems were expensive because they required special Lisp processors to run. However, the software now operates on many machines, including those from Sun Microsystems and Apple Computer. Porting the ICAD system to other machines made it possible for companies such as McDonnell Douglas to integrate the system into their products.

Larry Rosenfeld, president of ICAD, draws a distinction between his product and expert systems. According to Rosenfeld, the ICAD system does not draw conclusions or infer anything about a design that is not specified by design criterion. Instead, the system is programmed to automatically apply specified engineering attributes. Plus, the system can perform routine engineering work: catalog searches, configuration, calculation, and iterative design optimization. This makes it a particularly useful tool in areas where there are distinct, hard and fast rules.

For instance, General Motors Engine Engineering Group may use McDonnell Douglas' implementation of ICAD for part design. According to Rosenfeld, parts would be designed in ICAD, then sent to McDonnell Douglas' Unigraphics package for drawings and NC work. Process plans for the designs, however, would be generated automatically by ICAD.

" We believe that by using the ICAD approach we'll be able to reduce time to market," says GM's Dick Wandmacher. GM now writes procedural programs that automate some design operations. But it would be simpler to write and maintain such programs in ICAD.

Kodak has used the ICAD system for mold design. In this application,

JOB SHOP FINDS BIGGER IS NOT NECESSARILY BETTER

In 1987, with a mainframe and almost 30 terminals attached to its machine tools, Synchronized Design and Development Inc. (SDD), Sterling Heights, MI, looked like a model, high-tech supplier to the auto industry. But the company, which does body panel design, was about to be crushed by its computer system.

SDD was discovering what many other companies doing computer NC work on CAD geometry already knew. That it was almost impossible for programmers to use NC software to produce logical cutter paths from the tiny geometric entities provided by CAD systems.

"We were having trouble doing cutter paths for large panels. The entity counts were so large that the computer system would slow down. The software was not designed to machine contoured, multiple surfaces," says SDD President Barry Figel.

"We had to machine little patches of a large surface separately — and then link all of these paths together to create complete tool paths on body panels. We also had to set up containment work allowances to prevent out mills from violating adjacent surfaces," adds Figel.

As a result, the company found its CAD system held hostage by its 10-person NC department. Designers could not get work done because cutter-path creation took up so much CPU time. The situation finally got so bad that SDD quit using its computers for NC work.

The firm also was saddled with extravagant maintenance costs for its mainframe system. Maintenance contracts ran over $300,000 a year; in-house systems specialists cost even more. And the company needed the maintenance, since the system was logging over 300 hours of downtime per month.

The situation reached a point where the CAD/CAM system, which was installed to improve productivity, was costing more than it was worth. That was when SDD started looking at replacing its mainframe. To Figel's surprise, it found a solution to its problems in a PC package.

"Micro Engineering Solutions' Solution 3000 has solved important practical problems in manufacturing," says Figel. For instance, the program automatically groups CAD geometry into surface entities. Each surface can be distinguished by color. In addition, surfaces can be placed on a separate layer in the program. (In this way, Solution 3000 works like a CAD program. Geometry can be isolated on many layers. The geometry is then overlaid to display the entire picture.)

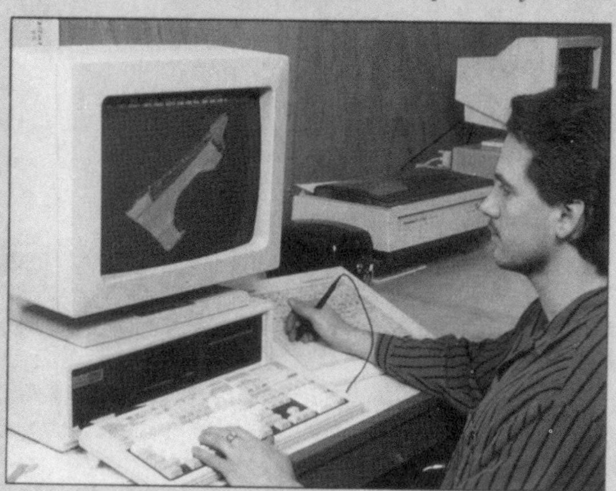

Small job shops are particularly vulnerable to the high costs of maintaining computer software and hardware. SDD cut maintenance costs by switching to PCs.

SDD also found geometry transferred through Solution 3000 translators required limited rework. This is important to the company, because much of the geometry it uses comes in on disk, magnetic tape, or over telephone lines. SDD now uses the IGES translators in the program as a filter to send data to its design group, which uses a different, workstation-based system.

According to CAD/CAM manager Brian Thibodeau, the PC system has considerably improved SDD's bottom line. Projects that once took 240 hours of programming work now take less than 100 hours. "And I know that the 100 hours can be reduced once we get everyone fully trained," says Thibodeau.

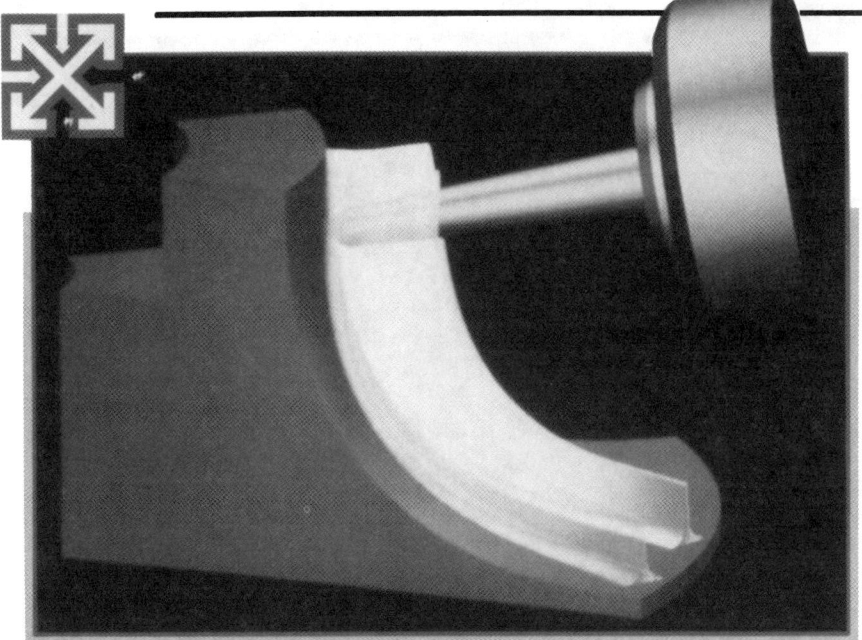

Graphics workstations let engineers more realistically simulate machining. For example, designers can use the computer's ability to rotate objects in real time to observe 5-axis machining from any angle. Strim 100 from Cisigraph Corp., Farmington Hills, MI, is shown on a Silcon Graphics Personal Iris computer.

the system uses design rules to create a mold for a part based on parameters entered by the designer. Kodak also may use ICAD to develop process plans and NC toolpaths. "We see ICAD being used to design a knowledge base for engineering and manufacturing, and then develop a planning model that ties both together," says Rosenfeld.

PCs bring down costs

There are over 50 PC-based NC programming systems available today, and many of them do the same work that was done in the past by mainframes and minicomputers. Though most systems are limited to 2 -axis machining, some offer 3 to 5-axis machining as well. (In 2-axis machining, the tool works primarily on the x and y axes. It can position itself in the z axis, but does not have the same freedom of movement found in 3 to 5-axis machining.)

PCs now contain fast microprocessors and can use over 1M byte of system memory. This means that they can do NC work faster than in the past, and can hold larger files. For the most part, however, software vendors are hamstrung by the 640k memory limitation imposed on them by the DOS operating system. Developers have found ways around the limit, which in the past prevented programs to operate on large 3D files involving 3 to 5-axis machining.

Vendors approach the 640k memory problem two ways. Companies like Point Control, Eugene, OR, and MCS use DOS extenders that allow programs to run in the protected mode of the 80286 and 80386 machine. Programs running in protected mode can access more than 640k. Therefore, large programs can reside entirely in main memory. This eliminates overlay swapping, where program routines are moved between the hard disk and main memory as needed.

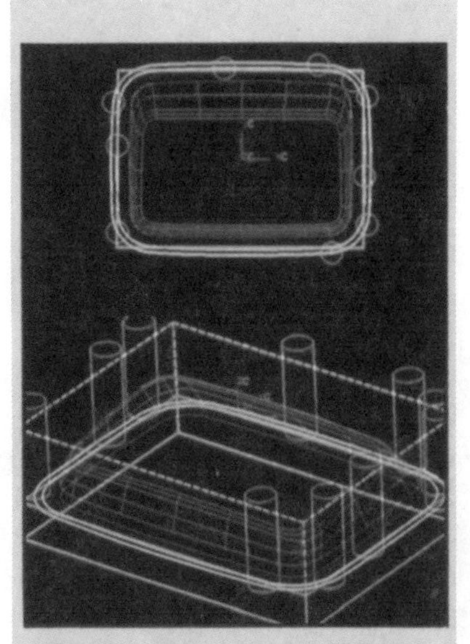

NC software from McDonnell Douglas allows simulation of 3-axis machining processes such as roughing. Newer software will use solid models rather than surface representations.

Another method is used by Micro Engineering Solutions (MES), Novi, MI. MES does not now use a DOS extender to speed program operation. However, it divides available RAM memory (usually between 3 and 4M bytes) into three parts — main memory, disk cache, and RAM disk. The first M byte of memory is divided into 640k of main memory and 512k of cache. The cache holds overlays from the hard disk. The rest of the available RAM holds the geometry database as it is developed by the user. Using memory like this eliminates some calls to the hard disk and speeds program operation.

Software developed for PCs runs differently than programs first developed for mainframes, and so can operate more efficiently. For example, the SmartCAM program from Point Control generates NC machine code directly from the graphics model, rather than through the APT or Compact II programming languages. Most older programs translate into APT or Compact II first, and then postprocess to machine code. SmartCAM recently was approved by systems integrator EDS for use inside General Motors Corp.

Many PC program developers cut their first teeth in the machine shop, which they feel gives them a distinct advantage over developers who got their start in computer programming. Their familiarity with the way machining is done allows them to write code more compactly. "Because we started out with dirty fingernails, we understand where production has to end up," says Bill Harris, president of MES.

But compact code does not necessarily mean trivial software. For example, Boeing Commercial Airplanes recently recommended Solution 3000 from MES to its suppliers because it accepts data formatted by Catia, Boeing's surface modeler. This means suppliers using Solution 3000 need not redevelop the mathematical surfaces they get from Boeing. "We believe we are 95% compatible with Catia," says Harris. ■

Reprinted from *Mechanical Engineering*, September 1988

A Closer Coupling

Design for manufacturing is quickly becoming one of the most important trends in industry today. As the pressure mounts to bring new products to the marketplace as quickly as possible, the ability to reduce expensive modifications during the manufacturing phase by concentrating on the initial design has obvious benefits in terms of time and money saved.

Design is the most important stage in the design-for-manufacturing process. This may seem strange when it is manufacturing that we are ultimately trying to improve. Nevertheless, a study by British Aerospace found that the first five percent of resources, expended during the preliminary design phase, typically determines 85 percent of the final cost and performance of a product.

But no matter how useful such sophisticated tools as surface and solid modelers are in the rapid development of the best possible design, they count for nothing if the product can't be manufactured or assembled. Before design for manufacturing can become a reality, production and assembly information will have to be incorporated into the CAD/CAM data base.

The manufacturing engineer is bound to ask whether the necessary machine tools are available and whether the designer's product can be assembled. If it cannot be made and assembled, it isn't a product. The more knowledge of production that is included in the CAD/CAM data base, the more efficiently the product can be manufactured. For instance, a design should not specify a tolerance of 0.0001 inch if the available tools can only machine to a tolerance of 0.001 inch.

Another important factor is the trend toward outsourcing: if the total cost of a product is to be reduced, it is essential that accurate and complete manufacturing information be communicated to the subcontractor in the right format. Moreover, a company may find it difficult to control its products if the parts it has outsourced need to be modified.

The incorporation of manufacturing information into the CAD/CAM data base has several implications. One is that the data base will have to deal with new types of information—not only topology and geometry, but also form features, tolerances, and text, all in machine-accessible format. Today, information about a product is stored in several forms—in drawings and words, on paper, in computers, and in physical models. None of these forms represents the sum total of information about the product's design and manufacture. The CAD/CAM system will have to store a complete model in electronic form.

One data structure capable of storing these large amounts of alphanumeric and relational information is a relational data base, which would also allow local data bases to be managed as part of a global data base. An industry-standard relational data base language, such as SQL, would provide ready access to the data base and also simplify the task of linking other data bases and proprietary applications to the CAD data base.

Feature-based modelers represent another step forward in data management. These modelers render into machine-accessible format all the information needed to manufacture a discrete component—mating relationships, part properties, form features, tolerances, surface finish. Taking into account the way a particular company manufactures components and the tolerances to which its parts can be produced, feature-based modeling allows designers to compare the manufacturing cost of different designs.

While feature-based modeling will help make design for manufacturing a reality, assembly modeling promises to improve the design of assemblies. Assembly modelers electronically represent such information as bills of materials, assembly drawings, and mating relationships. The relationship of parts, represented in terms of a tree structure, can be defined at the preliminary design stage. The information can be used in decisions regarding not only product design, but also certain manufacturing processes, such as scheduling and task allocation.

The ultimate goal of design for assembly is to have an expert system that, knowing the relationship between parts, can check designs to see if they can be made with fewer parts and can verify mating relationships automatically.

Interactive finite element modeling and analysis techniques can greatly accelerate the creation and evaluation of design alternatives. Such analysis tools can predict how designs in development will stand up to real-life loads, vibrations, and deflections. Using them can help to ensure that the manufactured product performs correctly and that time is not wasted on a product that will eventually be discarded.

Finally, simultaneous engineering will go a long way toward furthering design for manufacturing. Its two underlying concepts are parallel development of the product and the manufacturing process, and the creation and evaluation of multiple product and process design alternatives. Simultaneous engineering gives manufacturing and design engineers a real opportunity to determine in advance, by working together, the least expensive way of manufacturing a product.

Although design and manufacturing are coming closer together, the two will never merge. Both will continue to be specialties requiring specific tools, but in the future they will be linked. With simultaneous engineering, each participant will exercise authority in his or her own area of specialization, but all will use the same data base. This common data base will ensure that separate jobs are done with reference to each other.

Design is a world of creativity and experimentation. Manufacturing is a world of rigor, in which the concerns are quantity, time, and prices. Simultaneous engineering acts as a valve between the two.—*Michael Smith, Computervision*

Reprinted from *Assembly Automation*, August 1988

DFA as a primary process decreases design deficiencies

A survey of papers and design-for-assembly systems shows where this discipline should be heading.

P.J.Sackett and A.E.K.Holbrook, The CIM Institute, Cranfield, UK.

THE worldwide introduction of automation to the assembly process has led to the exposure of deficiencies in product design. Traditionally, production engineering considerations for complex engineering products have been left to the end of the design process. This is too late: 80% of a product's manufacturing costs are committed at the conceptual design stage[1].

Design-for-assembly methods have been developed that introduce the demands of production engineering after the functional design phase and not as a correction technique for completely designed and tested products. The value of these techniques is already evident from the current proliferation of technical literature that describes design case-studies. In the past, product cost savings between 15% and 70% have resulted from analysis and redesign.

The development of design for assembly stems back to elementary work studies carried out in 1911. These established repeatable operation times for manual tasks. The Systematic Approach developed by the General Electric Company was used to relate the value of product design to assembly times. In 1977 a programme of research was initiated at the University of Massachusetts, USA, and the University of Salford, UK, to develop a classification and coding system for manual handling and both manual and automatic assembly, that would form the basis of a systematic method for design analysis[2]. From the results of this programme, a number of design-for-assembly methods and analysis procedures have evolved.

The rapid rate of growth of this subject since 1980 has resulted in several different approaches being taken.

Current design-for-assembly methods

A survey of commercially-available systems and literature published over the last eight years has led to categorisation of design-for-assembly techniques into four groups:

- specific assembly operation theories (insertion and tangling);
- unstructured rules and concepts;
- procedural methods of applying rules;
- expert/knowledge-based systems.

The work in each category is described and discussed. An assessment of each method with regard to its economic effectiveness, value and applicability to a range of product sectors is included.

Specific assembly operation theories

Design-for-assembly concepts are frequently presented in the form of rules and guidelines. The specific assembly operation theories concentrate on a single design rule, usually analysing component geometry for its suitability to one aspect of assembly. So far, only two types of operation have been investigated: these are component design for insertion, and a theory to avoid the tangling of parts, the latter being suited to improved picking of parts from bins[3].

All the techniques are of a highly mathematical nature, converting the shape of a part into a value of assemblability, the value being stated in terms of geometric preference or statistical representation. For example, insertion envelopes are mathematically related to the geometry of the parts being joined.

Each method addresses only one design rule, and is therefore not able to satisfy a company's total design-for-assembly (DFA) requirements. These theories are suited to forming part of a more complete design system.

Unstructured DFA rules and concepts

The majority of existing design-for-assembly work falls into this category. Literature of this type presents collections of design rules and concepts which are derived from product structures and features, providing good inherent assembly characteristics.

It should be noted that, while each rule is correct and beneficial in its own right, any two rules can and often do provide conflicting advice. For instance, the objective of reducing the number of parts in a product may lead to two parts being formed as one; however, this may cause handling and other difficulties. In this case, only one rule can be applied: the one providing the greater cost reduction, yet still facilitating product functionality.

Another problem is that the sheer quantity of design-for-assembly rules available makes it hard to remember them all and to know which one to apply in a specific context.

Simple sets of rules have been constructed which effectively overcome the memory problem; one example, a ten-rule summary, is shown in Fig. 1[4]. But such sets of rules lack specific principles of application. For this reason, systematic methods of applying design rules are preferred[5].

Procedural methods of applying rules

These are systematic approaches to the task of design for assembly using written or software-guided procedures. The procedural methods can be separated into two principal types: spreadsheet analyses and rule-based systems.

Numerous spreadsheet approaches exist, all of which operate on the same principle. A considerable amount of time is required to assess a number of assemblability factors for each part of a product. These then have to be entered on the spreadsheet and summed to give a total assembly time. A description of the procedural techniques, their suitable applications, availability, ease of use and cost, follows.

The rule-based systems are no more than computer-assisted ways of presenting design rules, formed into a structure of assembly-related subjects.

- Minimise the number of parts in an assembly.
  Reason for parts being separate:
  (a) parts must move relative to each other,
  (b) different materials are required to give different properties,
  (c) for maintenance and replacement reasons,
  (d) to enable assembly on other parts,
  (e) manufacturing process constraints.
- Standardise a product's function and style.
- Standardise the use of components wherever possible.
- Provide chamfers and other location features for easy insertion.
- Design components that are easy to handle.
- Design product for minimum directions of insertion.
- Where possible, components should be grouped into sub-assemblies.
- Try to design parts that can be orientated rapidly.
- Design parts for stability during assembly.
- Try to minimise product weight.

Fig. 1. Summary of design rules

Expert /knowledge-based systems

These systems are the most advanced in the field of DFA, yet not necessarily the most effective at reducing assembly cost. Most of these systems are written in a declarative programming language such as Prolog or LISP, or an Expert System (ES) shell like Savoir. These facilitate easy representation of heuristic reasoning.

Each system runs through a lengthy phase of questioning the designer for product data needed for the analysis. The output of these systems gives specific advice on where a product or component would benefit from a change to its design.

To reduce clerical input to DFA software systems, graphical analysis of computer-aided design (CAD) drawings has been used in some cases.

Discussion

Some of the advantages of using procedure or knowledge-based design systems have been outlined. The systems are all based on a common three-part method, comprising a survey, analysis and output.

The survey consists of a stage of gathering data concerning the product, such as: component geometry, the parts list, materials, production volumes and costs. This data is acquired through either entry onto forms or by answering a set of assembly-related questions, and is passed to the analysis function.

The analysis stage is achieved in one of two ways. Firstly, the spreadsheet methods use estimated operation times (often converted to indices or costs) for feeding, grasping, manipulating and inserting, to calculate total assembly times for each part and a whole assembly. Secondly, the data procured from questions is used as knowledge for an inference engine that combines the precedents to generate assembly-orientated design advice. These systems, based on logical structures of rules, are commonly known as expert systems.

Stage three, the output from design-for-assembly methods, is limited by the quantity and types of data that have been entered. The spreadsheet methods do not operate with sufficient data to generate redesign advice, and are restricted to displaying assembly operation times and costs. A typical spreadsheet layout is shown in Fig. 2. In some cases, a design efficiency is calculated from a theoretical optimum design. When analysis of a product reveals a low efficiency, redesign is recommended for components bearing high assembly times.

By definition, an expert system should be able to demonstrate a level of expertise. The general DFA expert systems do not achieve this, giving little more advice than can be deduced from the alternative spreadsheet methods. However, the expert systems that are limited to a few design rules or a narrow field of engineering achieve a worthwhile level of advice, specific to the product being analysed.

The design-for-assembly techniques currently favoured by industry are almost exclusively based on procedural methods, and suited to mechanical and electromechanical products. Of the 14 systems that have been reviewed, 80% use a spreadsheet to quantify the assemblability of products; the remaining methods are rule-based.

Fig. 3 shows a table of the commercially-used DFA systems and comments on their individual features. Most of the systems restrict their application to product design only; however, the more complete systems such as 'ASSEMBLY'[6], 'Assembly Analysis and Line Balancing'[7], and the 'Productivity Estimation Method (PEM)'[8], also address system design and economics.

A survey carried out in West Germany revealed that, despite the well-known cost savings resulting from design for assembly, only five companies have formally adopted a systematic method. The lack of popularity arises from the time and effort necessary to analyse a product. In order to reduce time and clerical effort, a number of these systems have been written as software packages and are available to the public (Fig. 3).

Component	Item no.	Assembly sequence	Quantity	Handling time (s)	Insertion time (s)	Operation time (s)	Operation cost (p)
base	A1005-036	1	1	1.95	1.50	3.45	.86
seal	A1005-037	2	1	2.45	1.50	3.95	.99
sleeve	A1005-038	3	1	1.80	1.50	3.30	.83
weight	A1005-039	4	1	2.45	1.50	3.95	.99
cover	A1005-040	5	1	1.50	1.50	3	.75
spring	A1005-041	6	1	1.43	1.50	2.93	.73
washer	A1005-041	7	1	2.18	2.50	4.68	1.17
screw	A1005-043	8	1	1.50	7	8.50	2.13
TOTAL	—	—	8	15.26	18.50	33.76	8.44

Fig. 2. DFA spreadsheet for oil distributor assembly (assembly cost 0.25pence/s).

System name	Institution/ company	Authors	System type	Data entry	Contents/ method	Output	Manual/ software	Product sectors	Hardware/ system	Shortcomings ease of use	Training	Address	Cost (£)
A Designer's Guide to Optimise the assemblability of product design	General Electric	X	spreadsheet	form filling and look-up charts	UMass Handlability + Hitachi AEM	index values + costs	M	mech. + electro-mech.	—	lengthy no redesign advice	Y	1285 Boston Ave. Bridgeport 06601 USA	X
AEM–Assemblability Evaluation Method	Hitachi Ltd	T. Ohashi T. Yano	spreadsheet	form filling	100 point index for insertion, orientation, part count	index relates to cost	M	mech.	—	no handling no redesign advice	N	292 Yoshida-cho Totsuka-ku Yokohama Japan	X
AID Assembly Insight for Designers	IBM	J. D. Griebel	spreadsheet	questions + look-up charts	penalties 0-99 for handling and insertion	index values	M, S	mech. + electro-mech.	IBM pc + compat.	poor manual only design criticism	X	General Products Division Tucson, Arizona, USA	X
ASSEMBLY	Katholieke Universiteit Leuven	D. De Winter K. Machiels	spreadsheet + analysis	questions + look-up charts	DFA Handbook + line balancing, equipment selection, economics	op. times + system costs	S	mech.	IBM pc + compat.	complete system no redesign advice	Y	Celestijnenlaan 300B, B-3030, Leuven Belgium	250000 Belgian Francs
Assembly Analysis and Line Balancing	University of Massachusetts	C. Poli R. Graves	spreadsheet	form filling	evaluation of grasp, manipulate, insert, line balance	op. times costs	M, S	mech.	Digital Rainbow, VAX system	fairly easy with software	N	Amherst, MA USA	X
Assembly-Orientated Product Design	Fraunhofer Institut fur Produktion	R. Bassler T. Schmaus	spreadsheet	questions + procedure	evaluates function, assembly cost, assembly procedure	highlights weak design	M, S	mech.	X	no CAD link, no redesign advice	N	12 Nobelstrasse D-7000 Stuttgart 80 Vaihingen West Germany	catalogue 50 DM
ASSYST-ASsembly SYSTem	Politechnico di Milano	F. Arpino R. Groppetti	spreadsheet	questions + form filling	Integration of product and assembly system design	op. and system costs	M, S	mech.	VAX 11/750 VMS v4.5	too many questions, no redesign advice	N	Milan, Italy	X
A Systematic Approach to Design for Assembly	Lucas Eng. + Systems Ltd.	B. Miles K. Swift	expert system	questions + look-up charts	sequential analysis of function, feeding gripping + fitting	cost related indices	M, S	mech. + electro-mech.	IBM pc + compat.	many questions optimum design is unclear	N	Shirley, Solihull West Midlands B90 4JJ, UK	not available
Design for Assembly A Designer's Handbook	Boothroyd + Dewhurst, Inc.	G. Boothroyd P. Dewhurst	spreadsheet	questions + look-up charts	evaluation of grasp, manipulate insert	op. times design efficiency	M, S	mech.	IBM pc + compat.	time consuming no redesign advice	N	56 Sheerman Lane, Amherst MA 01002 USA	Handbook 18.00
DFA course	Smallpiece Trust	—	—	—	DFA Handbook	—	—	—	—	—	Y	Smallpiece House 27 Newbold Terrace E Leamington Spa UK	170.00
PEM-Productivity Estimation Method	Toshiba	K. Takahashi	spreadsheet	form filling look-up charts	estimates % assemblability 'E' + line balancing	'E' + system arrangement	M	mech. + electro-mech.	—	lots of form filling and no redesign advice	X	Toshiba Corporation Japan	X
Product and System Design for Robot Assembly	Robotic Assembly Consultants	R. Davisson A. Redford	spreadsheet	questions + look-up charts	revised form of DFA Handbook + more feeders	op. times design efficiency	S	mech.	IBM pc + Amstrad 1512	easy to use no redesign advice	Y	12 Harbury Ave. Ainsdale Southport PR8 2TA, UK	1500.00
Product Design for Economic Manufacture	Institution of Production Engineers	J. Corbett	guidelines	design subject	presentation of rules and figures	rules + figures	M	mech.	—	rules made clear, no procedure	Y	Rochester House 66 Little Ealing Lane, London, UK	12.00
—	Nooter Corp.	K. J. Korane	rule based expert system	chamber specification	applicaton of rules for flange design	geometric values	S	flanges	X OPS5 Fortran	specific to flanges only	N	Nooter Corporation USA	X

An 'X' indicates information not yet obtained

Fig. 3. Evaluation of commercially-used design for assembly systems.

However, considerable time is still required to enter product data into these system types. The DFA form-filling exercises demand more time and effort of designers than they are prepared to commit.

Late application of design constraints and DFA techniques results in an unnecessarily lengthy reiteration of the whole design process. The 'ADAM' and 'Assembly-Oriented Product Design' systems overcome this problem by dispersing product analysis throughout the design process[9,10]. This substantially reduces product development timescales; however, the volume of data is still prohibitive.

This attitude of designers to time-consuming DFA techniques could be overcome by having a permanent team dedicated to product analysis. However, the common pride of a designer in his work may make him unwilling to accept recommendations for change, from a separate (rival?) team. A better approach would be to provide the means for designers to appreciate and account for the needs of production during the initial design of a product.

The current trend in research should be to provide assistance to the designer throughout the application of design-for-assembly techniques. Advice-generating expert systems or knowledge-based systems have been used in eight of the nine research systems being reviewed. Fig. 4 outlines the contents of these systems and their methods of data input and output.

Of the three principal phases of each system — data input, analysis and data output — the method of analysis is common to all. The gathered data is processed through a logical mechanism to generate assembly-orientated design advice. The need for specific advice demands a comprehensive set of data to be entered: standard design and production engineering information such as standard parts and machine-

System name	Institution/ company	Authors	System type	Data entry	Contents/ method	Output	Manual/ software	Product sectors	Hardware/ system	Shortcomings ease of use	Training	Address	Cost (£)
ADAM- Assisted Design for Assembly and Manufacture	Cranfield Institute of Technology	A. Holbrook	knowledge-based system	solid model + product data	minimising parts, rationalisation, insertion + guidelines	verbal advice parts list	S	3D mech. electro-mech.	IBM CATIA + pc	easy to use, uses solid models	Y	CIM Institute Cranfield, Beds MK43 0AL, UK	700 00
Assembly-Orientated Product Design	Fraunhofer Institute for Manufacturing	R. D. Schraft	rule presentation	product design	list of rules for product, sub-assembly parts	rules + charts	S	mech.	CAD	not specific no procedure	N	IPA, Stuttgart West Germany	X
DACON- Design for Assembly CONsultation system	Hull University	K. Swift	rule based expert system	CAD analysis + questions	design for handling and orientation by Prolog expert system	redraws part	S	simple parts, high vol	Appollo Domain, PAFEC Dogs	very easy to use, only 2 design rules	N	Dept. of Engineering Design + Manufacture Cottingham Road Hull	X
Design for Assembly Optimal Suggestion Expert System	University of Michigan	M. J. Jakiela Papalambros	rule based expert system	questions	simplifies charts in DFA Handbook to imply design rules	rules + verbal advice	S	mech.	IBM pc Prolog	over simplified, too general	N	University of Michigan, USA	X
Effects on Product Design of Components Assembled by Automatic/Robotic Assembly	Cranfield Institute of Technology	D. Jackson	rule presentation	design subject selection	classification of 62 design rules into assembly subjects	rules	M, S	mech	Commodore PET	too many rules, no procedure	N	CRAG, Cranfield Beds, MK43 0AL, UK	X
Expert System for Design and Assembly of Mechanical Parts	Academy of Science and Technology	A. Nazif F. Saleh	rule based expert system	input shaft parameters	strength and fit shaft joints	shaft size tolerances	S	shafts	micro computer + graphics	specific to shaft design	N	Cairo Unviersity Egypt	X
HERNEX	Cranfield Institute of Technology	J. Hernani	rule based expert system	2D drawing + questions	design for orientation + assembly advice	verbal advice	S	2D parts mech.	IBM pc SAVOIR	too many questions, too general	N	CRAG, Cranfield Bedford, UK	X
—	Cranfield Institute of Technology	Trichelot	rule based expert system	questions	HERNEX rewritten in Prolog	verbal advice	S	mech.	IBM pc Prolog	too many questions too general	N	CRAG, Cranfield Bedford, UK	X
—	Heriot-Watt University	I. Aitchison	rule based expert system	product specification	design of assemblies by specifying product function	product design	S	mech	X Prolog	not yet linked to CAD	N	Dept. of Computer Science, Edinburgh University, UK	X

An 'X' indicates information not yet obtained

Fig. 4. Evaluation of design for assembly research systems

tool capabilities. This data can be stored on company databases and need not be entered for every analysis.

Questions concerning the product being analysed can often be reduced to enquiries about component geometry and parts-list data. Three systems have been developed that automatically extract a high proportion of the necessary data from a computer-aided design system, see Fig. 4. However, the present level of product definition resident on CAD systems is not in itself sufficient to allow DFA analysis.

The early graphical analysis systems (DACON and HERNEX) were restricted to applying handling and orientation rules to simple, high-volume components. Whilst these systems were good at reducing the effort demanded of the user, they addressed only a small band of the assembly analysis spectrum. The most recent system, 'Assisted Design for Assembly and Manufacture' (ADAM), moves into the field of complex assemblies and is applicable to low- and high-volume parts.

The supported design rules encompass minimising part count; component rationalisation; and design for insertion through using a solid-model representation of the assembly.

These research systems have significantly reduced the demands on the system user and provide assistance down the redesign path, even offering redrafting of parts in two systems[9,11].

Conclusion

All the current systems of design for assembly can be effective in reducing assembly cost.

Currently, the companies commited to DFA are almost exclusively those using or considering automation. This should not be the case, as the benefits to manual assembly from considerate design practices, though not essential to production, are also significant.

The most popular systems use a spreadsheet to evaluate a product from its constituent assembly operation times. Three levels of company commitment to design for assembly have been identified and are outlined as follows:

•Collections of assembly-orientated rules distributed to design and manufacturing engineers;
•Operation of an existing DFA system — manual or software-based;
•Developing advanced data capture and analysis techniques to improve effectiveness, design assistance and ease of use in a customised application.

Whatever method is used, integration of the design and production engineering departments should be encouraged from an early design stage. The ongoing education of both these departments in the company's manufacturing capabilities, the availability of new technology and good design practice, is facilitated by the active adoption of design-for-assembly philosophies. This is best achieved by a systemised approach. □

References

1. D.C. Gossard, 'Designing for Assembly: Research Issues. Computer Integrated Assembly.' *CAM-I's 15th annual meeting and technical conference, Oct.14-15, 1986,* San Antonio, Texas, USA.
2. A.H. Redford, K.G. Swift, R. Howie, 'Product Design For Automated Assembly,' *Assembly Automation, 2nd International Conference,* 1981.
3. J. Haeusler, 'Design for Assembly - State of the Art,' *Assembly Automation, 2nd International Conference,* 1981.
4. A.E.K. Holbrook and P.J. Sackett, 'Design for Assembly - Guidelines for Product Design,' *Assembly Automation, Seminar Notes, 9th International Conference,* London, 1988.
5. P.J. Sackett and A.E.K. Holbrook, 'Designing with automation in mind,' *Institution of Engineering Designer's Official Reference Book and Buyers' Guide,* 1986/87, p. 535.
6. D. DeWinter, D. Reynaerts, S. DeCorte and K. Machiels, 'Assembly', Katholieke University, Leuven, Belgium.
7. C. Poli, R. Graves and R. Gropetti, 'Rating products for ease of assembly,' *Machine Design,* August 21, 1986, pp. 79-84.
8. K. Takahashi, K. Senba, 'Design for automatic assembly,' *Assembly Automation, 7th International Conference,* Zurich, 4-6 Feb, 1986.
9. A.E.K. Holbrook and P.J. Sackett, 'Positive design advice for high-precision, robotically-assembled product', *Assembly Automation, Seminar Notes, 9th International Conference,* London, 1988.
10. R. Bassler and T. Schmaus, 'Procedure for assembly-oriented product design', *Assembly Automation, Seminar Notes, 9th International Conference,* London, 1988.
11. K.G. Swift, *Knowledge-Based Design for Manufacture,* Kogan Page Ltd., 120 Pentonville Road, London, UK, 1987.

Reprinted from *Assembly Engineering*, May 1989

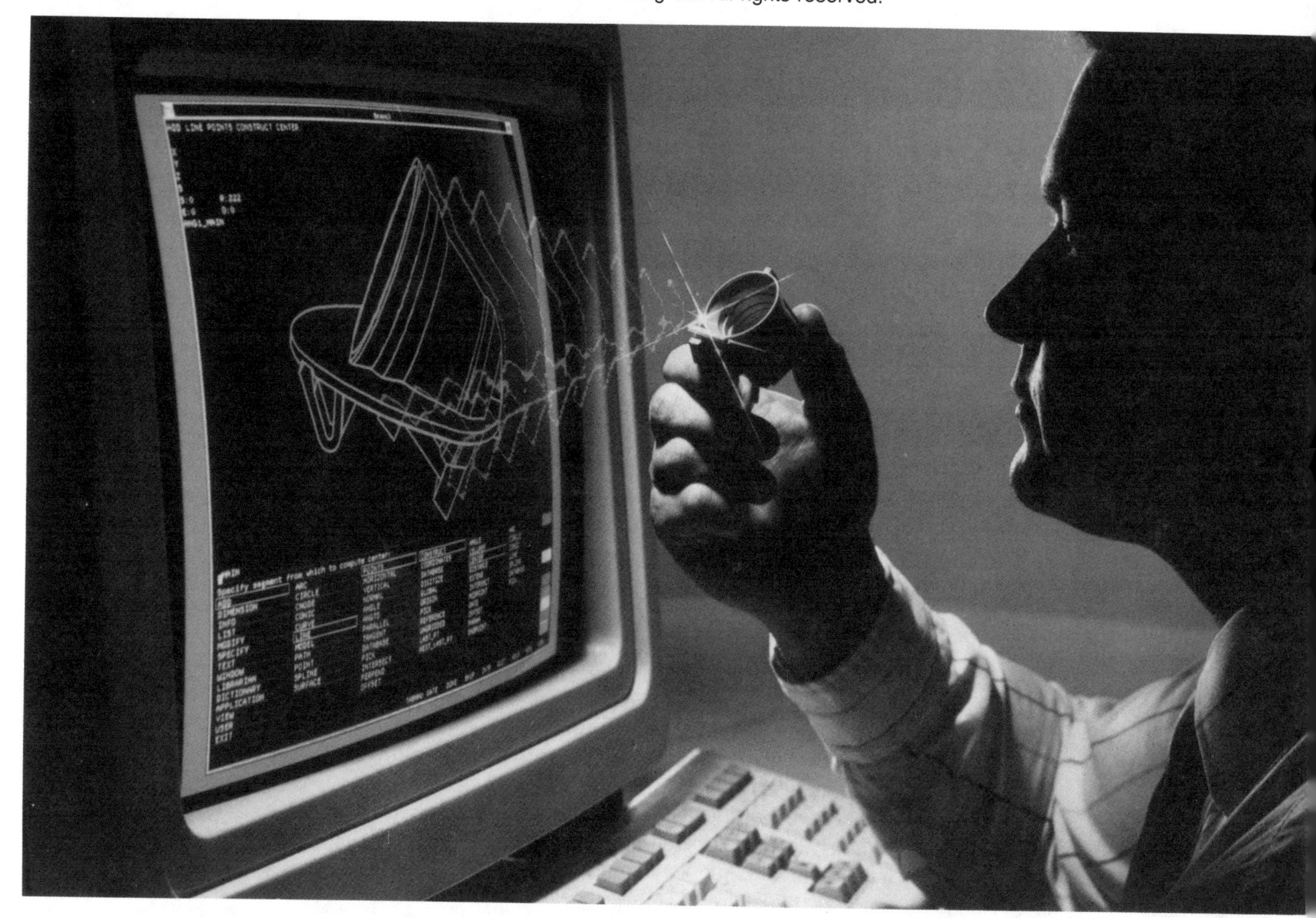

ITW Fastex

DFA Promises and Delivers

The design-for-assembly "revolution" continues. Here is an overview of its principles, and a sampling of applications where it has worked.

LINDA K. SCHUCH
Associate Editor

In addition to reliability and maintainability, producibility should be a major design criterion for most products. If it can't be produced—or assembled—efficiently, a product is not well designed.

Traditionally, producibility has focused on manufacturing efficiency and capability, as well as ensuring the timely availability of materials and equipment. Recently, however, the concept has broadened and now includes quality, flexibility, and responsiveness.

"Quality includes performance, features, reliability, conformance, durability, serviceability, aesthetics, and perceived quality," says Dr. Henry Stoll, manager of design for manufacture, Industrial Technology Institute (Ann Arbor, MI). "Flexibility implies the ability to manufacture to customer order in any sequence and in lot sizes of one or more.

"And responsiveness means quickly and easily bringing a product to market or adapting it to changing customer needs, new materials, and new or improved technologies. Doing it right means not incurring major cost or timing consequences."

Dr. Stoll lists the following as desirable design goals:

- family of products derived from one maximum, all-encompassing design
- all models produced on one production facility
- set-up time for changes reduced to minutes
- planned migration path to future product generations and technologies
- short design cycle and quick response time to market
- price below requested target

How do manufacturers meet these

demanding goals? By considering producibility very early in product development. DFA practitioners do this long before making concept decisions and other irreversible product and process choices. "The ultimate goal is smooth transition from product development to assembly," says Dr. Stoll, "followed by continuous improvement and uninterrupted production of a quality product."

Bart Huthwaite, director, Institute for Competitive Design (Rochester, MI), agrees. He believes that the seeds of all costs are directly related to product design. "Once a product passes through its development stages, the cost of making changes escalates," he notes. "Although product design typically involves only 5% of total product cost, it can directly impact 70-95% of your manufacturing operations. It drives part, processing, quality, and overhead costs."

Huthwaite uses the term *design for simplicity* along with design for assembly. He says it means reduced material, labor, and overhead costs; easier assembly; a smoother manufacturing process; and better quality. Product launches are faster and less painful, and capital equipment costs are lower. "Simplify first and automate later, if you have to at all," he advises.

Huthwaite urges companies to apply DFA methodology in teams: "Such an approach encourages participation by all members, product and process designers alike, technical and nontechnical members also. The result is an array of innovative solutions to design problems." He also warns that it won't be easy. Group problem-solving is not deeply ingrained in Western culture. Individualism is the norm; however, it must take a back seat to team efforts and accomplishments.

Dr. Stoll and Huthwaite have both compiled lists of DFA principles. Their guidelines will help you not only understand the methodology, but also realize its benefits.

Design for a minimum number of parts. Fewer parts mean less of everything needed to manufacture a product. Nonexistent parts cost nothing to make, handle, orient, inventory, purchase, inspect, or service.

Eliminate extra parts by incorporating their functions into multifunctional parts. For example, combine mating or contacting parts that don't move independently of each other into an integral part by considering alternative materials or net-shape manufacturing processes.

Develop a modular design. Create desired functional variants from basic and variation modules. Satisfy common functional requirements with standardized modules, allowing increased production of identical parts. Isolate technologies or functions that are likely to change in a separate, adaptive module. Standardize interfaces to maximize interchangeability.

Modular design offers several advantages. Engineering updates are brought on-line sooner and at lower cost. More product options are possible, and product lines can be broadened at reduced cost. Testing can happen well before final assembly, which enables isolating and solving problems without major disruptions. Disassembly and field service are also easier.

Don't fight gravity. Minimize the number of assembly directions. Top-down (Z-axis) is best; it means simple robots and insertion tooling, with gravity actually helping assembly.

Designing against gravity means expensive fixtures and increased clamping costs. Instead, develop a sound base component on which to build. Make it a fixture to help identify, position, and retain the top-down parts that follow it during assembly (Figure 1). Exploit the fact that gravity is free.

Reduce processing surfaces. Design so all assembly is finished on one surface before moving to the next (Figure 2). Multiple processing surfaces mean wasted motions and time. From an automation standpoint, they increase fixture and equipment costs.

Assemble in the open. Provide a clear view and easy access during

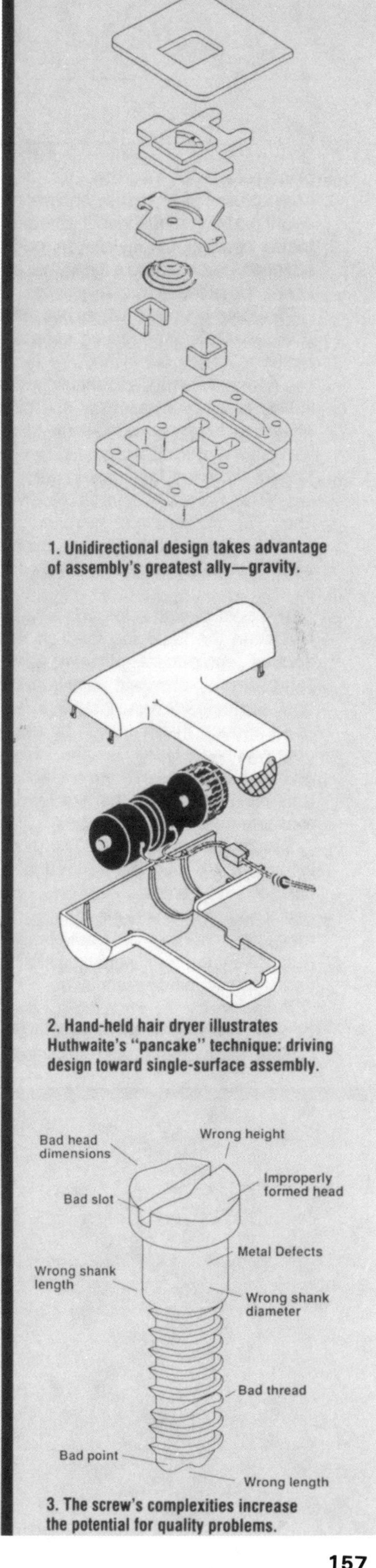

1. Unidirectional design takes advantage of assembly's greatest ally—gravity.

2. Hand-held hair dryer illustrates Huthwaite's "pancake" technique: driving design toward single-surface assembly.

3. The screw's complexities increase the potential for quality problems.

placement and insertion, for both manual and automatic operations. Avoid parts or sequences that require tactile sensing for installation or inspection. Such "blind" assembly exposes the process to quality risks.

Eliminate fasteners where possible. Avoid designs that require separate fasteners. They are difficult to feed, can cause jamming, and require monitoring. In manual assembly, the cost of driving a screw can be six times the cost of the fastener. Huthwaite targets threaded fasteners in particular: "They are like spoiled children. They do so little yet demand so much." He contends that the simple screw is actually very complex (Figure 3).

The best design approach is to incorporate the fastening function into another component. Snap-fit parts, for example, eliminate screwdriving and torquing. When fasteners are used, reduce their variation by using common head types, lengths, sizes, and materials. Select from a rationalized list of standards that are readily available from several suppliers.

Design for part identity. The more symmetrical a part, the easier it is to handle and orient; positions are easier to achieve and maintain. Symmetry also reduces assembly quality risks. The need for it increases significantly with high-rate automation.

If symmetry is impossible, then over-emphasize the asymmetry. Design components with easily recognized and exaggerated external features to make alignment foolproof. Don't rely on human judgment or repeatability. A slot, for example, may be difficult to detect; a pin or chamfer can help orient it.

Optimize part handling. Avoid flexible parts wherever possible; they are difficult to automate economically. Maintain part orientation from the point of manufacture to the point of assembly. For example, automatically unload plastic parts to a palletized tray. Another approach is to link parts together as part of a web or package in a tube or magazine.

Avoid part designs that nest, tangle, stick together, are slippery, or require careful handling. Add flames or ribs to parts with locking angles; design in barriers to prevent parts from interconnecting; and close loops and ends on springs. For robotic or automatic handling, provide symmetrical vertical surfaces to simplify gripper design.

Design for easy part mating. Part misalignment and tolerance stackup occur when components produced by different processes (e.g., stamping, injection molding, casting, and machining) from different vendors are mated. Compliance must be designed into both the product and the production process. Techniques include generous chamfers on both mating parts, adequate guide surfaces, and specs that require part consistency.

Provide nesting features. Although parts should not nest during handling or feeding, self-locking features are desirable during final assembly. Nested parts do not have to be repositioned or clamped before final assembly. Design projections, indentations, or other surface features that maintain the orientation of the parts already in place.

Develop a reliable, durable design. Avoid alternatives that require specialized or unusual operator skills or training. Eliminate cluttered, chaotic, or confused installations of wires, hydraulic lines, and other interconnecting components. Avoid designs that are sensitive to uncertainties like weld strength, bolt preload, lubrication condition, or voltage fluctuation.

When possible, design assemblies with components in a low-energy state so only external work produces a change. An assembled snap fitting, for example, is in a low-energy state, but a preloaded fastener is in a high-energy state.

Spending more time at the design stage and applying these principles have brought dramatic results to many companies. IBM, for example, designed its Proprinter using DFA techniques (*AE,* "Design for Assembly in Action," Jan '87, pg 64, and Figure 4). Springs, fasteners, and adjustments were eliminated. Part counts and operations required were squeezed down to 32 steps, with total assembly completed in 3 min.

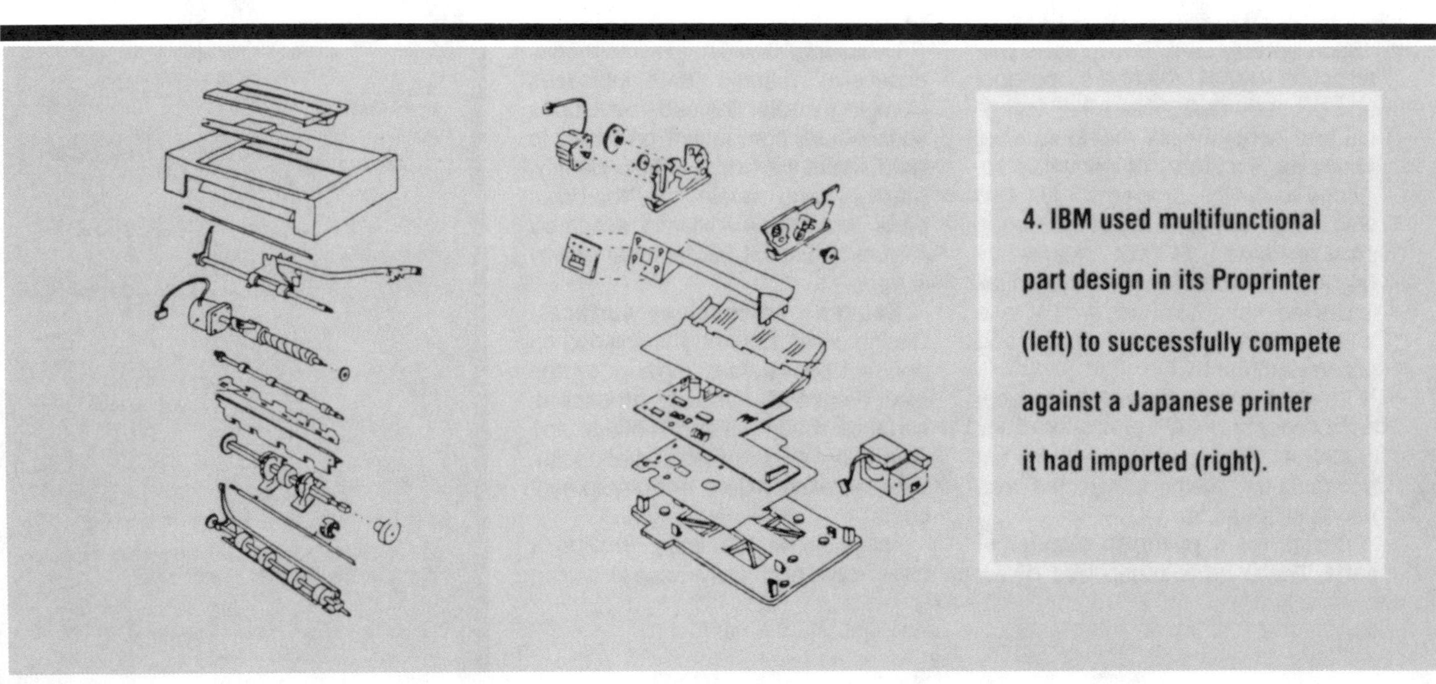

4. IBM used multifunctional part design in its Proprinter (left) to successfully compete against a Japanese printer it had imported (right).

One of IBM's competitors still requires 185 assembly operations and takes 30 min to assemble its product.

At Ford, Huthwaite trained over 4500 engineers in DFA and participated in the Team Taurus design group. He reports cost savings of $700 per car by designing for simplicity. He also cites NCR Corp, where the redesigned 2760 terminal boasts 85% fewer parts, 75% quicker assembly, and close to $1.1 million in total savings. (*AE,* "NCR Cashes In on Design for Assembly," Sept '88, pg 15).

In the appliance industry, builders don't totally redesign products as often as in other markets. But companies still apply DFA principles to solve design problems. ITW Fastex (Des Plaines, IL) has helped appliance manufacturers address many challenges. For example, the firm worked with a range maker to design new oven broiler glides. Engineers replaced multiple-element, mounting systems with one-piece, easy-to-install, thermoplastic parts.

The track-type glide snaps into a prepared hole in the range chassis, replacing a rivet-mounted metal roller assembly. No tools are required, and the installer doesn't need to orient the symmetrical part. A large bearing surface meshes with the guide channels in the broiler drawer and supports the sliding weight of the unit. And, yes, the heat-stabilized nylon remains tough in the broiler environment.

Another innovation is a one-piece, snap-in terminal block. The modular design replaces several components and eliminates labor-intensive assembly. The product also incorporates an integral strain-relief feature to eliminate yet another component. John Rachow, Appliance Products Group, reports that Fastex is working with several major appliance builders on such cost-reducing projects.

General Electric's laundry division faced a problem with the light socket inside its dryer drum. In addition to its initial costly installation, the socket caused service problems. Fastex worked with GE engineers to design a new one that snaps into the existing metal shroud. It allows easy installation in the plant and easy replacement in the field.

As a bonus, the heat-stabilized nylon socket meets GE's requirement of accepting a candelabra screw-base bulb in place of a more expensive bayonet style. No changes to the existing reflection shroud were necessary.

Finally, another GE-Fastex joint effort resulted in an improved design. Metal anchors had been used to hold cantilevered shelf brackets to the inside of refrigeration liners. Installation was labor-intensive, and the anchors often allowed foam insulation to leak into the refrigerator, causing odor and taste problems.

Engineers were challenged to develop an anchor that would perform better, with the added stipulation that it had to accommodate robotic equipment. An anchor and an assembly system were finalized after two years of development.

Robots now Spiralweld™ plastic anchors to the back side of the refrigerator liner to hold the brackets in place. Spiralwelding is a bonding method that spins plastic fasteners against a plastic panel until friction causes the similar materials to melt and fuse. In this application, the anchors reinforce load-bearing areas of the thin panel, supporting the weight of attached components. The solution improved both product quality and assembly efficiency.

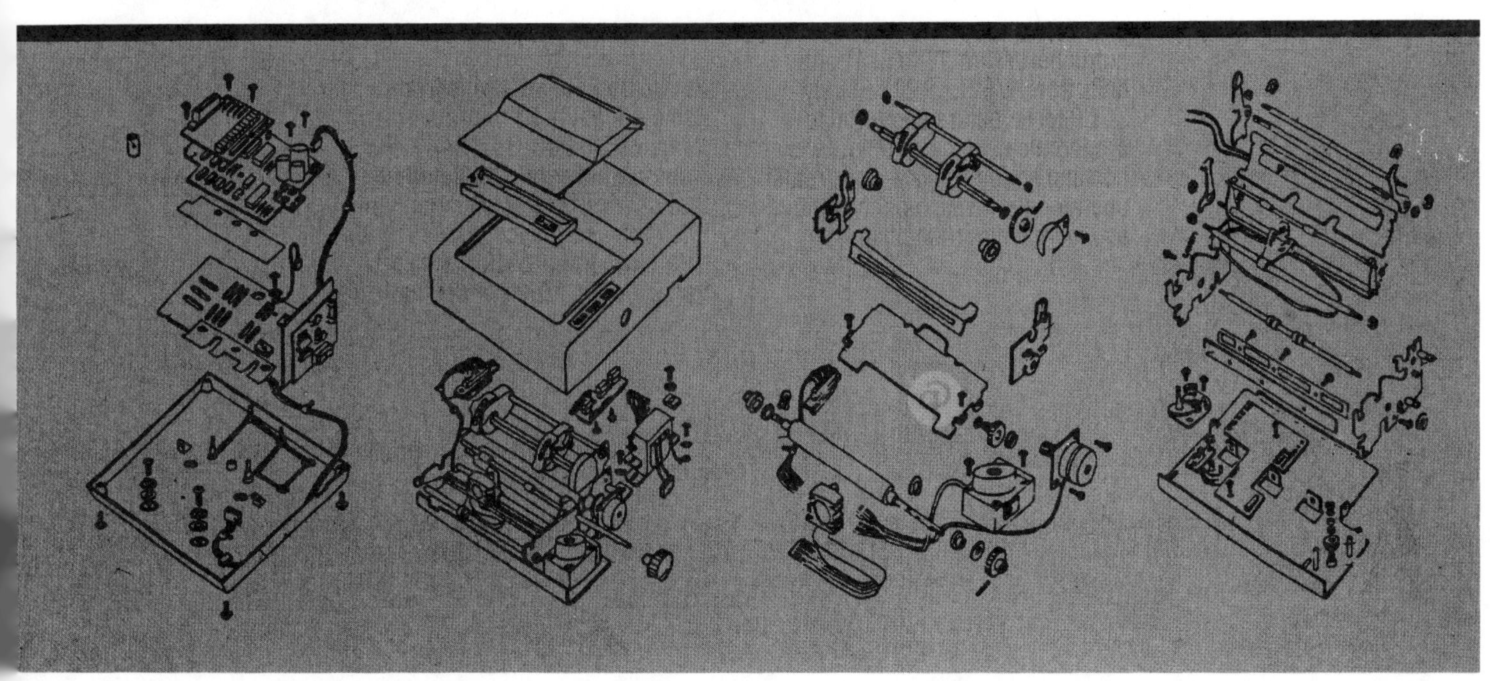

Presented at CASA/SME AUTOFACT '89 Conference, October 1989

Simultaneous Engineering

By David P. St. Charles
Valisy Corporation

Simultaneous engineering is not a technology issue. It is a people and communication issue. Why has simultaneous engineering emerged only recently as a dialog once dominated by CIM? Because after the progress made making specific functions work better -- for instance, improving design with CAD, improving inspection with coordinate measuring machines -- we have found that the remaining problems are not with how we execute these functions. The problems are with how we think of them in the context of a total process.

In the traditional linear process, the design department generates an engineering document. It then goes to manufacturing, where the engineers there figure out how to make the part. The drawback with this linear process is that the design may be unnecessarily difficult, time-consuming or expensive to manufacture or inspect. We say "unnecessarily" because sometimes two different designs for essentially the same part are equally good in terms of part functionality, yet distinctly unequal in terms of how quickly, cheaply and reliably it can be produced.

Simultaneous engineering is different because it is a non-linear process. Simultaneous engineering means involving manufacturing and quality engineers in the design engineering stage. The idea is to get everyone's input before the design document is finalized. This way, design engineers are exposed to what it will take to get a design through manufacturing and onto the assembly line. And they learn what they can do to get an equally good design there faster, more reliably, for less money. There are two specific benefits to this dialog:

- Manufacturing operations are more productive because the resource expenditure per part is minimized by using the lowest impact design. It is estimated that 80 percent of the cost of manufacturing a part is fixed in the design stage.

- Manufacturing operations are more productive because design documents are not sent back for engineering changes due to manufacturability or inspectability problems. Engineering changes cost an average of $2,000 per part and the average part normally goes through twenty changes in its lifetime.

Perhaps the biggest problem in implementing simultaneous engineering, once you get the agreement to implement it, is that engineers in design, manufacturing and quality assurance do not speak the same language.

One reason major authorities such as Daratech and Dataquest have identified Valisys software as a significant tool for implementing simultanous engineering is that Valisys goes a long way to providing a common language for all these engineering disciplines.

It does this by providing information to each function -- design, manufacturing and quality assurance -- that is relevant to their concerns -- manufacturability, inspectability, etc. This information derives directly from the original design database.

Many people think that a common engineering database exists (or at least has been defined). But nothing that you could truly call "common" has emerged. Current engineering databases share design geometry, but that is about all. After that, the NC database, the inspection database, etc., are only superficially linked. Something in the NC database could be inconsistent with the design database, without anybody being able to tell the difference. A situation like this is hardly consistent with the concept of simultaneous engineering. In fact, one would hope to use simultaneous engineering to correct a problem like this.

Valisys is not a database per se. But it might look like one, because it makes the engineering database more robust. The information Valisys adds lets engineers see the connection between different databases. This is what provides the common language referenced earlier. It makes it possible for design, manufacturing and quality engineers to know they are working with the same information.

Here is an example. By looking at a CAD model, most manufacturing engineers cannot determine the exact relationship between the features on the drawing. This information is important because with it, the manufacturing engineer can design his tooling accordingly.

Using Valisys, the manufacturing engineer would not have to interpret abstract drawings to see the relationship between features. Valisys would show him this relationship in simple graphics and text -- automatically. The design engineer's only role would be to specify the perpendicularity with a drafting symbol.

Now that the manufacturing engineer has built the right fixture, there is less deviation between parts because getting the correct relationship between features is no longer a hit-or-miss proposition. Consequently, the percentage of high quality parts goes up dramatically.

What about inspection of the part? How can you ensure the part will be accurately inspected without resulting in unnecessary engineering changes? The Valisys design checking function tells the designer if he has made an error in the specifications on his drawing. For example, if the designer says he wants a particular feature to be perpendicular, but not what it should be perpendicular to, Valisys will tell him to specify.

Even further, not every quality engineer takes the time to discuss with a design engineer what features of the part need to be measured to tell whether or not the design specifications were met in production. With Valisys, the designer's specifications are automatically made part of the engineering database in the form of inspection requirements.

Assess for yourself what the cost would be if we failed to inspect this part properly -- and passed half of the bad parts produced? Or if we scrapped 25 percent of the good parts, because we were not using the proper inspection criteria.

These are not all the capabilities of Valisys, but they are key examples of the kind of support Valisys gives to the simultaneous engineering process. Following is a summary of this process:

First, if a company wants to implement simultaneous engineering, you can see that Valisys will help shoulder the engineers' load. In effect, Valisys adds to the engineering database some of the dialog that takes place between disciplines in the simultaneous engineering process. Valisys catches mistakes or highlights issues that otherwise might not come out.

Second, Valisys provides a common language for different kinds of engineers, because it presents key design, manufacturing and quality assurance information in easy to understand graphic form.

Third, Valisys helps individual engineers do their jobs better, because they now have more information to work from. Whether they talk to one another or not, each engineer works with the same information every other engineer is using.

When we put the advantages of Valisys to work in a simultaneous engineering environment, the big picture benefits emerge. We reduce the number of expensive engineering changes, we reduce per-part manufacturing costs, scrap goes down and part quality goes up, and with Valisys we do all this with fewer man hours involved, saving engineering time starting with drafting all the way through final inspection.

DOUGLAS AIRCRAFT COMPANY -- CASE STUDY

We are discussing some big dollar savings here, so that deserves some substantiation. Here is a case study from Douglas Aircraft Company in Southern California.

Douglas recently completed an installation with Valisys and has decided to further implement Valisys in a number of machining and inspection centers within their manufacturing facilities. Douglas has a facility in Torrance where parts are machined and inspected. These finished parts are then passed off to the Long Beach facility for assembly. There, they make the MD-11 and MD-80 commercial jets as well as the C-17 military transports. As you can imagine, the factory floor area in both plants is enormous.

To implement Valisys, Douglas assembled a team of personnel including CAD/CAM engineers, CMM programmers, quality and tooling people. This team thoroughly evaluted Valisys within the Douglas organization. The evaluation process involved a pilot installation in one CMM room to determine whether Valisys could function as an inspection tool for Douglas. They were initially interested in the ability of Valisys to create, run and analyze inspection paths, working from a CAD model.

In charge of the team was Jeff Duke, Senior Systems Engineer, CAM Systems. Summing up the findings of the team, Duke said, "The installation showed us that we could save time and money using Valisys both in manufacturing and inspection and also in the design of parts."

"Hard tools" or "check fixtures" are an important issue at Douglas, specifically, the time and cost involved in manufacturing and maintaining them. Rick Hutton, Section Manager, Tool Design is responsible for creating these tools from engineering drawings. Many of these tools bear a price tag of up to $1 million each.

But Hutton feels the Valisys "Softgauge" function can help eliminate much of this cost. In a recent conversation with Hutton, he said, "Valisys gives us the ability to produce a soft tool vs. a hard tool. Valisys will allow us to take a one-time model and keep updating it electronically and use a CMM to produce the results that we need without any of the hard tooling. The soft tool that Valisys represents is a milestone which I feel can cut down our costs by at least 80 percent."

Also on the subject of tooling, Wes Robinson, Branch Manager, Tool Design, said, "Valisys is another step toward eliminating hard tooling that is not necessary. I think it can eliminate probably 80 to 90 percent of hard tools, especially some of the very complex tools that are difficult to build."

Besides reducing tooling costs, the Douglas team cited Valisys for its ability to reduce CMM proofing time. Proofing is a task performed on the CMM to determine the optimum inspection path. The operator must continually run and rerun the path, making changes when necessary, until it runs correctly.

Jeff Wood, Manager of Quality Engineering, said, "We're looking at a reduction of CMM time. One reason is Valisys automatically defines the inspection path and characteristics to be checked. The operator doesn't have to define his own inspection path, and this can save a considerable amount of machine time."

After obtaining these results, Douglas has decided to implement Valisys in a variety of machining and inspection centers. They are currently in the process of writing an implementation plan, and expect to put it in operation over the remainder of this year.

Regarding the Valisys implementation, Duke said, "CAM systems has been contacted by the other components at Douglas such as the Huntington Beach Astronautics Division, the Culver City Helicoptor Division and also the Mesa/Phoenix Area Helicoptor Division. They were interested in learning what Valisys could do for them to improve their quality and manufacturing capabilities. We are currently working on implementation plans for all of these sites plus other."

Other members of the evaluation team at Douglas have expressed their feelings on the Valisys organization:

Jeff Wood said, "I expect Valisys to grow with us, to be able to recognize our needs and be able to adapt to us. I think our needs are similar to the industry, so I feel our partnership is going to be beneficial to Douglas as well as Valisys. To this point, technology hasn't put enough emphasis on inspection programs and now, with Valisys we're making that step forward."

And Joyce Baba, CAD/CAM Senior Principal Consultant said, "I don't think I can say enough about what Valisys can do for Douglas at this point. It can save time and money, and I think all companies are looking for a systems that can do that."

As a result of the Valisys installation to date, Douglas sees big dollar savings ahead -- savings big enough to pay for the software in a very short period of time.

Presented at SME Simultaneous Engineering Conference, November 1988

Simultaneous Engineering in the Conceptual Design Phase

By Henry W. Stoll
Industrial Technology Institute

It is now widely recognized that doing design for effective cost, quality, and delivery requires a simultaneous engineering approach. Recognizing this is only the first step, however. The pivotal second step, actually doing it within the constraints imposed by the organizational structures, operational practices, and specialization oriented culture of today's large corporations, is the challenge. One approach is to provide the design team with effective and easy to use design for manufacture (DFM) methods which require and promote cooperation and team work between design and manufacturing.

One area where we see this approach being particularly effective is in the conceptual design phase. In this phase, the design is fluid, hardware is still remote, few numbers of people are involved, and constraints and interactions have not become tight. Hence it is in this stage that the design team has the most latitude to explore alternative possibilities and therefore the greatest opportunity to identify the inherently best product and process concept for manufacture. However, because of the design freedom available and the importance of functionality, producibility and other manufacturing considerations are easily deferred or overlooked during this phase. It is precisely for this reason that disciplined methods are needed in the conceptual design phase to help insure producibility of the final product.

In this paper, we report on work we are doing at the Industrial Technology Institute (ITI) to develop DFM methodologies for conceptual design. In particular, we describe a five-step procedure for integrated product/process design which is intended to help encourage parallel development and evaluation of several alternative product and process concepts.

TRIAL DFM METHODOLOGIES

We have developed several preliminary DFM methodologies for use in conceptual design and are currently field testing these ideas in various ways. One approach we are trying is to offer a two or three day workshop on-site at a client company where we train practicing engineers in new DFM methodologies and then apply them to a design problem of current interest or concern to the company. This allows us to evaluate the usefulness of the methodology while simultaneously helping the client company to address an existing design need in innovative and new ways. DFM methodologies which are currently being tested, evaluated, and improved in this manner include:

- **Manufacturing Envelope:** A formalized method for capturing a company's current/existing production capability, experience, and best practices; developing a list of preferred methods of fabrication, assembly, material handling, and inspection; standardizing and rationalizing tooling geometries, etc.; and presenting this information to the designer in a form which can be readily applied in a new product design. The manufacturing envelope provides guidance and producibility goals to the design team and acts as an early warning mechanism for producibility related problems.

- **Producibility Goals Method:** A systematic methodology for developing specific producibility goals for a particular design task or project. The methodology also provides a mechanism for insuring producibility by "red-flagging" areas of non-conformance and facilitates the definition of a "global" producibility measure which can be used to maximize producibility of the design for a given set of production circumstances (e.g., the manufacturing envelope).

- **Robustness Assessment:** A systematic technique for assessing the robustness of a proposed product concept and process plan in the early stages of design.

- **Parts Reduction Assessment:** A systematic technique for analyzing various product structures and design alternatives to define the practical design alternative having the least number of parts with the least number of processing operations.

- **Integrated Product/Process Design Method:** An iterative analysis-redesign for manufacture methodology based on precisely stated design principles and guidelines. The method consists of five steps: (1) specify detail design goals; (2) optimize current product and process (if appropriate) following prescribed procedures; (3) synthesize and optimize new product and process concepts following prescribed procedures; (4) evaluate, combine, refine into one or more best designs; and (5) select one or more best designs for further development.

All of these methodologies are supported by worksheets, checklists, etc. Some are being further developed as PC based tools. Also, as a result of our contract research program, we have been able to substantially improve the trial methodologies that we have tested so far as well as identify some possible new ideas and approaches.

At present, none of the trial methodologies are in widespread use. However, our experience so far in the experimental DFM workshops indicates that they all have potential for providing needed discipline for simultaneous engineering during the conceptual design stage. A key issue in their use is that management recognize the value in using them and that their use be given the time and commitment required.

The "Integrated Product/Process Design Method" is a general design procedure which provides a systematic approach to conceptual design of the product and process and a structure for applying other DFM methods and tools such as DFM Guidelines, Design for Assembly, and the other trial methodologies listed above. The rest of this paper is devoted to a more detail discussion of this method.

INTEGRATED PRODUCT/PROCESS DESIGN METHOD

Two major goals of simultaneous engineering are parallel development of the product and the manufacturing process, and the creation and evaluation of multiple product and process design alternatives. The "Integrated Product/Process Design Method" is a systematic, five step procedure (Figure 1) for identifying alternative product and process concepts which we have developed as a disciplined approach for achieving these goals. The method starts with a comprehensive set of product and manufacturing design goals and converges toward one or more fully integrated and coordinated product/process design concepts. The basic activities addressed in each step are described in the following paragraphs.

Step 1. Prepare by reviewing the functional requirements and constraints of the design problem, listing the advantages and disadvantages of the current product design and manufacturing methods, and develop producibility design goals using design for manufacture (DFM) methodologies such as manufacturing envelope and producibility goals method. If appropriate, other competitive benchmarks should also be developed during this step by "reverse engineering" leading competitor products.

Step 2. Optimize the current product design and production methods to the fullest extent possible using the principles of DFM and methodologies such as parts reduction assessment and robustness assessment. The results of this step should represent the greatest improvement possible in the present design while still staying within the bounds of current practice and experience. This improved design should be retained as a viable design alternative and possibly used as the baseline against which new product and process concepts are judged.

Step 3. Building on steps 1 and 2, identify innovative new product and process concepts which are inherently easier to manufacture. Depending on the particular circumstances involved, use one of the following design strategies:

- **Product Driven Design.** First identify a new product concept and then develop an appropriate manufacturing plan to produce it. Simplify and optimize the resultant product concept and manufacturing plan as a coordinated product/process system using DFM principles and guidelines (Figure 2). Repeat to generate as many alternative product/process concepts as possible. This strategy should be selected when the product is not seriously constrained by any major manufacturing consideration and/or where several acceptable alternative manufacturing approaches are possible.

- **Process Driven Design.** First propose a manufacturing concept and then develop an appropriate product concept which can be manufactured in the proposed innovative new way. Simplify and optimize the resultant product/process concept by eliminating and/or coordinating product and process interactions and simplifying design details and manufacturing operations (Figure 2). Repeat to generate as many alternative concepts as possible. This approach should be selected when manufacturing constraints dominate the design or when significant manufacturing simplifications, economies, or quality improvements can be achieved by designing the product to be manufactured in a particular way.

- **Ideal Driven Design.** This strategy is recommended when the product and/or process driven design strategies fail because the desired product concept and preferred method of manufacture are incompatible for one reason or another. The procedure in this case is to first identify the "important few" problems or issues which stand between the proposed "ideal" product concept and a preferred or "ideal" method of manufacture and then to define DFM solutions which eliminate the incompatibilities without unduly compromising either the product concept or the desired method of manufacture. In broad terms, this involves the following four step process:

    1. Formulate the IDEAL product and manufacturing concepts.

    2. Identify BARRIERS that block the way.

    3. Determine CRITICAL ISSUES that form the heart of the problem.

    4. FOCUS on the critical issues, all else is a chain of technicalities.

In performing step 3, the design team should seek to identify as many alternative product/process concepts as possible. The probability of identifying and selecting the "best" concept is greatly improved when there are many possibilities to select from. This is because the greater the number of ideas, the more likely the best idea will be included.

Step 4. Evaluate, combine, and refine the results of step 3 into one or more "best" concepts. In doing this, the design team should consider all of the goals of the design including technical, manufacturing, and marketing needs and capability. When many concept variants exist, one approach is to first group the alternatives together according to similarities in solution principle, layout, form, etc. Each grouping is then carefully evaluated, the best features from the various alternatives are combined, and the less desirable features eliminated. The objective is to narrow the field to one preferred concept for each major group and to optimize the preferred concepts as much as possible with respect to the design goals.

Step 5. Select one or more "best" designs from the list of preferred concepts developed in step 4 and the optimized version of the current design developed in step 2. It is recommended that this selection follow rigorous practices such as those used in "value engineering". Relative merits and costs of each alternative should be carefully scrutinized and objectively evaluated before a selection is made. If a clear decision between two competing concepts is not apparent, it is recommended that both concepts be further developed until the proper choice can be made apparent.

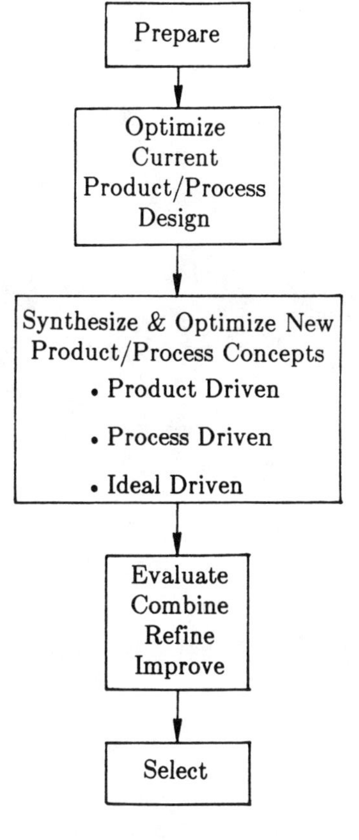

Figure 1: Simple, Five-Step Process

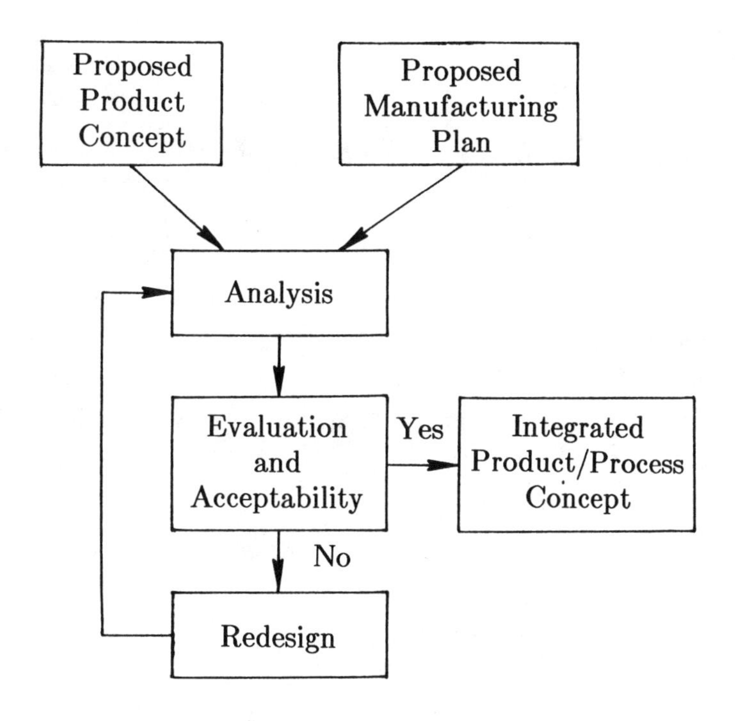

Figure 2: Analysis-Redesign Model

PRELIMINARY OBSERVATIONS

We have been using the Integrated Product/Process Design Method in our contract research and in various DFM workshops presented on-site to client companies. Use of the method in workshops has ranged from a brief overview given as part of a Design for Assembly workshop to an hour discussion given as part of a day long briefing on DFM to using the method as the main theme for an advanced DFM workshop. We have used the method in our contract research to perform conceptual design and as a test-bed for developing and evaluating various approaches to "producibility measurement" which we are currently investigating. Based on this experience, we are able to offer several preliminary observations regarding the method and its use which are summarized in the following paragraphs.

In every case we have been able to achieve the producibility design goals which were delineated as part of step 1 of the method. This lends great support to our contention that producibility, like functionality, reliability, and maintainability, can be achieved if quantified design goals are set at the beginning of the development project. Typical producibility goals that might be set in step 1 include:

- Design for top down assembly
- Develop a "club sandwich" construction
- Design for modular assembly
- Avoid separate fasteners. If a separate fastener must be used, use only 1/2 inch long #6 sheet metal screws.
- Parts count efficiency to be greater than 0.7
- Producibility rating for injection molded parts to be greater than 0.8

Step 2 is a critical preparation step for step 3 in that it requires the design team to carefully and thoroughly examine the current product and process design, decide what is good about it and what is bad about it, optimize it as much as possible, and as a result, form a clear vision of what the new product and process design ought to be. We have also found that this is an ideal time to study leading competitor products, again to provide a clear vision of what is required for the new design to be superior in the market place and to provide a jumping off point for innovation.

In step 3, most design teams start out with the product driven design strategy. However, as they begin to develop conceptual ideas, many teams evolve to the process driven design strategy because they find that it is easier to design the product with less ambiguity if they know beforehand the assembly sequence, testing process, material handling scheme, etc. By proposing the manufacturing process first, the best answer to questions such as how the product should be modularized, how optional features should be structured, how testing should be accomplished, and what needs to be designed into the product to simplify fixturing, tooling, number and type of operations, testability, etc. become immediately clear. Finally, as the various product/process design alternatives become better defined, the strategy tends to shift from "process driven design" to "ideal driven design" in order to resolve conflicts and iron out incompatibilities between the best manufacturing plan and the best product concept.

There seems to be a general tendency for design teams to quickly home in on their favorite product/process concept and once they have what they feel is the best concept, there is a resistance to developing other alternative concepts. We feel that a great advantage of the method is provided by step 4 where several alternative concepts are traded-off and the best features or aspects of each are combined and refined into even better concepts. In every case where there has been several viable product/process concepts to choose from, we have found that step 4 invariably leads to significant improvements. Hence, to reap the full benefits offered by the method, management should expect that several alternative product/process concepts be identified and should provide the necessary time and resources required to properly perform steps 3 and 4.

We see our approach of involving industry in the development of DFM methodologies such as the Integrated Product/Process Design Method by offering on-site workshops which focus on particular product and process needs as being a "win-win" approach for both the company and ITI. The company benefits by being exposed to these new ideas and we gain important feedback regarding the validity and usefulness of the methodologies. In all of our experiences so far, innovative new product and process concepts have also been identified, making the small cost of the workshop well worth while. Because DFM methodologies require close cooperation between product design and manufacturing, another benefit that we have seen for the client company is a building and strengthening of ties between design and manufacturing and an improved awareness of the power of the design for manufacture approach. Ultimately, improved communication and cooperation between design and manufacturing is the greatest benefit that any DFM methodology can hope to provide.

CONCLUSION

Identifying the best product and process concept for manufacture is pivotal to new product development success. It is now widely agreed that simultaneous engineering gives manufacturing and design engineers a real opportunity to determine in advance, by working together, the least expensive way of manufacturing a product. But simultaneous engineering doesn't just happen, it requires a disciplined approach. The "Integrated Product/Process Design Method" represents a positive step in this direction.

Reprinted from *Manufacturing Engineering*, December 1987

People in R&D and manufacturing face different needs, have different personalities, and work to different standards. Here are some thoughts on how to bridge between them

Worlds Apart—Bridging R&D and Manufacturing

By Robert Szakonyi
George Washington University

Research and development people are different from manufacturing people. To characterize R&D people and manufacturing people as living in different worlds, however, is an understatement. The truth is, *R&D personnel and manufacturing personnel in a company may have almost nothing in common except that they work for the same company.*

For example, a senior manufacturing manager—who had once been in R&D—related his company's experience as though his company had never taken a project from R&D to manufacturing. He had 25 years of experience and knew that the problems in the transition from R&D to production have been covered in journals and discussed at conferences for many more than those 25 years. Yet when asked, "Why are you reinventing the wheel as if nothing had ever been done in your company before?", he answered: "Because with each sunrise, the problems of carrying out the transition crop up again."

R&D lives in a world of change, manufacturing in a world of discipline. With planned change and training, the two worlds can be bridged.

If with each sunrise a new effort must be made to solve problems between R&D and manufacturing, something very fundamental must be involved. What are those fundamentally different worlds of design people and manufacturing people?

The world of R&D is a world of *change.* This is not to say that design engineers do not want stable employment, a stable laboratory environment, or the stability of benefits from their companies. They do, of course. Change, however, is embodied in their work. The everyday questions of their work are: What is a new idea for a product? How can we improve the manufacturing process controls?

On the other hand, manufacturing marches to a different drummer: Keep the costs down. Meet the schedule. There is only one way to meet those demands—*discipline in production.* Change is disruptive. In manufacturing, there is almost no better way to increase costs and cause delays than to inject change without preceding the change with exhaustive planning.

Given the different goals of R&D people and manufacturing people, the laboratory and the manufacturing plant are understandably managed differently.

In the laboratory, failure is common. R&D, by definition, is risky. R&D managers understand this. If they are not pressured themselves, they do not penalize people who have done good technical work that by happenstance does not pan out. These people, moreover, generate new concepts; they do not necessarily implement them. Once they have developed a concept as far as they think they should (though perhaps not as far as they really might), they are anxious to begin a new project.

On the other hand, in the manufacturing plant, failure is disastrous. Failure means very high costs, and it usually means a deviation from schedule. As a result, manufacturing managers tend to run a tight ship. Potential improvements or new concepts that have not been shaken down and proven may mean a major failure in the plant. The manufacturing people responsible for that failure, should it happen, will be penalized severely.

Given the difference between the R&D and the manufacturing worlds, what can be done to effect a harmonious transition between them? Based on discussions with more than 1200 R&D and manufacturing managers at 180 companies—124 in the US and 56 in Europe—three solutions present themselves: (1) establish a buyer-seller relationship between the two areas; (2) soften the changes that new products and manufacturing technologies generate; and, finally, (3) create a new discipline in production.

What is a buyer-seller relationship between manufacturing people and design people? It is a *deal* like the deals made between the company as a whole and its outside customers.

Frequently in companies it is forgotten that deals have to be continuously negotiated between functions. People in functional areas often think, "My group has an assigned mission. I know what I am supposed to do. If I do my job well, I help the company. So what is the problem?"

The problem is the difference in worlds between functional groups. If a deal is not struck, the different worlds of R&D and manufacturing will split apart again overnight.

The suggested deal is based on matching the needs and capabilities of R&D with those of manufacturing. Putting together the deal calls for skills needed for any deal—shrewd organizational maneuvering, insight into the most opportune time to make offers, and the sense not to force the other negotiator into unnecessary risks. The question each has to ask the other is, "How can what I do—or refrain from doing—make your work more effective?"

More than just an interest in making a deal between R&D people and manufacturing people is needed. Company operations have to change so that there is a solid basis for a deal.

First, consider changes that can be accomplished with the R&D people. Softening change in this way does not mean that a company should hold back either technology or change. But a company should manage change in a workable way so that the change is more livable to manufacturing people, who need a stable environment in which to produce.

The first step in managing change is to educate design engineering about production

Educate R&D about production. The first step in managing change is to educate the primary creators of that change—design engineers—about production. Among the approaches that can be taken are these:

- Develop courses, possibly taught by manufacturing personnel, on designing for producibility, particularly for young design engineers
- Send R&D people to plant start-ups to observe
- Establish a manufacturing group within R&D to provide technical input into design during project development
- Develop courses on statistical controls for R&D people so that they may speak more knowledgeably with manufacturing people about day-to-day variables
- Create time-limited but significant jobs within manufacturing so that R&D people can get hands-on experience in manufacturing.

Improve the transition. Once efforts such as these are in place to educate R&D people about production, the company can address the problems of carrying out the transition from R&D to manufacturing. Because this subject has been discussed at length for decades, the subject can be limited to three issues here: a management goal to carry out the transition and two stumbling blocks in the way of reaching that goal.

First, the goal. One way of looking at the transition from R&D to production is to focus on the stages by which a new design is released to manufacturing. In an ideal world, manufacturing people would begin preparing the production of a new product (or implementing a new manufacturing technology) long before a design is completely released. R&D people would therefore get feedback about the producibility or validity of their ideas long before the final design stage. In such a transition, there would be a regular exchange of information between R&D people and manufacturing people. In this ideal world, the concept would be brought to production smoothly through regular producibility discussions at each stage—in rough form, at a more advanced stage, when nearly complete, and finally at validation and release.

But, reality poses at least two stumbling blocks to reaching this goal. The first is the issue of accountability. Who is in charge at each of the various stages in the transition process? Who determines what the new product should be, or who determines the new manufacturing process? After all, R&D people provide the concept, but the manufacturing people are going to have to produce the new product or make the new manufacturing process work.

The second stumbling block is that of execution. Who provides the resources at each of the stages in transition? Who provides the skills? For example, during one project, 2 ½ years of work were required to bring a new manufacturing technology successfully up to speed after it had arrived on the plant floor. After the first six months, the new technology had been brought roughly 33% up to speed. At that point, most of the R&D people were assigned to other projects. A few, however, stayed because they were committed to the technology. If these had left as well before the 2 ½ years were up, the new manufacturing technology probably would have been rejected because of its poor performance. At times, therefore, the transition between concept and execution may be overwhelmed by the deep drain on skills and resources required.

What can a company do to make the transition from R&D to manufacturing smooth? The first thing is to see the problems clearly and to acknowledge the size of the stumbling blocks. The problems need to be addressed continually because they will not go away.

A company can try a number of procedures to ensure that the problems are dealt with effectively in the laboratory and in the plant. Colocating key R&D people with key manufacturing people is one way of doing this. Creating a hybrid design/manufacturing group that keeps one foot in the lab and one foot in the plant is another. No way is foolproof, of course, because the success of any procedure ultimately depends on the company and on the people a company assigns to these positions. Given that R&D people and manufacturing people live in different worlds, there is no way to guarantee they will not split apart—except to continually face the problems between them.

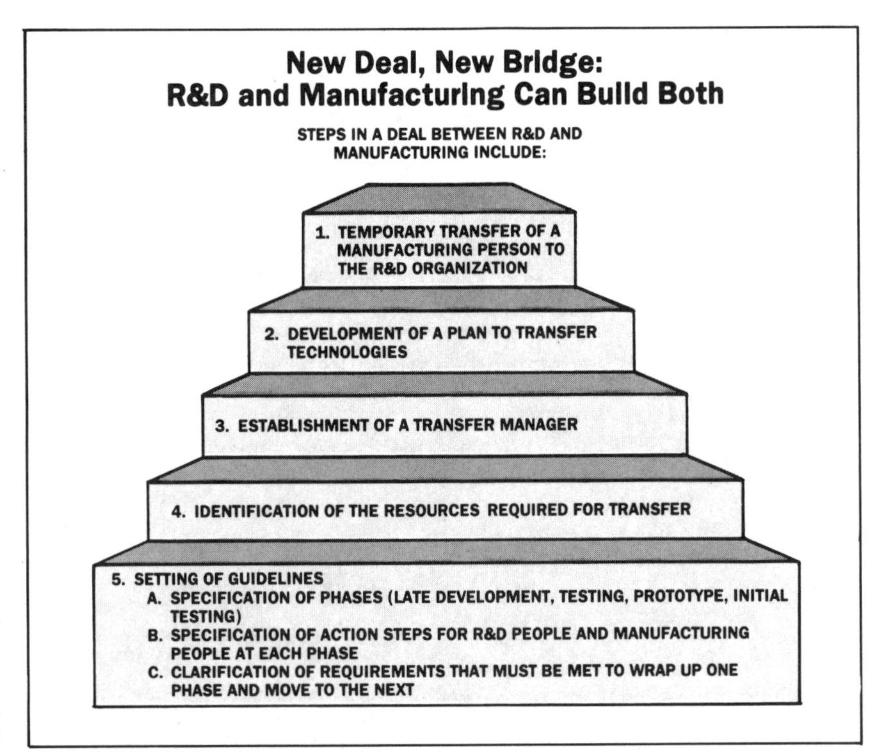

In addition to softening technological change, many companies need to create a new discipline in production. Traditional disciplines in production that focus solely on schedules and costs no longer work.

Schedules and costs cannot be forgotten in the new discipline, but they must be set in a broader context of other potentially more important goals to be met. An interesting benefit is that schedule and cost goals are more often met when pursued in context than when they are pursued for their own sake. A new discipline in production assures that the organization not only meets traditional goals, but also meets goals that it has traditionally neglected.

Upgrade training. One of the most neglected goals in manufacturing is the training of manufacturing personnel. A neglect both of technical training and of management training has helped make them perceive technology as a threat.

Manufacturing people need to know more about the equipment they use, about the equipment they *could* use, and about the manufacturing processes in which they play a role. Almost universally, manufacturing people would benefit from a better understanding of the dimensions and tolerances of the parts they make and the potentials of flexible manufacturing. Better technical training could help manufacturing people understand various types of in-process quality control to catch mistakes almost immediately.

Better management training of manufacturing people could bring them up to speed on participative management or just-in-time materials management. Traditional management—simply ordering subordinates what to do, stockpiling large inventories to escape scheduling problems, and the like—is no longer adequate in manufacturing organizations.

Moreover, when manufacturing people do stop perceiving technology or new manufacturing systems as a threat (because they have learned to understand them and have participated in implementing them), then manufacturing people will welcome improvements.

The new manager. Developing manufacturing managers of a "new school" must happen hand in hand with upgrading the training of manufacturing people. Unless manufacturing managers encourage and reward those manufacturing people who improve their equipment operation or improve work flows, people will go back to old ways.

What is the "new school" of manufacturing managers? Most importantly, they are managers who do not concentrate solely on schedule and cost goals and who do not manage like an autocrat.

It is not that manufacturing managers of the new school do not push to get schedule and cost goals met. But they think about other things as well: producing quality products, managing materials well, making needed capital investments, and planning future manufacturing operations. Technology is not a problem for them, but a tool for getting the productivity increases that they desire.

A manufacturing manager of the new school understands that people are as important a resource as capital equipment, materials, and inventories. Rather than giving commands and expecting simple obedience, he or she encourages manufacturing people to develop better ways to meet schedule and cost goals. Further, the new manager fosters the teaming of plant managers, supervisors, and line managers so that they work not only toward getting each member's quotas met on time and within cost, but also toward coordinating operations across manufacturing.

Toward the goal. Managers manage according to the systems currently in operation. To get from where we are now to the new school, something more is needed—a rethinking of manufacturing operations so that a whole new set of standards and expectations are used in evaluating manufacturing operations. That rethinking will create a new discipline in manufacturing.

But creating a new discipline will not be easy. There are two great hurdles in the way.

The first hurdle is conceptualizing the problem. The ultimate rationale behind rethinking manufacturing operations is to develop procedures for meeting schedule and cost goals more effectively. What, of all the innumerable things going on in a plant, is the plant manager going to change to cut scheduled time for production by 20% and costs by 40%? Without a conceptual model of the entire manufacturing process, a plant manager will not be able to ask the right questions and certainly will not be able to achieve either goal.

The second hurdle is the emotional and organizational problems caused by stripping away the old ways to evaluate what really functions effectively in a company's manufacturing operations. Many of the old ways of doing things were developed to disguise problems or to buffer personal and organizational conflicts—sometimes at the price of delayed schedules and cost overruns. To create a new discipline in production, those problems and conflicts have to be faced.

Key criterion. Once the hurdles are overcome, the central element of a new discipline in production can be understood. That fundamental element, the ultimate criterion by which to evaluate all manufacturing operations, is that the operations make the company more competitive.

Flowing from that ultimate criterion are a number of questions such as these:

- How effective is the manufacturing organization's use of materials, and how well does it plan for its materials needs?
- Does the organization use technologies at least as advanced as those of the company's competitors?
- Has the optimal size of the factory been analyzed, and have appropriate adjustments been made in capital investments, equipment layout, and authority lines?
- Have the work assignments in manufacturing been analyzed and identified so that everyone understands the work flow?
- Has flexible manufacturing been investigated, especially when a family of similar parts needs to be produced?

The list could go on, of course. The important point here is not whether any list of questions is inclusive, but that such a list of questions probes the implicit assumptions underlying the traditional discipline of manufacturing. No longer can building up inventories, minimizing capital investment, defining jobs narrowly, and constructing larger and larger plants be automatically defined as better by manufacturing people. If competitive pressures require a manufacturing organization to change, then manufacturing people have to think differently, strategically.

A strategy aimed at encouraging technological change and creating a new discipline in production consists of individual plans of action along a number of fronts. Upgrading the training of manufacturing people, integrating computer systems, improving quality, increasing automation, instilling participative management, upgrading capital equipment, improving the management of materials—all these need action. More importantly, a strategy requires a set of priorities by which to determine how hard each of the individual plans of action should be carried out along its own front.

A new manufacturing strategy, therefore, embodies manufacturing managers' decisions after they have rethought their company's manufacturing operations. After managers have developed a conceptual model of how the manufacturing process works in the company and after they have overcome the emotional and organizational problems brought to light by stripping away the old ways of doing things, they will be able to work directly to implement new product manufacturing skills and technologies. ■

Presented at the SME Simultaneous Engineering Conference: Making It Work
June 1989

Synchronous Engineering: Implementation of Modern Techniques

By Quentin C. Turtle
Technology Management Group

INTRODUCTION

The successful commercialization of new products depends very heavily on two traditionally diverse organizations working closely together. In the future, it will be ever more important for the R&D and manufacturing people to work more closely together than they have in the past. No company can afford to develop any given new product more than once. In fact, your company cannot afford to develop the next new product more than once.

Your R&D personnel probably perform in an environment that stimulates free thinking, creativity and innovation. They probably are not so concerned with the downstream company requirements of product engineering and manufacturing engineering. In fact, an attempt has traditionally been made to free the creative inventor's mind from final development, engineering and manufacturing activities and requirements.

On the other hand, manufacturing requirements demand a high degree of detailed documentation and methodical planning to satisfy customer order requirements and to manufacture products and inventory them in response to forecasts.

Linking these two diverse organizations together toward the common company goal of successful new product introduction, which includes the business plan, requires changes in behavioral patterns in both organizations.

The solution lies in a collective understanding of the total body of technology representing the new product, understanding the total product development cycle in your company, understanding factors leading to failure and success and properly structuring the technology transfer program. The solution lies in developing a new style of company-wide, new product management, coupling R&D, Marketing, Manufacturing and Finance without inhibiting creativity and always addressing the end users' requirements in product performance and cost as well as time schedule. The solution must be found - it must be found - and implemented because only then can your company compete well. Only then can it survive in the domestic and global markets.

CHALLENGES AND SOLUTIONS

A key question when structuring the program for transferring new product technology from R&D into Manufacturing is when to involve the responsible manufacturing personnel in the development program. A survey of companies has revealed that the successful ones attribute their success to addressing new product issues in the following order: manufacturing; development and then research where, traditionally, most companies have addressed research; development and then manufacturing. The successful companies study manufacturing, design and research in that order on purpose. Overall company objectives are better served if there exists that kind of plan. Research, development and design are then carried out with the requirements of manufacturing in mind - for highest quality and lowest cost, including development cost.

Turning now, specifically, to the discussion of synchronous engineering, let us examine some issues.

Many products are labor intensive. Companies have, as an objective, minimized outside-purchased material because, if everything were purchased, then the company becomes an assembly house and the competitive edge is lost. Many organizations can assemble. On the other hand, if a new product is structured so that the company adds significant value, based on intellectual property and proprietary rights, then the company is more competitive. Consequently, the cost of labor is significant, and reducing the cost of labor is highly desirable. Therefore, providing a design which is easier to manufacture and assemble should always be a primary goal.

Secondly, discipline of a company-wide group working together for the express purpose of developing a given product for lower cost manufacture and greater reliability and acceptance results immediately in a long lasting camraderie that does not otherwise exist. When a primary goal at the outset is to establish teamwork in a group consisting of research and development engineers, manufacturing engineers, manufacturing assemblers, purchasing and marketing people, then a closer liaison is established. This liaison is maintained throughout the whole product development cycle into the manufacturing cycle and, in fact, for the life of the product.

Ease of manufacture, reduction of parts and cost, increased quality and reliability and ease of training is designed into the new product from the very outset.

Manufacturing processes are better specified and are designed in earlier through intra-group communications and feedback. As a result, the time to Manufacturing and time to the market are reduced. The development cost is therefore less because later design changes to incorporate optimum manufacturing processes are unnecessary; they are brought to mind early and the initial engineering design is consequently one which leads to these optimum processes.

Another great advantage of the manufacturing engineers' participation is that estimates of the manufacturing cost are more accurate. The manufacturing process and methods engineers are fully knowledgeable of the product and feel a part ownership of the product early on. Consequently, their knowledge and interest in estimating product cost and the resulting accuracy are better.

A very significant subset of simultaneous engineering is the design for manufacturability. A case study will be presented later in this paper.

Returning now to understanding the total product development cycle in the company, there are several key points to be made.

The development cycle really begins with product concept. It ends with the product being manufactured smoothly on a timely basis and with many satisfied customers.

The technology transfer begins with the involvement of key manufacturing personnel early in the development program. It begins with the integration of key departments including manufacturing, quality control, production control, materials management and marketing personnel all integrated into one company team. Who decides when these people begin to be involved and to what extent? The answer is one which must result in the product being designed only once. If R&D designs the product without manufacturing involvement, the manufacturing engineers will most likely make significant changes for tooling, process and assembly requirements. Your company cannot afford to develop the product more than once. The answer is that once research has been completed and the development phase begins, let the manufacturing people decide when they wish to become involved. Because of the pressures from their responsibilities, they will want to become involved early in the development cycle.

The technology transfer program must be based on a formalized, disciplined approach. There must be a detailed work plan with realistic schedules made by experienced people. Then there must be a commitment to schedule by everyone having responsibility and control. The plan, therefore, must have credibility to the point where there is agreement by all on the team and everyone "buys in." Schedules are then kept by careful planning and rigorous follow through in order to meet the business plan which, of course, transcends all other considerations.

Two disciplines for accomplishing the above, in addition to those that have served you well in the past, are two that you may not be familiar with.

Kepner-Tregoe is a discipline including potential problem analysis and prevention including a four-point situation appraisal. Kepner-Tregoe also includes a five-step quantitative decision analysis procedure. It is a proven, useful technique for managing product development and technology transfer.

Robert H. Shaffer and Associates have developed a technique for breakthrough. A concept for identifying and selecting small, well-defined, manageable tasks has been implemented. A very structured plan is written for carrying out a first task as a first step. Then a second one is scheduled and so forth. The Shaffer technique has been a highly successful one. Many companies are now employing it.

Of utmost importance in the overall technology transfer plan is the selection and provision of an effective champion. Studies have shown that most products successfully commercialized have each had a product champion. There has been an individual who, while the team has been very active and successful, has had, head and shoulders above anyone else, an intense, innate desire that the product be successful. He then must pursue the product commercialization vigorously. Another comment regarding the champion will be made later.

Secondly and very importantly, there must be top management support for the pursuit of the team and the champion.

How does the company of people interested in and responsible for new product success win top management support? There must be a business plan with valid forecasts and favorable, attractive return - it must be at least as good, if not better, than any other current product return on investment projection in order to command the resources required to meet the plan.

As an illustration of proper design for manufacturability and assembly, a case study will now be cited.

CASE STUDY

The writer has recently designed a column gage, an instrument for measuring a process variable and which has a column display. In the course of the development program on this product a major cost reduction was effected in a structured, disciplined manner.

It was extremely important during the development of the column gage to provide specified functional and performance characteristics, equally important to control cost.

The design challenge was to improve manufacturability and reduce the projected product cost, starting from a prototype generated in the early phases of development. The challenge, then, was to reduce the number of components in general, and to minimize the number of fasteners and other parts directly related to assembly. Of prime importance was the large liquid crystal display and its subassembly; this operator interface was the major focus in the product development efforts.

It was felt that a major change in the packaging concept was required, along with a complete re-examination of all the components and their respective needs. There were many subassemblies and fasteners, and it was determined that the number of each could be significantly reduced. The housing was previously a molded plastic design. It was thought that fasteners could be eliminated if aluminum extrusions were used and slip fits designed in. At the same time, this would provide a rugged enclosure that was much more protective of the circuitry against static. Also, the product would be virtually radiation free.

A team consisting of the project manager, the assigned manufacturing engineer, a designer, an assembler and the marketing product manager was formed, and the design for greater manufacturability was underway.

Design-for-manufacturability-and-assembly (DFMA) techniques developed by G. Boothroyd and P. Dewhurst were used. Software developed by Boothroyd-Dewhurst was loaded into a PC and the team went through the complete program -- responding to the prompts and often discussing the quantified costs and efficiencies of alternative design approaches. This led to a high performance product, yet one that exhibited a significantly simpler combination of components and means of assembly.

The fundamental objective was to reduce to a minimum the number of components, thereby improving cost and reliability, and providing a design that was extremely easy to manufacture and assemble.

The DFMA software program addresses each and every component. The breadboard, or preliminary test prototype, is disassembled or at least viewed as an exploded view of the product. One can imagine a repair parts diagram or parts replacement pictorial.

The software screen presentations ask the questions about each individual component as it is added to the assembly: a) is this part necessarily of different material from all other parts in the assembly?; b) does this part move (slide or rotate) relative to all other parts in the operation or use of the product?; c) must this part be removable because otherwise parts that satisfy the criteria cannot be assembled?

The key to making the product easier to manufacture and assemble is the answer to the questions. For example, if the answers to a, b and c above are all negative, then the particular part under study is either unnecessary or the part might be combined with others. By considering each component in this manner, a significant number of them can be eliminated.

Although the analysis is not limited with regard to the type of component, a rich area of exploitation in this study is that of separate fasteners. Separate fasteners are even undesirable. They are difficult to handle and time-consuming to install. It is preferable to use snap fits, sliding locks in extrusions or molded hinges.

The product resulting from the analysis is one with significantly fewer components and one which is less costly to manufacture in terms of both material and labor.

Major emphasis must also be placed on modularization. A goal in manufacturing is to build major subassemblies, test them and then store these product subassemblies that are now of guaranteed quality, so that they will be ready for final assembly in response to specific customer orders.

In the case of an electronic instrument, such as the column gage--with input/output controls and connectors, a signal processing section, a power supply and a housing -- one can identify minimum interconnection interfaces and then modularize these interfaces.

For example, it is possible to find minimum interconnection boundaries in the circuitry so that each major functional circuit section can be located on a separate circuit board. Also, the front panel and its components, including controls and display, can usually be handled as a single subassembly module. The rear panel with its connectors can be identified as another module. Each of these can be subassembled and tested by QC because each has been defined and designed as a functional subsection of the overall product.

A mechanical or electromechanical product can be analyzed and designed in a manner parallel to that described above for the electronic product. The procedure then is to go through the whole product in a structured manner, following a progressive step-by-step procedure.

All the DFMA questions must be answered during this phase. The secondary outcome of this is that everyone on the team observes the product becoming easier to manufacture. The design engineer, the manufacturing engineer, the production foreperson, the product manager and the QC engineer all experience the considerable benefit. A strong camaraderie is also established as a result of the team working together during the product development cycle. This is an immensely important result.

The seasoned, professional engineer realizes that the product that best meets all the users' requirements is the simplest product. Likewise, the product that meets all the functional requirements and, in fact, excels over the competition in features and functions, while at the same time consisting of the simplest combination of components, is always the most desired product.

The component count in the new column gage was reduced by twenty-five percent over the original prototype using quantitative DFMA design techniques. The material cost was reduced by twenty percent, the assembly time, by fifteen percent.

The development project was then completed and a mean-time-before-failure analysis on the resulting product indicated an MTBF of 18,600 hours.

The Intelligent Column has been on the market now for several months, with favorable results.

ADDITIONAL FACTORS

Some additional key factors in the overall issues of simultaneous engineering are now in order to discuss.

What are the principal requirements for people managing simultaneous engineering activities? The three most important qualities are: personality, intelligence and a penchant for hard work. A successful manager must have these three. In most failures, it has been found that there is a deficiency in one of these.

The central issue in technology transfer is primarily a people issue. The team must consist of proper representatives of various functions and adequate training is a requirement. Leadership styles must be addressed. Also, of critical importance is the characteristic of the champion. Do not assign an individual as champion because he happens to be available. Assign your best qualified individual. If his supervisor does not want to provide this individual, the supervisor is remiss and should be convinced of the importance of this matter.

Another issue is the selection of materials and components for a new product. Are these to be selected by the research people or by the engineering people? This is a critical question in simultaneous engineering. As an example, some companies hope to compete with the newest integrated circuits to exploit the technical features in capturing the market, but these companies are troubled with lower start-up volumes and, therefore, higher cost. Other companies work with the second to last generation of components at high volume to perfect productivity. These components are already debugged. The cost is lower because of trouble-free, high volume supply. Through careful study and implementation, manufacturing processes and cost can be perfected incrementally, one step at a time, eating away at competition.

The customers' needs must always be kept in focus. The prototype should be introduced to the customer group, a group of 'customer friends'. The prototype, as it is developed, is looped through the customers until they like it - until it is a product they would surely buy - only then is it developed. To protect confidentiality, this may have to be done through a focus group.

Remember to limit product complexity. Time and cost are not linear with complexity; they are exponential. Keep new product features to a minimum in any one release. There must be some, to be sure, but question everyone of them. Maximize parts reusability, employ new technology well, emphasize flexible design -- move quickly to market.

As a test, ask yourself the following questions: Can you say that competitors have lower cost to develop; Can you simplify the product design; Were past failures due to too much technological change; Can you have more standard components or more upgradeable designs? If you answered yes to more than one of the above, you have trouble.

CLOSING

The primary elements in transferring new product technology from R&D into Manufacturing then, are the people, the technology and the transfer implementation plan. These are the primary elements in simultaneous engineering. Selection of the team and properly integrating them on a timely basis is of paramount importance. Multi-functional involvement early on and providing for a sense of ownership are valuable. Making certain that the technology is, in fact, developed to a point where it can be released and, then, training people in this technology are necessary. A written plan, attractive to all, as well as a detailed, scheduled work plan are required. The R&D principal must take the initiative in providing this support as the company cannot leave the top management/R&D partnership to chance.

Since people comprise the primary issue, key individuals must understand the total product development cycle in your company. It is they who effectively implement the technology transfer plan.

ABOUT THE AUTHOR

Quentin C. Turtle is a consultant in the field of technology transfer, project management and R&D productivity. He has over twenty-five years experience in industry in new product development and commercialization.

He served with the Esterline Corporation as Vice-President of Research and Development of their Federal Products Corporation subsidiary for ten years and was in charge of all technology development and transfer. Federal Products markets quality control gages and instrumentation.

Previously, Dr. Turtle was Vice-President of Engineering and Manufacturing at International Data Sciences, a principal supplier of instrumentation in the telecommunications field. Prior to that he was employed by General Signal Corporation developing primary and secondary instrumentation for process control.

He has done consulting work for IBM and the Harris Corporation.

He is now President of Technology Management Group in Cranston,RI.

Reprinted from *Assembly Engineering*, March 1986

Applying Design for Assembly Principles

Robert Waterbury
Senior Editor

Design for assembly requires more than simply applying a few principles. It entails subjective evaluations and human adjustments, too.

Companies that seek to increase production efficiency by automating the assembly of established product designs often dig themselves a costly hole. They frequently end up purchasing and installing expensive capital equipment unnecessarily. Why? Because the product probably wasn't designed to optimize assembly; especially automated assembly. If more time were spent in rethinking and redesigning the product for automated assembly, expensive and/or unnecessary capital expenditures could often be avoided, and assembly efficiency improved. Another example of working smarter; not harder.

Speaking at a recent Assembly Technology Expo, General Electric project manager Gerard Hock explained that GE got involved in assembly evaluation techniques out of necessity in 1980. A study conducted at that time showed that the efficiency of GE's assembly labor force was about 50%. Translated into dollars, it meant that $350 million a year was being wasted due to inefficient assembly. And most of the deficiencies, he noted, could be corrected by the proper application of design for assembly techniques.

DFA Evaluation

Design for assembly, Hock explained, produces simpler designs by generally reducing the number of parts, and utilizing less complex assembly operations. It improves the design engineering/manufacturing assembly interface by fostering better teamwork, and providing for early resolution of process, tooling, and design problems. It also promotes better quality and higher reliability through the use of simpler, easier to assemble designs.

In designing for assembly, all mechanical and electromechanical parts and subassemblies should be evaluated first of all for ease of assembly. This includes analysis of the plane(s) of insertion, methods of insertion, methods of joining, fixturing requirements, and assembly rotations, if required. Each of these criteria should be assigned a penalty factor relative to the difficulty it poses to an automated assembly station. Furthermore, each part or subassembly is assigned a combined penalty rating based upon the total number of factors complicating the assembly operations.

By combining the scores of each part or subassembly, it is possible to assign an overall rating for a given product design. This rating, in effect, indicates how easy or difficult it is to assemble that design in relation to other similar designs. The table included as part of this article provides one simplified approach to rating product designs.

The starting point for any design analysis is obviously an existing design. It may exist as a rough sketch, finished design, working model or prototype. The more tangible the design form, the more detailed the analysis can become. In general, however, there are three basic questions that must be answered about a specific part or subassembly:

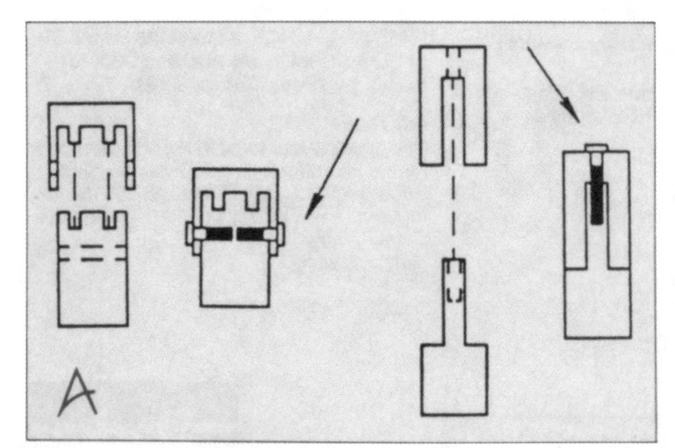

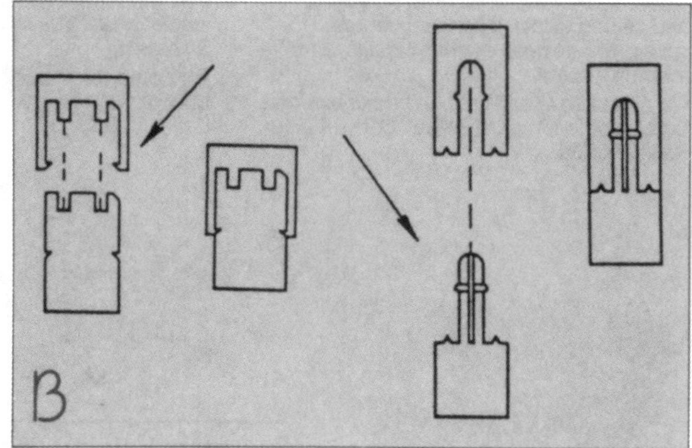

Where appropriate, parts that require fasteners (A) can be replaced by redesigns that use press-fit or snap-fit locking mechanisms (B).

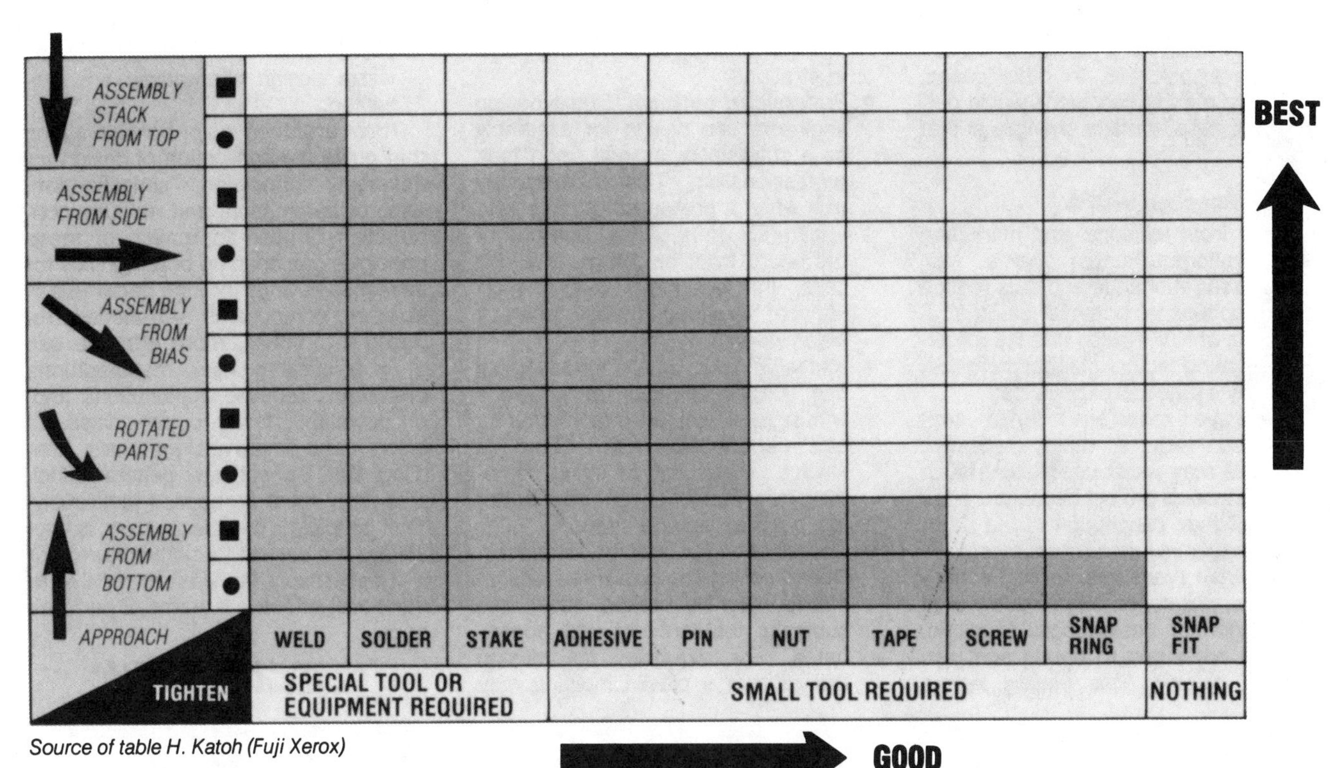

Source of table H. Katoh (Fuji Xerox)

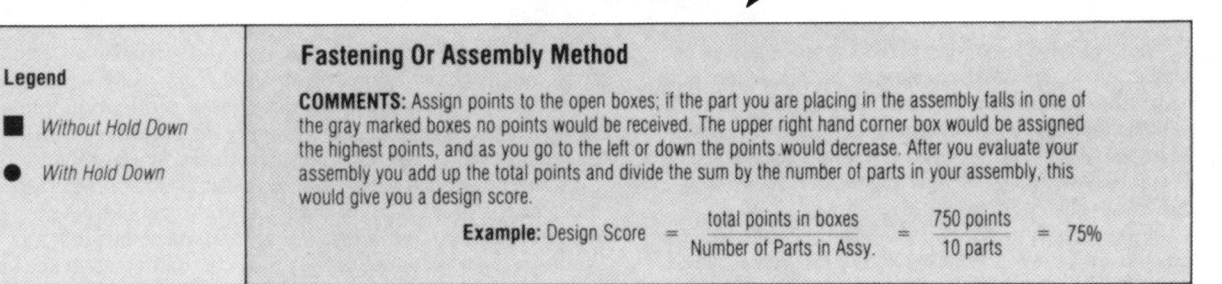

Legend

■ *Without Hold Down*

● *With Hold Down*

Fastening Or Assembly Method

COMMENTS: Assign points to the open boxes; if the part you are placing in the assembly falls in one of the gray marked boxes no points would be received. The upper right hand corner box would be assigned the highest points, and as you go to the left or down the points would decrease. After you evaluate your assembly you add up the total points and divide the sum by the number of parts in your assembly, this would give you a design score.

Example: Design Score $= \frac{\text{total points in boxes}}{\text{Number of Parts in Assy.}} = \frac{\text{750 points}}{\text{10 parts}} = 75\%$

- Can it be eliminated?
- What is the cost to bring it to the proper position and orientation for assembly?
- What is the actual cost of assembly, itself?

The objective in eliminating parts is to optimize "busy" parts (those that perform a number of useful functions) and get rid of "lazy" parts (those that serve little or no functional purpose). In addition, there are three principal questions that help determine whether or not a part or its function can be eliminated:

- Does it move relative to other parts?
- Is it made from a different material?
- Does it have to be removed for service?

Any "yes" answer to these three questions probably means that the part cannot be eliminated. All "no" answers, on the other hand, probably mean that there is no need for that part, although there may still be a need for the function it performs. Oftentimes, this function can be transferred to another, more essential part.

Design for assembly (DFA) evaluation shows the consequences of a design far ahead in the production cycle. Product design is also the key to subsequent parts feeding problems. The best time to assess parts for the ease with which they can be supplied and oriented for assembly is at the very earliest design stages—before they are combined into subassemblies. Specific DFA evaluation principles are described in detail in "Design for Producibility—The Road to Higher Productivity," June 1982; "Computer Aided Design for Assembly," June 1983; and "Designing Parts for Automated Assembly," February 1985 issue of Assembly Engineering.

Case studies show that application of these principles can greatly reduce actual assembly costs. The N.V. Philips Co. in Eindhoven, Netherlands, applied design for assembly principles to about 30 products over a two year period. The products included a car radio tuner, portable iron, video camera, and telephone. Results showed that it was possible to reduce the number of parts and subassemblies in each by anywhere from 20 to 80%. Likewise, the assembly time and tooling costs could be reduced by similar percentages.

Taking into account the materials, direct costs, and other charges, such savings could theoretically be translated into 15 to 20% price reductions in the final product. In reality, however, cost savings closer to 5 or 6% were more commonly achieved. The differences in the theoretical and actual figures are attributable to the fact that not all of the design modifications that are possible are necessarily practical. Some, in fact, were judged to be too risky to incorporate. In other cases, prevailing market conditions would perhaps not support major changes at that time.

Human Reaction to DFA

Aside from technical and marketing considerations, human inertia also tends to resist change and thus reduce the magnitude of potential savings. J. Boorsma of N.V. Philips lists the following mitigating factors gleaned from that company's practical experience:

- *Extensive redesigns.* When confronted with a major redesign, people may resist undertaking such an extensive project because it does not fit their conception of the product and/or its immediate market.
- *Schedule pressures.* Project schedule commitments, often made well in advance of design considerations, exert pressures to create and produce quickly. This usually leaves precious little time for design analyses and subsequent modifications.
- *Recurring design frustrations.* Having successfully created a design concept that meets engineering requirements, the designer is naturally reluctant to start over again. He would prefer not to retrace his efforts, with all of its attendant frustrations.
- *Professional criticism.* Some design engineers see design for assembly as a criticism or assault upon their professionalism. This is especially true when it comes from those who apparently do not know one end of the board from the other. In which case, the designer's ready retort may be: "We tried that before, and it didn't work."
- *Redesign risks.* Design for assembly can sometimes lead to redesigns with greater inherent manufacturing and performance risk. There is always greater risk in trying something new that just might do a better job, but is as yet unproven.
- *Coordination of design subtasks.* Due to scheduling pressures, design efforts may be broken down into subtasks that are not well coordinated. For example, the design drawings of a plastic molding may be released before other metal components are finalized for production. Rare, indeed, are the times when a total design can be evaulated before making any production commitments.
- *Design disagreements.* Sometimes there are simply irreconcilable differences when two or more equally viable design alternatives are presented.

There are three major considerations that guide the application of design for assembly principles: simplification, ease of automation, and modular construction. Proper application of these principles can open up opportunities for increased savings through better use of standard components or materials, reduced inventories, more effective use of sophisticated new manufacturing processes, reduced malfunctions and/or downtime, fewer opportunities for errors, and improved materials handling. But, as with any general principles, they must be applied judiciously. The simplest, one-piece part is not always the easiest to manufacture; nor is it necessarily the least expensive or most cost-effective in terms of performance.

Presented at SME Machining Systems Conference, May 1989

Analysis of Production Line Efficiency from the Viewpoint of a Japanese Production Engineer

By Takuro Yamada

Mitsubishi Heavy Industries, Ltd.

1. PREFACE

Simultaneous engineering between automobile companies and machine tool builders has become of interest recently in the US. From the beginning of a new automobile production project, automobile manufacturers' engineers and machine tool engineers are members of the same project team as cooperative planners for the smooth start up of production. The purpose of this engineering is to establish mutual technical understanding, to shorten delivery, and to improve investment efficiency.

When a user decides all the specifications from a user's viewpoint, or a machine tool manufacturer estimates the facility from only his perspective, both sides often are unsatisfied with each other's ideas. Simultaneous engineering is a good solution to this problem.

Why has simultaneous engineering become popular so quickly in the US? One reason is that the difference in production line efficiency between the US and Japan has been recognized. With increased cooperation between US and Japanese automobile manufacturers and surveys of Japanese lines, managements of US makers have recognized this difference in efficiency.

US maufacturers have been interested in observing the Japanese method for developing a new model car. This method is the same as simultaneous engineering.

It seems strange to the author that the US automobile industries (which have a long history and much experience) are suffering from lower levels of production efficiency than those in Japan, and are forced to plan more expensive production lines which have excess over capacity to cover the lower efficiency.

In this paper, the author will discuss this difference from a Japanese engineer's viewpoint based on his business experiences in the US.

High speed metal cutting with high power and highly rigid machinery is not dealt with in this paper, it being another technological genre.

2. THE DEFINITION OF PRODUCTION EFFICIENCY OF A METAL CUTTING FACILITY

There are four different indexes of production efficiency of a metal cutting facility. In this paper, each efficiency is defined, but E1, E2 and E3 represent the major points outlined.

(1) Productivity ratio: E1

$$E1 = \left(\frac{\text{Chip making time in a machine cycle}}{\text{Machine cycle time}}\right) \times 100\%$$

E1 shows the effective machining time ratio in a machine cycle when a machine runs under normal conditions. E1 is an important index in short cycle machining or in mass production such as in a transfer machine.

(2) Up-time ratio and/or machine efficiency: E2

$$E2 = \left(\frac{\text{Total machine hours} - \text{nonproductive machine hours}}{\text{Total machine hours}}\right) \times 100\%$$

E2 is an important consideration for production engineers. The numerator is effective machining hours in total machining hours, i.e., machine downtime and machining hours to make rejected parts are removed from the total machine hours. E2 is the ratio of hours of effective part production to total machine running hours.

(3) Overall line efficiency: E3

$$E3 = \left(\frac{\text{Effective production hours}}{\text{Total running hours of a line}}\right) \times 100\%$$

E3 is similar to E2. This can deal with the effectiveness of a buffer station between machines.

(4) *Stock Removability: E4*

This index shows the capability of a machine expressed in terms of the amount of stock removed in a given unit of time. A highly rigid machine is evaluated using this index. Of course, this index will be influenced by many factors, i.e., machining process, machinability of material, heat treatment, rigidity of a part, tool materials, etc.

It is difficult to compare machine capability under the same conditions and proper machining conditions. This index is not dealt with in this paper. In Japan, there is not as strong a market demand for high stock removal rates as there is in the US. In Japan, the first priorities of most production engineers in automobile industries are quality, limiting space, and conserving energy. The use of excess horsepower is considered useless.

3. *PRODUCTIVITY RATIO: E1*

3-1 *Invalid Machine Time*

Invalid machine time in a machine cycle refers to the following:

(1) *Parts handling time*

Parts transferring, loading/unloading, pallet changing (APC), parts locating, and clamping/unclamping times.

(2) *Tool rapid approaching and retracting times including air-cut time.*

The control of acceleration and deceleration rates is most important to decrease these times. CNC systems can precisely control the machine movements, decreasing these times to less than 1 second.

(3) *Tool changing (ATC) and attachment changing (AAC) times*

When combination machining by one set-up is used in a machining center, these times effect E1. A newly developed mechanical drive mechanism of a turret lathe can index tools in 0.5 seconds, raising E1 effectively in short cycle turning operations.

3-2 *Target of E1*

E1 is most important when the cycle time of a transfer machine becomes shorter. A mechanical parts transfer mechanism and CNC feed unit can decrease the invalid time of a transfer machine to less than 6 seconds, so the efficiency E1 can be expected to be over 80% when the machining cycle time is 30 seconds. But, the efficiency E1 decreases to 70% when the cycle time is 20 seconds. Figure 1 shows the relationship of machining cycle time and E1.

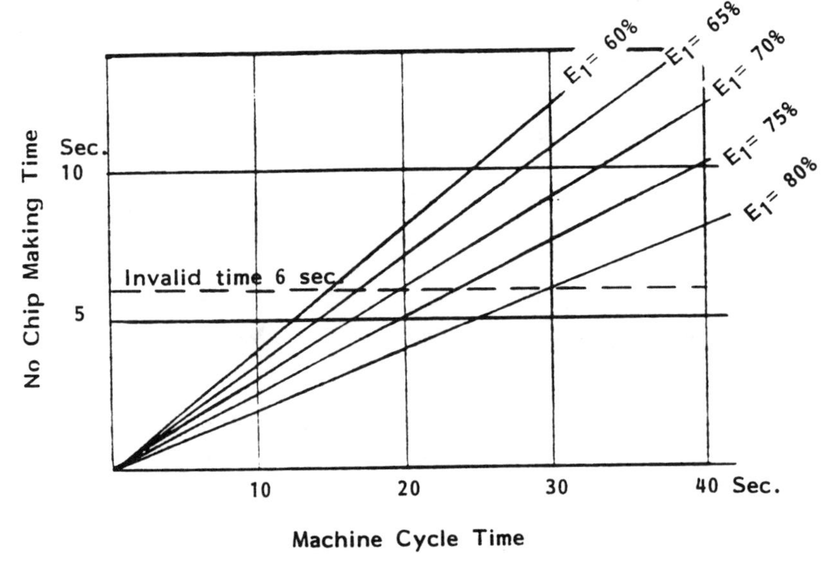

Figure 1 E1 of a Transfer Machine

When the transfer distance is 1200-1400 mm, a transfer bar serves as the link mechanism. For over 1400 mm transfer distance, a ball screw driven by a servo motor is used. In either, to increase E1, compact design is essential. If the transfer distance is over 2000 mm, a high level E1 can never be achieved.

A minimum target value of E1 is 75% in the case of conventional and flexible transfer machines. E1 of machining centers in FMS can be expected to be only 20-40% because of long ATC and APC times. Minimum cycle time of a machining center to achieve good E1 is over 30 minutes. If the cycle time decreases to less than 10 minutes, E1 decreases to less than 20% as shown in Figure 2. At the Machine Tool Show in Japan last year, machining centers equipped with fast APC (functioning within 10 seconds) and turning-centers equipped with an auxiliary spindle presented impressive demonstrations of decreased invalid machining time.

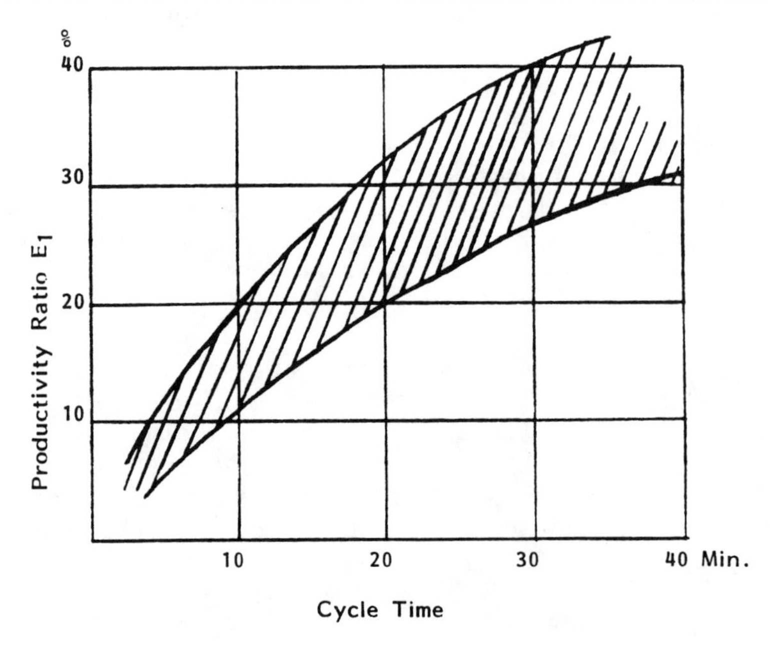

Figure 2 E1 of a Machining Center

3-3 Increasing E1

Shortening strokes, increasing acceleration rate by using light weight components, and minimizing moving elements are the solutions to increasing E1. In the case of short cycle times, decreasing machining time by only 0.1 seconds strongly effects E1. Design concept innovation including new material adoption makes a considerable impact on E1.

In this regard, Japanese customers are extremely concerned with the conservation of energy. For example, a user's engineer who strictly evaluates the layout distance of machines said, "One meter is one second for an operator to walk." He evaluates machine size by time units.

Thus, Japanese machine tool builders, because of demands of their market and their backgrounds, have realized a higher level of efficiency E1, productivity ratio, than US manufacturers.

4. UP-TIME RATIO/MACHINE EFFICIENCY: E2

4-1 Downtime of a Machine

This point is of concern to engineers throughout the world and has become more complicated recently because flexibility of machine tools has been strongly requested by users.

The downtime of a machine is grouped in three different categories.

(1) Scheduled stops

Changeover, maintenance, production changes, tool changes, measurements, adjustments, etc.

(2) Loss

Making inferior parts.

(3) Unscheduled stops

Tool breakage, unscheduled tool changes, machine failures, unscheduled changeover of the machine, labor performance, etc.

Another grouping is also possible, i.e., machine stops and human stops.

These stops of a machine all cause a shortage of planned production quantity. Scheduled stops and losses can be estimated when production engineers make the production planning. To allow for unscheduled downtime, additional marginal quantity must be estimated in the machine capability. So estimating E2 is an important consideration when determining a machine's capability to satisfy a scheduled production quantity. If estimated E2 is too low, the planned capability of a machine must include excess capacity compared with scheduled production quantity, and requires excess investment by a machine tool user.

4-2 Target of E2

The target of E2 of a transfer machine of Mitsubishi Heavy Industries is usually E2 $\geq$ 60-90%. A transfer machine is run with conservative cutting conditions to keep stable tool lives and a machining center is run with aggressive conditions to increase productivity. The latter is provided with many cutting conditions monitoring systems to avoid tool breakage and stores spare tools in ATC magazines. Changeover of a machine cycle is not difficult if CNC is employed. Pallets and work holding fixtures are important parts of the cycle. Usually, additional sets of spare pallets and fixtures are provided to do changeover in parallel with production to avoid excess changeover time.

Direct feed transfer (free transfer) is usually difficult unless the parts design is completely standardized and serialized. In the past, a transfer machine for cylinder blocks of 2, 3, 4, and 6 cylinders and another one for cylinder heads of 2, 3, 4, and 6 of direct injection and combustion chamber types were run on direct feed system machines. In any case, proper limitation of part models and sizes for easy changeover of tooling helps maintain a high E2 of any kind of machine tool including machining centers.

Total avoidance of unscheduled machine downs is impossible; a more realistic goal is the prevention of accidents by using a proper monitoring system.

Most causes of unscheduled downs of a machine are minor and can be easily remedied in a few minutes, so a slight improvement of little imperfections through a quality circle is very effective.

The most basic way to improve E2 is to use the simplest functions and mechanisms. The use of established technology only is essential to increase the reliability of the total machine.

4-3 Comparison of a Synchronous System and a Non-Synchronous System

The reliability of these two systems has often been compared. The yield of a ten station synchronous transfer line system with an average up-time reliability of 98% for the transfer mechanism and each station, would be 0.98^{11} =80%. On the other hand, a non-synchronous system still has 98% reliability because it is not time-locked. Is this analysis correct? This estimate contains several contradictions because of the assumed conditions of comparison as follows.

(1) Synchronous Transfer Machine and FMS

The productivity of these two production systems is so different it is difficult to compare them.

For a FMS to have the same productive capacity of a transfer machine, at least two to three times the number of machining centers must be used compared with the number of machining units in a transfer machine. The difference of E1 is 75% for the first as compared to 20-40% for the second. Also, a transfer machine usually has multiple spindle head stations.

The machining units in a transfer machine are only sharing a simplified operation with a simple function. Machining centers in an FMS have multi-functions including ATC and APC and, moreover, have only one spindle. Machining units with a multiple spindle heads usually travel in only one axis. A machining center moves through multiple cycles in three axes to do the same operation. There is also a great difference in the transfer of parts. The comparison of these two different systems using the same number of machining units and same up-time reliability estimate is, in the author's opinion, nonsense.

(2) Synchronous Transfer Machines and Non-Synchronous Transfer Machines

If there are two different transfer machines which have the same productivity and the same number of machining units, it is still difficult to decide which has the higher reliability or E2. In this case, the difference is only in the transfer mechanism.

As mentioned above, the unscheduled downtimes of a transfer machine are of short duration. In such a case, the non-synchronous transfer machine has advantages over the synchronous. To avoid time-lock among all the machining stations, more than 10 pallets must be stored between stations making quick pallet transfers difficult as compared to a synchronous transfer system. In the synchronous transfer system, one controller controls all pallets. But a non-synchronous transfer system has many distributed pallet controlling mechanisms and their controllers.

The up-time reliability of a non-synchronous transfer mechanism may be inferior to a synchronous system or at best equal. If the up-time reliability of these two transfer machines shows a clear difference, the author must assume there are other factors which influence the reliability difference.

4-4 Machinability and Efficiency

Tool life and chip disposal are definite factors by which to judge manufacturing systems, e.g., stand alone machines, FMSs, and transfer machines. High efficiency can be expected on aluminum and cast iron, but not on forged steel and ductile iron.

Aluminum die cast parts have good machinability and minimum stock removal leading to the highest efficiency -- in many cases over 80%. The only notable drawback is deformation of workpieces which causes accuracy problems.

On the contrary, forged steel parts have poor machinability, hardness fluctuations, and stock removal fluctuations. Also, chip disposal is usually difficult. As a result of these factors, an average of only 60% efficiency can be expected in the case of forged parts. The extent to which a system user can control the quality level of materials determines the system's efficiency. Japanese metal suppliers have succeeded in raising the quality control level to the highest in the world permitting Japanese automobile industries the highest level of efficiency and stable tool lives.

4-5 Tool Management and Efficiency

Tool management is as important as material control to the efficiency of a manufacturing system. Recently, almost all turning tools and boring tools have indexable inserts, making tool management in this area easier.

The most important function of tool management is the regrinding of high speed steel tools and tungsten carbide brazed tools, i.e., drills, taps, reamers, broaches, etc. Generally speaking, when a user regrinds a tool, the tool life becomes shorter than the original tool or a new tool. This reflects the technical difference of tool grinding between a tool maker and a tool user. In many cases, the user's regrinder cannot completely remove the worn part of a tool (for example, it is very difficult to remove completely the worn part on the margins of a reamer or broaching). To regrind the correct shape of the web thinning of a drill is also a difficult job for a user's tool room. The fluctuation of tool life is often caused by such incomplete regrinding of tools.

A tap is not as expensive as other tools, so it may be used as a throwaway tool.

TiN coated tools, which have become more popular, must be carefully reground to be used effectively.

4-6 Other Factors

A system's efficiency is influenced by many other factors.

The power source seriously affects efficiency. In Japan, there are zones of 50 Hz and 60 Hz. A CNC grinding machine which was equipped with a US controller had no problems in the 60 Hz zone, but in the 50 Hz zone, almost all of the machines had troubles. A similar experience occurred in Wisconsin where a machining center had serious troubles during the night shift, but in the daytime had no problems. In this case, it was finally discovered that the surge voltage of the power source switching at midnight caused the troubles of the CNC controller.

Of course, the skill and knowledge level of workers and maintenance people make an important difference in the efficiency of the system.

Formerly, in Japan, people treasured the life style of an employee in a company, so managers normally expected to have enough educated subordinates. But, recently, younger workers transfer more frequently (as in the US) and in the near future, many Japanese managers will experience a shortage of skilled people. Recently, as the result of international vocational training competitions shows, the workers' skills in South Korea, Taiwan, Hong Kong, etc., have surpassed Japan in many industries. This means the skill level of Japan's workers has reached a transition period when the efficiency level, now supported by skilled workers, will lower without new technological developments.

5. OVERALL LINE EFFICIENCY: E3

Overall line efficiency is basically the same as efficiency E2, but, in this section, the effectiveness of buffer stations in a production line must be discussed.

About 20 years ago, buffer stations were recognized as essential in a line to avoid mutual interference of downtimes. Automated large capacity buffers were distributed between transfer machines in a line. Or, in a small batch production line, many cartboxes or trays were placed between machines to stock parts.

Many engineers questioned the effectiveness of these large size buffers and started to analyze them theoretically and in computer simulations. Their studies of buffers were aimed at the avoidance of downtimes by optimizing the buffer size in a line. For this purpose, many computer line simulation programs have been developed and it is even now very effective to use these programs on existing lines to determine the source of a bottleneck.

Soon after, the market needs suddenly changed to production of more models in a mass-production line. In such a technical environment, the concept of the "Kanban production method" of Toyota Motors became popular in Japan and quickly spread worldwide.

This concept eliminated buffers in a line and demanded higher reliability of machines and easier maintenance of them. It meant that if a machine went down in a line which had no buffers between machines, the total production line stopped. So the manager in charge of a line made desperate efforts to maintain the line to avoid confusion and requested the purchase of only excellent machines. Of course, at the beginning, all Japanese industries experienced many serious troubles, but gradually, years, all newly installed production lines completely eliminated buffers between machines and, instead included flexibility to quickly changeover the total line. The inventory of parts has been reduced. Today the standard idea of automobile industries is to stock only a half day's volume of machined parts between the machine shop and the assembly shop.

Thus, the success of the "Kanban production method" directly affected the management concept of production line efficiency.

6. IMPROVEMENT OF PRODUCTION LINE EFFICIENCY

6-1 Improvement Steps

The improvement of production line efficiency must take a long term view including planning, introduction, and post installation of a line. It is difficult to state the priority of which improvement step is the most important.

In Japan, top management generally regards the planning step as the most critical to avoid the investment risk. This step is earlier than the starting time of simultaneous engineering in the US. At the post installation step, quality circle activity is used effectively in Japan. Thus, the difference of the line efficiency between Japan and the US is based on fundamental differences of management styles resulting from cultural differences of both countries.

6-2 Improvement Outline for Planning

This outline reflects the managing concept of top management of companies. It is not dominated by the present market condition (which fluctuates at all times), but looks forward several to ten years to future prosperity of the company.

(1) Standardization of product design

No one doubts the effectiveness of design standardization in concept, but in practice it is very difficult to realize in a company. At the origin of a new product, the standardization of design is realized by engineers, but usually during the product life cycle, the concept changes little by little. Finally, when a new production system is planned and cutting tools and fixtures are arranged, people are surprised at the enormous numbers of kinds of them. And, it is too late to consolidate them and decrease the number for reasons such as supplying maintenance parts. Usually a parts maker will produce only standard parts, but customers strongly request diverse specifications. The author had this experience with a US customer who carried many numbers of tools which had only slight variations in diameter.

Figure 3 shows an example of how the standardization of products effectively reduced the number of cutting tools when MHI Kyoto plant introduced an FMS. In this case the machined parts are prismatic components for machine tools. Standardized thread hole depth and arrangement can eliminate much input data from NC part programs. Hole locations are specified only by the pitch circle diameter and the hole pattern location.

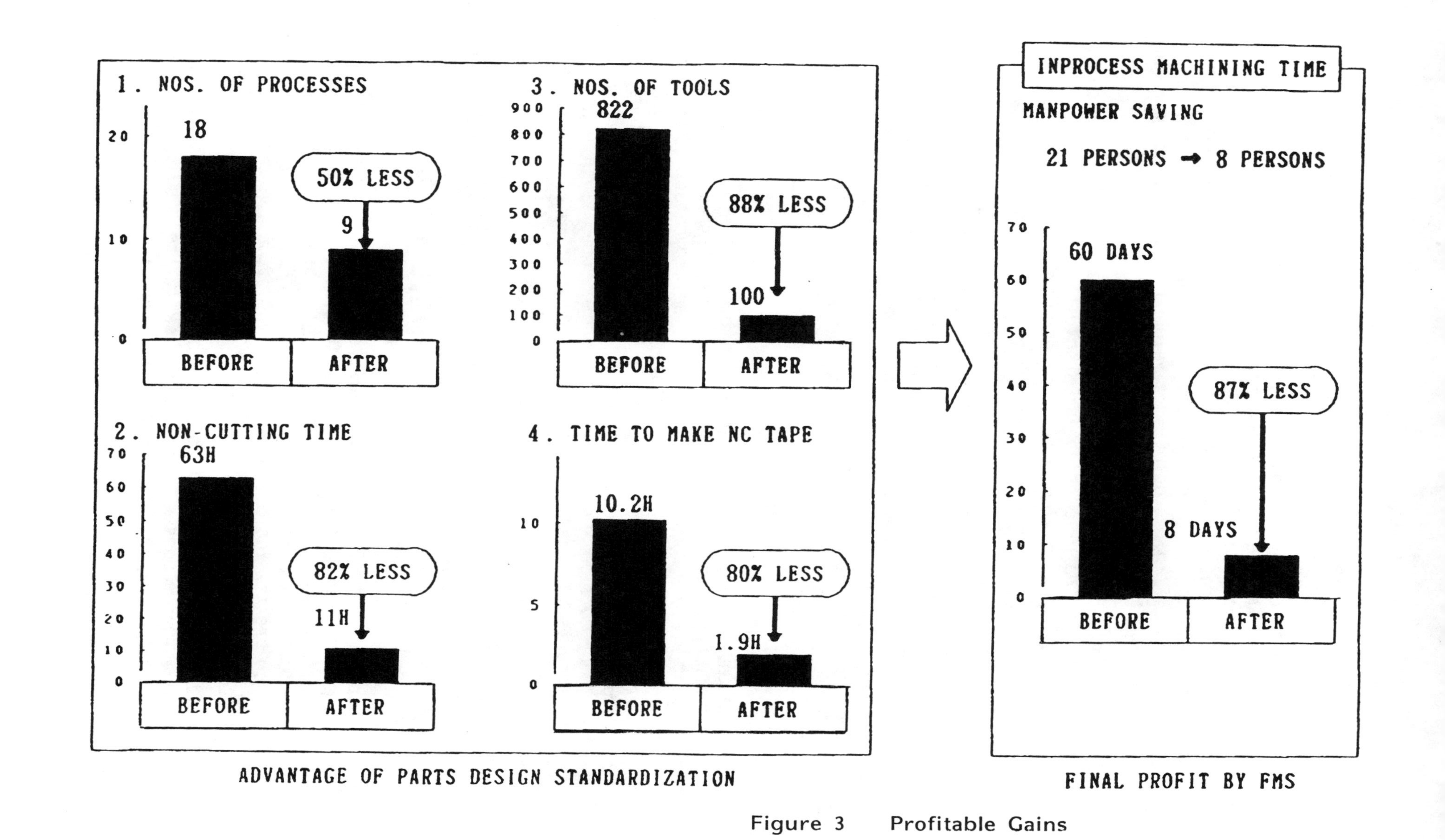

Figure 3 Profitable Gains

In the design of automobile parts, if the designer recognizes the minimum spindle distance required in a multispindle head and arranges hole locations greater than this minimum distance, fewer machining stations will be needed in a production line. This will result in higher line efficiency. Slight design changes of inclination of oil holes or off-set value of holes may be more easily accommodated in the machining process.

In spite of these advantages, standardization of parts design through inside discussion is often difficult, so simultaneous engineering involving outside members is more effective.

The choice of standard parts is also a good way to avoid trouble, but users have a tendency toward special specifications. Users would be well-advised to understand the advantages of standard parts of the machine tool manufacturer. In the same way, Japanese people are used to choosing the chef's recommendations, but the US people like to specify the details of cooking by themselves. It is also seen in the different approach to social living behaviors of the individual and his sense of obligation to the community in both countries.

(2) Education of the Employee

In Japan, generally speaking, the top management of companies are striving to raise the educational level of their employees. In the US, there are many excellent management education courses for managers in universities, but it seems to the author not so popular to give workers the opportunity to attend educational training courses at the company's expense, or to hear an educational lecture by outside speakers during working time as in Japan. "An enterprise depends on people" is a popular management motto in Japan.

Educational efforts, of course, raise the worker's motivation and, therefore, raise the line efficiency. To foreigners, the Japanese appear to be working too hard without enjoying living life, but the author will emphasize that people are working autonomously, self-motivated within their shops.

Quality control, preventive maintenance, etc., are difficult to implant in a shop without habitual education of all workers. The practical application of them to solve problems cannot be expected without a high level of workers' motivation. QC, ZD, PM, etc., which are raising production line efficiency in Japan originated in the US.

Still, there is a clear difference of efficiency in both countries. The author supposes that is most likely the result of differences in educational efforts of top management.

6-3 Improvement Outline at the Introduction of a Line

(1) Simplification of a System's Functions

Although it is certain the reliability of a simple machine has higher efficiency than a complicated one, many machine tool users are requesting higher machine efficiency and more complicated functions at the same time. Machine tool builders must explain to their customers more earnestly the benefits of a simple machine if they want high level efficiency.

People expect that FMS is the best solution to accomplish changes JIT (Just In Time). Still, no one (makers or users) discusses the obvious disadvantage of lowering the productive efficiency of an FMS with JIT. For increasing the productivity of an FMS, a user must avoid excess changeover of it for JIT.

To use a FMS producing over hundreds of different workpieces including one or two per year quantity of maintenance parts with over several hundred different cutting tools, the customer must prepare a very complicated changeover schedule and be prepared to accept lower productivity than from a system which produces only up to ten different workpieces using less than fifty different cutting tools.

According to the author's experiences, Japanese customers have a tendency to accept maker's proposals to raise a system's efficiency, but in the US, usually customers will not compromise the functions of their first imagination.

A research report comparing the FMS efficiency of the US and Japan noted that FMSs in Japan have higher productivity and efficiency than the US. The author supposes the difference is caused by difference in the number of specifications and functions.

(2) Simultaneous Engineering

Simultaneous engineering is very effective in the development of a production line of higher reliability by combining the mutual technical experiences and understanding of users and makers. Machine tool engineers must understand the design concepts of the user's products and propose ideas of design modification which improve productivity from the viewpoint of machine tool designers. Product designers must keep their ideas modest enough to allow for simple machining and design standardization.

In some cases, it is important to give a limitation of kinds, shapes, and sizes of machined workpieces for increasing system efficiency.

In regards to software in particular, the development and modification are not only expensive, but also difficult to completely avoid "bugs". The use of existing standard software programs is the best way to bring about early start-up of the project. If the use of existing software is impossible, it is a good suggestion to separate system software and management software to avoid the latter confusing the former. MHI's FFMS (Free Flow Manufacturing System) is a unique manufacturing system. See Figure 4. It does not use a computer to control system movement and results in significantly higher reliability of the system.

(3) Parts and Tools Control

The quality of incoming parts and tools directly affects the system efficiency.

Stock removal, dimensional accuracy, hardness of incoming parts, and tool regrinding conditions affect tool life and machining accuracy. A successful user of a system usually can manage these items. For example, a user who has sufficient experience and confidence in FMS does not require a tool monitoring system because he has complete data of tool life when using standard cutting conditions on his products. On the other hand, a user who requests special part programs, tooling, and many monitoring systems should be viewed with caution by a machine tool builder. In other words, the efficiency of a system depends on not only reliability of the system itself, but also on the user's process engineering capability.

(4) JIT and Production Scheduling

The minimum inventory is a very important managing concept for an enterprise to meet a bewildering change of market situations. However, JIT delivery can be accomplished as the final result of total plant logistics. The front part and/or machining shops must be controlled by proper strategic management of delivery time and quantity.

From the author's experience, a manufacturing system shows the highest efficiency when enough incoming parts are accumulated at the loading station. If the supply to the loading station is curtailed sharply (as with JIT), efficiency scarcely reaches the highest level. The reason for this difference is psychological tension which is experienced by the operator of the system.

The production manager must consider the psychological and mental outlook of workers and correctly judge the merit and demerits of JIT management.

Figure 4 Free Flow Manufacturing System (FFMS)

(5) Mutual Trust Between a User and Makers

The author also believes that the efficiency of a system is decided basically by the first acceptance test of the user at the manufacturer's site. The system, of course, must satisfy the machining accuracy and productivity specifications which were decided when the order was placed with the manufacturer. Usually the user requests, additionally, the change of many items before installation, for example, changing the location of switch boxes, the addition of monitoring lamps, changing safety guards, etc.

These changes occur mostly because of errors in ordering specifications or incomplete communication of the specifications between planners, operators, and maintenance people of the user. As a precautionary measure, a good suggestion is to hold early meetings of all members or at least of planning engineers, operators, maintenance people, purchasing people, and manufacturer's engineers. This meeting will clarify the specifications and the terms of the acceptance test. At least, the operators and maintenance people will feel pleased with the specifications decided upon by themselves and will approach the acceptance test optimistically. A system which is accepted with the user's satisfaction will produce at higher efficiency in the user's plant than a machine accepted with many changing items.

A typical example occurred about twenty years ago in the US. Four gear cutting machines were purchased by a user. The machines were installed in two gear shops with two machines in each. The machines in one shop were smoothly accepted, but the others experienced problems in the first run-off test at the user's site. After about one year passed, the shop manager who accepted the latter machines said to the other manager who accepted the former two, "The machines have too many troubles and cause too many headaches." The other manager proposed, "I am satisfied with the machines, so, if you agree, we will exchange yours for those of another maker." They changed machines. The author was informed of this news from the dealer and visited the user after several months. The manager who had all four machines said, "I was completely satisfied with all the machines. They have no problems." This was the manager who trusted the machines from the start and the machines met his expectations.

People say the human relationship between a manager and his subordinates has an important effect on the motivation and productivity of his shop. The author believes, after many experiences, that a machine tool more precisely obeys the operator's mind than that of his boss.

If simultaneous engineering cannot ensure the mutual trust between a user and a maker, it is of no value and only a waste of time and money.

6-4 Improvement Outline -- Post Installation

(1) Quality Circle Activities

No manufacturing systems can be perfect even if a knowledgeable engineer did the engineering and a skillful worker did the assembly. Only operators and maintenance people can find slight imperfections and improve them. The most effective improvement of these imperfections may be accomplished through daily activity of quality circles. A system's efficiency may be increased up to 10% this way.

On the contrary, when a maker tries to repair these imperfections, they will be ineffective if their ideas are based solely on information from the user's manager without looking at the machines. Effective improvements will be successful only when they are based on practical experience with the system.

(2) Controlling Materials and Incoming Parts

The quality control of incoming parts has progressed remarkably. It seems better to have less stock removal; however, it is very difficult to decide how much because too little may cause difficulty in location and (too often) makes them too hard to maintain good machinability.

The composition of steel parts affects hardness and machinability. All steels have a proper hardness range for good tool life and chip making. When the efficiency of a system is unsatisfactory, not only machine tools, fixtures, and cutting tools, but also materials and dimensions of incoming parts must be checked.

(3) Periodic Inspection of Facility

The maintenance of a machine also has a strong effect on efficiency. A knowledge of the history of previous problems is essential to raise efficiency.

In machine control systems, reliability and the ability to maintain the circuits have progressed remarkably. CNC and PLC have proper diagnostics for easy maintenance. Monitoring systems protect machine tool damages during a long, unmanned operating.

In more recent years, young workers prefer to study software, so fewer people study hardware. Therefore, in Japan, the number of skilled hardware maintenance engineers is decreasing. It is important to educate excellent hardware engineers to expect highly efficient production.

7. CONCLUSION

This paper has tried to present ideas for increasing the productive efficiency of machine tools and production lines from the viewpoint of an experienced Japanese Systems Engineer.

Production engineers must analyze the efficiency of machine tools and production lines from both view points: theoretical and practical.

Usually people like to discuss the theoretical side only, as in reliability analysis or computer simulation. These theoretical analyses are, of course, essential to plan a new project or to improve an existing facility. In this case, the analyst must never commit the error of making imperfect assumptions which are mostly based on inexperienced prejudice and visionary supposition. Before trying the theoretical analysis, engineers must go to the machine site and watch precisely how operators are dealing with the machine. No matter how precise the analyses are, the process engineers cannot calculate the effect of the workers' motivation.

So engineers must try practical
improvement plans including total management to raise the line efficiency in conjunction with the theoretical approach. In particular, management which enhances workers' motivation can increase the line efficiency at least several and maybe up to ten percent.

The relationship between a user and a maker influences the efficiency.

The man/machine relationship may make a considerable difference in many cases. Productive efficiency is the combination of man/machine systems.

For engineers to increase efficiency by even 0.1 percent, they must approach the goal from many directions.

As a Japanese Systems Engineer, the author will continue to try to discover more ways to reduce production cost. He also sincerely hopes that excellent American engineers will develop new American systems management which will increase efficiency in the near future.

Presented at the Spring National Design Engineering
Show & Conference, March 7-10, 1988, McCormick Place North, Chicago, IL

Analyzing Product Assemblability Merit

By Carl F. Zorowski
North Carolina State University

Historically, designers have concerned themselves primarily with creating designs that fulfilled the functional requirements of new product specifications. Fabrication and assembly of the resulting design then became the responsibility of the manufacturing engineer. All too often this process lead to the need for significant redesign of the product by manufacturing to accomodate fabrication and assembly facility and cost constraints. Todays competitive marketplace with its requirements for shorter lead times, greater productivity, and improved quality all at lower cost dictates a pressing need for integrating product design into the total manufacturing process and life cycle of the product. The increasing order of magnitude of cost required to implement design changes in the fabrication, assembly, and field use stages of the life cycle dramatically illustrates the importance and impact of early design decisions in product development. The level of attention given by the product designer to the needs and requirements of the manufacturing process by which the product will be produced and the environment in which it will be used will dictate to a large degree the competitive "success" the product will enjoy in the marketplace.

To specifically assist product designers in developing designs that are more compatable to the assembly needs of modern manufacturing process technology a microcomputer software tool called Product Design Merit (PDM) was developed. The tool provides designers with a systematic method for analyzing a given design and calculating a numerical merit rating for the assemblability of the product. The process of merit assignment rewards design alternatives which incorporate principles of design for ease of assembly and penalizes designs that include potentially redundant or unnecessary parts. By using PDM alternative design configurations can be systematically evaluated and numerically compared on a common basis for assemblability. Through the application of this merit rating tool the designer becomes more aware and sensitive to the impact of design decisions on the assemblability of a product and is assisted in identifying those areas where redesign can improve ease of assembly. This in turn facilitates the integration of the design process into an important aspect of the total manufacturing effort.

The PDM Analysis Tool

The analysis segment of the PDM software tool consists of two main components. The first is based on the three primary functions required in any assembly process: (1) feeding the part to the assembly, (2) inserting the part into the assembly, and (3) fastening or capturing the part in the assembly. All of these functions can be accomplished through a variety of means dependent on the design of the part under consideration. For example, Table I lists ten options by which a part might be captured or fastened in an assembly. These vary from a simple, low cost, easy to implement "slip fit" to a significantly complex and more costly to automate method like "welding". In PDM a part designed to be slip fit into an assembly and require no additional fastening is given a high merit rating for that assembly function. At the other extreme if the part had to be fastened by welding in would receive a low merit rating. The order in which the options are listed in Table I, from the "easiest" to the "hardest", as viewed from assemblability together with the merit values assigned each method represent subjective decisions on the part of the author and are fixed in the program. In a similar fashion PDM provides the designer with a fixed selection of options for the feeding and insertion functions. Again representing the subjective judgement of the author these are listed from the highest to the lowest merit ratings as viewed from assemblability in Table II and Table III.

In using PDM to evaluate the assemblability of a product the designer is required to make these three assembly function option choices for each part as it is sequentially added to the product. The software assigns the pre-selected merit figures to each option choice and calculates a combined assemblability merit for each part. When all parts have been analyzed the program determines an average combined assemblability merit figure for the product. The details of how these calculations are performed in the software are presented in reference [1]. The assigned merit figures in the program vary for "100" as the best to "0" for the poorest as contributing to "ease of assembly".

The second component of the analysis segment of PDM consists of a part redundancy criteria evaluation. The designer is asked the following three questions about each part: (1) does it move relative to adjacent parts, (2) do adjacent parts need to be made of a different material, and (3) does the part need to be separate to permit assembly or disassembly? A "no" answer to all three questions recognizes that there is a high probability that the part can be eliminated through redesign. Elimination of extranious parts always improves assemblability. The existence of potentially redundant parts in a design , numerically penalizes the final assembly merit calculated by PDM for the total product. Details of the penalty calculation can be obtained from reference [1].

PDM User Features

PDM is user friendly, menu driven, and interactive. Subroutines offered and selected by the user through the menu include: performing a new product analysis, reviewing a summary of results, reviewing part input data, editing input data, saving a parts data file, retreiving a parts data file, printing out hard copy results, calling up merit figure definitions, and terminating the program. Help screens are also included to assist the user interpret and understand the option choices offered for the three assembly functions. Results of an analysis are immediately available providing a summarized overview of the design/assembly evaluation and a numerical assembly merit for the product. These results can be used to quickly identify areas of potential redesign that could result in improved assemblability.

Example Applications

Presented in Figures 1 and 2 are the original and reconfigured designs for a pneumatic pressure sensor studied by Boothroyd and Dewhurst [2] using their DFA, Design for Assembly, analysis technique. The objectives of DFA are similar to those of PDM: to provide (1) means of quantifying assemblability, (2) identification of potentially redundant parts, and (3) recognition of redesign opportunities that would lead to assemblability improvement. Applicaiton of PDM to the design in Figure 1 generated the results in Table IV. These results indicate that the screws and piston stop are redundant and can be eliminated by redesigning the cover into a snap fit plastic molding with the stop included as shown in Figure 2. Also the piston feeding could be improved from manual to machine with the addition of a handling stud on the top surface. With these changes the average combined merit of 79 for all parts and assembly merit of 45 for the original product were improved to 92 for both merit values as shown in Table V for the redesigned sensor; a significant improvement in the assemblability.

A second example of product redesign, following analysis with PDM, that resulted in dramatic improvement in assemblability is the household fan illustrated in Figure 3 and first reported in reference [3]. The original design contained 46 part and 3 subassemblies in the total product. Application of PDM resulted in an average combined merit of 81 but an assembly merit of only 28. The analysis identified a significant number of potentially redundant parts and strongly suggested that the subassemblies might be eliminated. The redesigned fan shown in Figure IV contains only 16 parts which along with other modifications improved the assembly merit to 84.

Table VI presents assemblability results of five common products using PDM and the evaluation of a redesign of the product suggested by the analysis of the original design. Some general characteristics of the results are to be noted. Significant differences between the average combined merit (for all parts) and the assembly merit (for the product) for the original design is indicative of the identification of a number of potentially redundant parts. This is illustrated in the first four examples reported. The elimination of these redundant parts through redesign will yield a marked improvement in the assembly merit even though the average combined merit does not increase greatly. In fact in example 2 the average combined merit actually decreased.

The fifth example presented in Table VI is different from the first four. Only one redundant part was identified. Its elimination would only slightly increase the original assembly merit. In this case redesign required special attention to the assembly functions associated with each part. This effort together with elimination of the unnecessary part resulted in a significant increase in the assembly merit.

It should also be noted in all the redesign results that the average combined merit is identical in all cases in Table VI to the assembly merit. This is indicative of the designer having satisfied himself that there are no more redundant parts. An analysis of the same redesign by another designer might well yield a different conclusion.

Care must be exercised in only comparing the relative numerical value of assembly merit for alternative designs of the same product in concluding whethter assemblability has been improved. The absolute values of the numbers generated have little inherent meaning themselves. A merit of 70 or better does not necessarily mean the product design "passes". Under no condition should assembly merit values be compared between products. Just because the automatic coffee maker has an assembly merit of 83 does not mean it is "easier" to assemble than the dry cell charger with its assembly merit of 77.

Limitations and Constraints of PDM

The choice and ordering of the three sets of assembly function options represent the subjective input of the author. Other listings of assembly function options, other rank orderings, and other assigned merit values might be chosen and be more appropriate for some specific design/manufacturing environment. In addition, the manner in which the combined merit ratings for each part and the assembly merit for the product is calculated is somewhat arbitrary. The present version of PDM does not permit the user to change any of these characteristics. All of these features represent limitations and constraints of PDM in its current form. A new version of PDM is now being developed which will permit the user to change the default option listings of the assembly functions and revise the numeric merit values assigned to each option choice. This will permit an emphasis or bias to be built into the analysis which might better suit a specific application environment.

Conclusion

It should be kept in mind that the overall purpose of PDM as an analysis tool is to make the designer more aware of the effect of early design decisions on assemblability, assist him in identifying those areas in which redesign may improve ease of assembly, and provide him with a common basis of analysis to make quantitative comparisms between product design alternatives. PDM will not, on the other hand, tell the designer what assembly functions to choose nor how to redesign the product or its constituent parts. Nevertheless, it has proven its ability, even in its present restricted form, to help improve the assemblability of existing product designs when applied as intended by designers willing to interpret and implement the guidance PDM provides.

Table 1

Fastening Function Options

1. Slip of slide fit
2. Snap or light press
3. Clip or snap ring
4. Rivet or stake
5. Glue or adhesive
6. Screw
7. Nut and bolt
8. Press fit
9. Solder
10. Weld

Table II

Insertion Options

1. Vertically down from top
2. Horizontally from side
3. Combination of veritcally down, horizontal, and/or at an angle.
4. Vertically down, horizontal or at an angle with a rotation
5. Combination of vertically down, horizontal, and/or at an angle with a rotation
6. Vertically up from bottom

Table III

Feeding Options

1. Vibratory bowl/ programmable feeder
2. Precision pallet/ slide tray/magazine
3. Standard pallet/ standard containers/ conveyor
4. Special handling/ applicators
5. Manual feeding

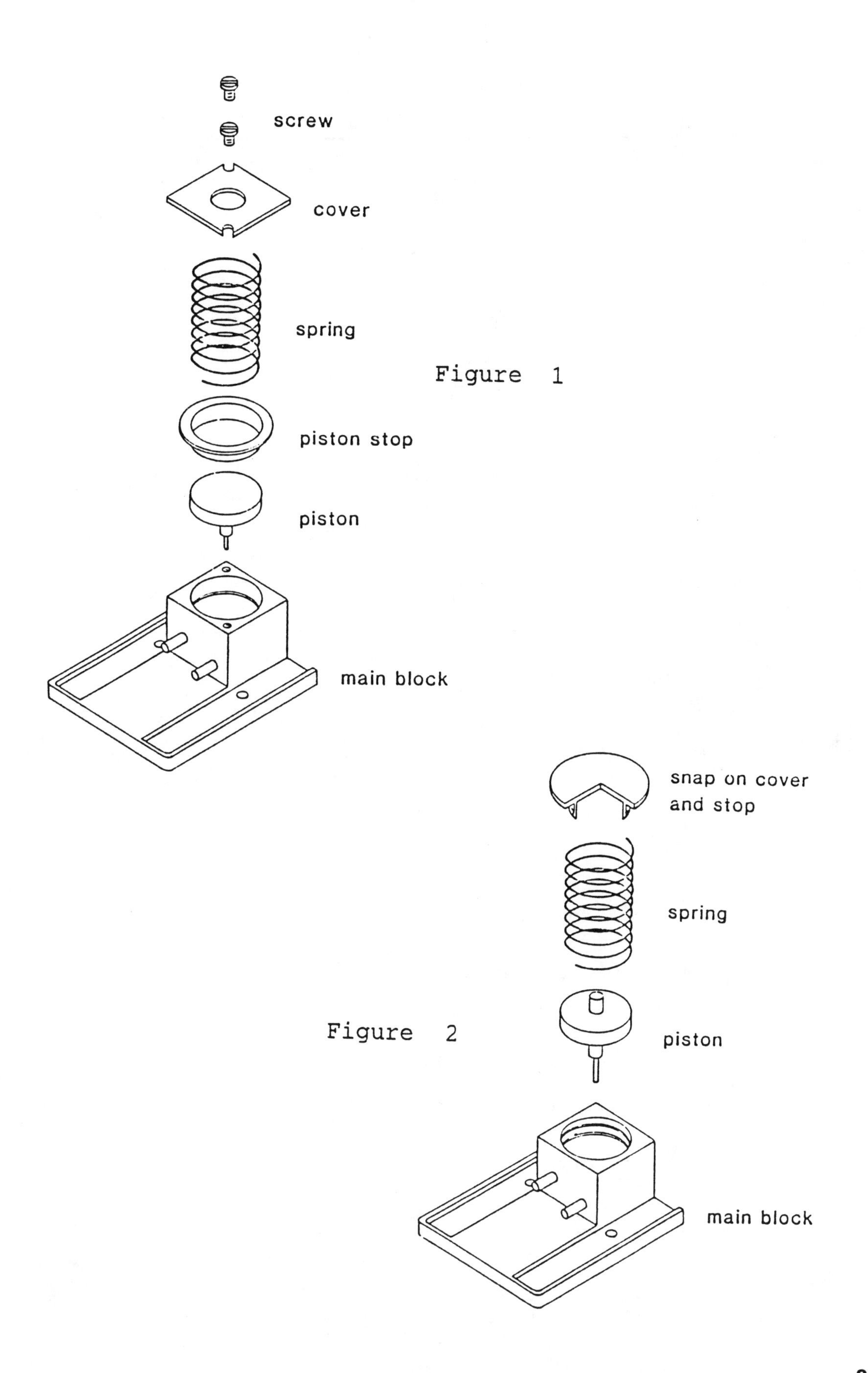

Figure 1

Figure 2

Table IV

Merit Analysis Results
Air Piston - Figure 1

Part No.	Part Name	Insertion Merit	Fastening Merit	Feeding Merit	Combined Merit	Redundant Part
1	Main Block	100	100	50	87	No
2	Piston	100	100	0	82	No
3	Piston Stop	100	89	100	96	Yes
4	Spring	100	100	67	90	No
5	Cover	92	89	0	74	No
6	Screw	100	44	0	63	Yes
7	Screw	100	44	0	63	Yes

Average Combined Merit: 79 Assembly Merit: 45

Table V

Merit Analysis Results
Air Piston - Figure 2

Part No.	Part Name	Insertion Merit	Fastening Merit	Feeding Merit	Combined Merit	Redundant Part
1	Main Block	100	100	50	87	No
2	Piston	100	100	83	95	No
3	Spring	100	100	83	95	No
4	Cover	100	89	83	91	No

Table VI

	Original Design	Redesign
1. Dry Cell Recharger		
Parts:	29	12
Avg. Combined Merit:	75	77
Assembly Merit:	65	77
2. Automatic Coffee Maker		
Parts:	31	12
Avg. Combined Merit:	84	83
Assembly Merit:	41	83
3. Hand-held Phone		
Parts:	40	17
Avg. Combined Merit:	73	81
Assembly Merit:	42	81
4. Electric Shaver		
Parts:	22	11
Avg. Combined Merit:	83	92
Assembly Merit:	41	92
5. Portable Trouble Light		
Parts:	12	11
Avg. Combined Merit:	57	74
Assembly Merit:	52	74

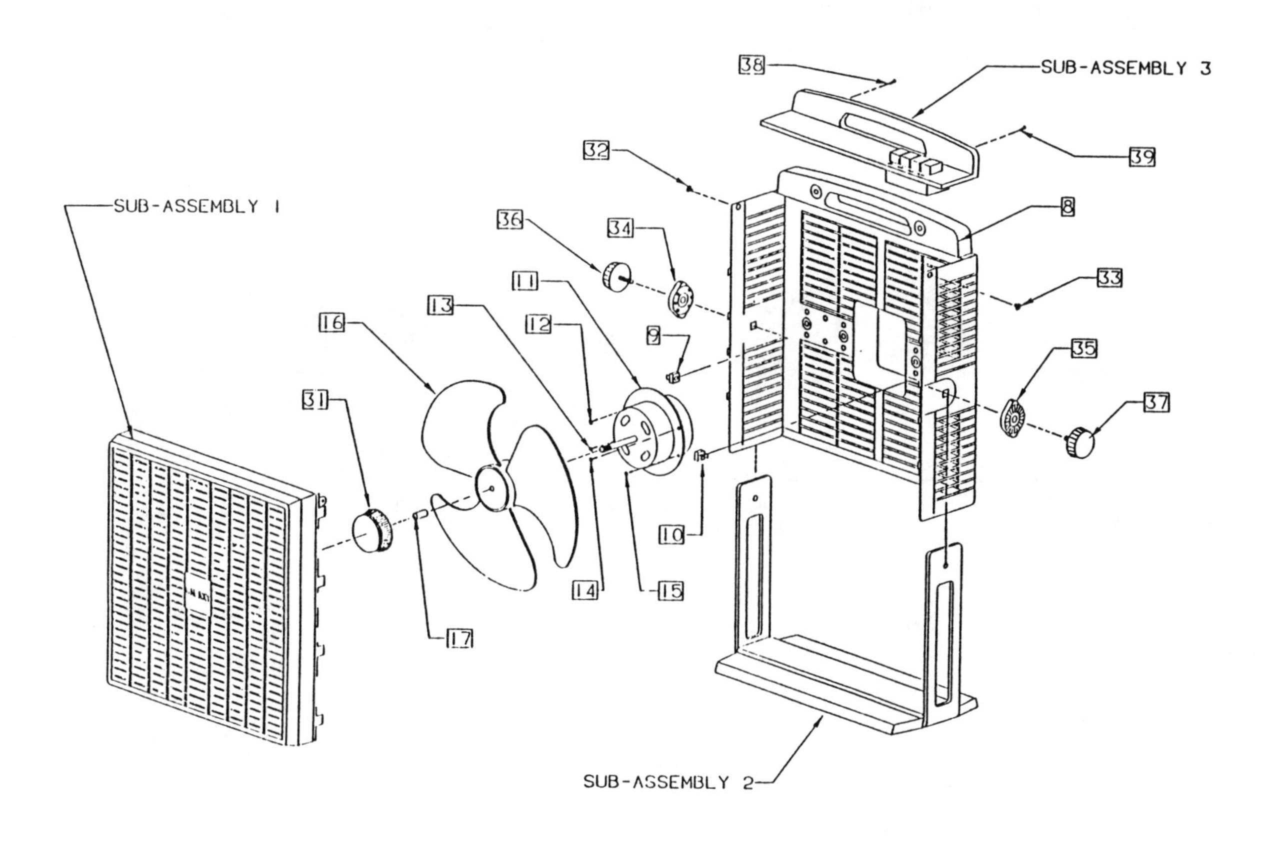

Figure 3

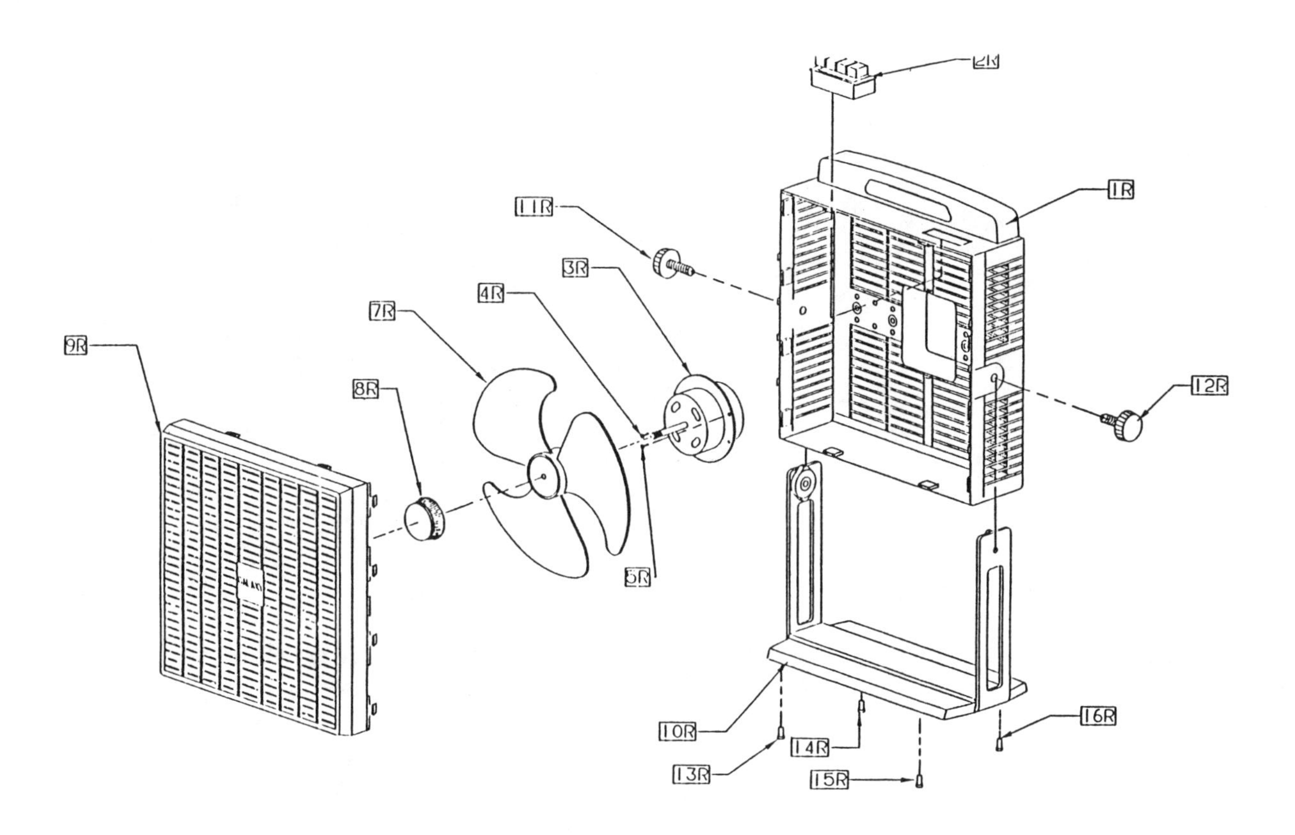

Figure 4

References

1. Zorowski, C. F., "PDM - A Product Assemblabilty Merit Analysis Tool", ASME Design Engineering Technical Conference - Oct. 1986 - Paper No. 86-DET-124

2. Boothroyd, G. and Dewhurst, P., Design For Assembly, A Designer's Handbook, Department of Mechanical Engineering, University of Massachusetts, Amherst, MA, 1983

3. Zorowski, C. F., "Product Design Merit: Relating Design to Ease of Assembly", Proceedings of the 8th International Conference on Assembly Automation, IFS(Publications) Ltd. UK, 1987, pp. 161-170

Dr. Zorowski is Director of the Integrated Manufacturing Systems Engineering Institute at NC State University. The IMSE Institute is a University/Industry cooperative graduate level education and research program offering an interdisciplinary masters degree in manufacturing systems and conducting basic and applied research in modern manufacturing systems technology. Prior to this position Dr. Zorowski served as Associate Dean for Engineering and Head of the Mechanical Engineering Department at NC State University.

CHAPTER 3

SIMULTANEOUS ENGINEERING ACCOMPLISHMENTS

Reprinted from *Manufacturing Engineering*, April 1988

How can DFA and DFM reduce manufacturing costs? Read on

Product Design for Manufacture and Assembly

By G. Boothroyd and P. Dewhurst
University of Rhode Island

Design for manufacture (DFM) means different things to different people. For the individual whose task is to consider the design of a single component, DFM means the avoidance of component features that are unnecessarily expensive to produce. Examples include the following:

- Specification of surfaces that are smoother than necessary on a machined component, necessitating additional finishing operations
- Specification of wide variations in the wall thickness of an injection-molded component
- Specification of too-small fillet radii in a forged component
- Specification of internal apertures too close to the bend line of a sheet metal component.

Alternatively, the DFM of a single component might involve minimizing material costs or making the optimum choice of materials and processes to achieve a particular result. For example, can the component be cold-headed and finish-machined rather than machined from bar stock? All of these considerations are important and can affect the cost of manufacture. They represent only the fine-tuning of costs, however, and by the time such considerations are made, the opportunities for significant savings may have been lost.

It is important to differentiate between component or part DFM and product DFM. The former represents only the fine-tuning process undertaken once the product form has been decided upon; the latter attacks the fundamental problem of the effect of product structure on total manufacturing costs.

The key to successful product DFM is product simplification through design for assembly (DFA). DFA techniques primarily aim to simplify the product structure so that assembly costs are reduced. Experience is showing, however, that the consequent reductions in part costs often far outweigh the assembly cost reductions. Even more important, the elimination of parts as a result of DFA has several secondary benefits more difficult to quantify, such as improved reliability and reduction in inventory and production-control costs. DFA, therefore, means much more than design to reduce assembly costs and, in fact, is

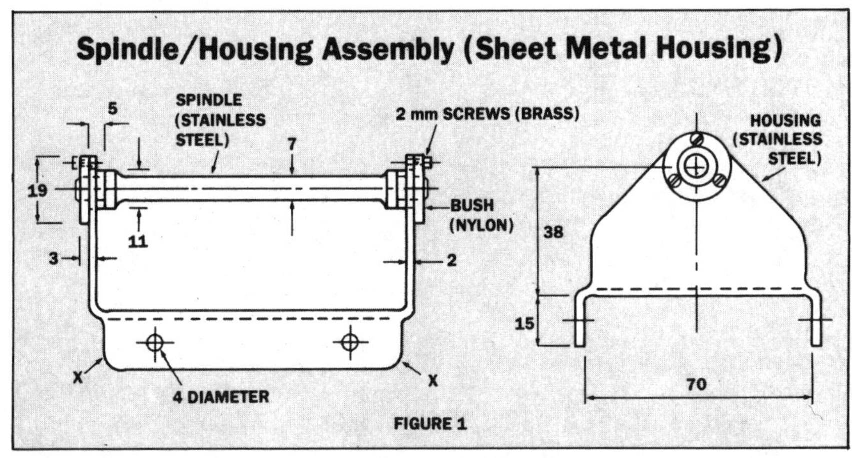

This spindle/housing assembly has 10 separate parts and an assembly efficiency rating of 7%.

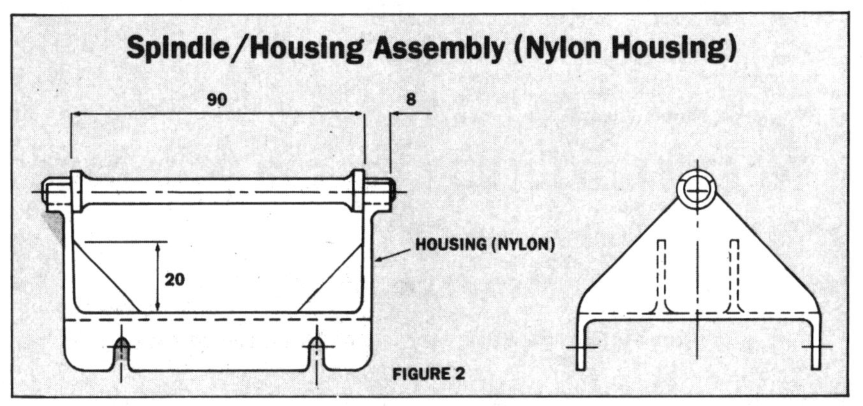

The two-part design, utilizing an injection-molded nylon housing, has an assembly efficiency rating of 93%.

DFA Goes How Far Back?

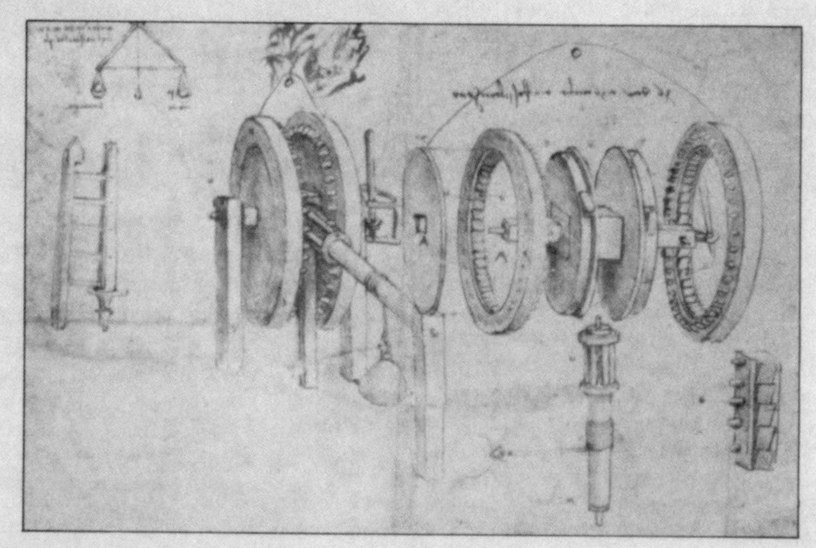

Design for assembly (DFA) concepts may reach back further than we suspect—perhaps to the late 15th century and Leonardo da Vinci.

This illustration, *folio 8 of the Codice Atlantico*, presents an assembled view (left) and an exploded view (right) of a mechanism for changing the rocking motion of an upright lever (right, both views) to the rotary motion of a shaft, so as to lift a heavy weight. The drawings, incidentally, are outstanding not only in their novelty of assembly—which gives us insight into early notions somewhat consistent with those of today's DFA principles—but also in their clarity, especially the exploded view.

In operation, the upright lever is rocked back and forth, and the stone suspended by a rope is wound upwards around the horizontal shaft. The rocking lever swings a square shaft, upon which are fixed two wheels, each having pawls in their outside edges, acting in opposite directions. The pawls engage ratchets in the bores of two outer rings. These ratchet wheels also have gear teeth engaging a common lantern gear on the final shaft.

When the operating lever is pushed one way, one of the pawls engages its ratchet wheel. Pushed the other way, the other pawl engages its ratchet wheel, but the shaft revolves in the same direction. ■

—Robin P. Bergstrom
Editor-in-Chief

central to the issue of product DFM. In other words, part DFM is the icing on the cake; product DFM through DFA *is* the cake.

DFA derives its name from a recognition of the need to consider assembly problems at the early stages of design; it therefore entails the analysis of both product and part design. For some years now, an assembly evaluation method (AEM) has been in use at Hitachi. In this proprietary method, commonly referred to as the Hitachi method[1], assembly element symbols are selected from a small array of possible choices. Combinations of the symbols then represent the complete assembly operation for a particular part. Penalty points associated with each symbol are substituted into an equation, resulting in a numerical rating for the design. The higher the rating, the better the design.

Another quantitative method, developed by the authors and known as the Boothroyd and Dewhurst method[2], involves two principal steps:

- The application of criteria to each part to determine whether, theoretically, it should be separate from all the other parts in the assembly
- An estimate of the handling and assembly costs for each part using the appropriate assembly process—manual, robotic, or high-speed automatic.

The first step, which involves minimizing the part count, is the most important. It guides the designer toward the kind of product simplification that can result in substantial savings in product costs. It also provides a basis for measuring the quality of a design from an assembly standpoint. During the second step, cost figures are generated that allow the designer to judge whether suggested design changes will result in meaningful savings in assembly cost.

The third quantitative method used in industry is the GE/Hitachi method[3], which is basically the Hitachi method with the Boothroyd and Dewhurst criteria for part-count reduction added.

For business reasons, companies are seldom prepared to release their manufacturing cost information. One reason is that many companies are not sufficiently confident about their costing procedures to want manufacturing costs made public for general discussion. In such an environment, designers will often not be informed of the cost of manufacturing the product they have been designing. Moreover, designers do not usually have the tools necessary to obtain immediate cost estimates relating to alternative product design schemes. Typically, a product will have been designed and detailed and a prototype manufactured before a manufacturing cost estimate is attempted. Unfortunately, by then it is too late. The opportunity to consider radically different product structures has been lost, and among those design alternatives might have been a version that is substantially less expensive to produce.

Currently, there is much interest in having product DFM and DFA techniques available on CAD/CAM systems. By the time a proposed product design has been sufficiently detailed to enter it into the CAD/CAM system, however, it is already too late to make radical changes. A CAD representation of a new product is an excellent vehicle for making effortless detail changes, such as moving holes and changing draft angles. But for considering product structure alternatives, such as the choice of several machined parts versus one die casting, a CAD system is not nearly as useful. These basic, fundamentally important decisions must be made at the early sketch stage in product design.

A conflict thus exists. On the one hand, the designer needs cost estimates as a basis for making sound decisions; on the other hand, the product design is not sufficiently firm to allow estimates to be made using currently available techniques. The means of overcoming this dilemma is another key to successful product DFM—namely, early cost estimating.

Many individuals are proposing rule-based or axiomatic methods to help designers achieve efficient designs. The axiomatic approach proposed by N.P. Suh and his co-workers[4] was based on attempts to identify common properties of successful designs. These common properties, such as how the proposed design satisfies the functional requirements, were then proposed as axioms of good design. Design axioms can thus be viewed as global product guidelines that can coexist with component guidelines for details such as hole spacings, fillet radii, and draft angles. However, axiomatic approaches have two major weaknesses when manufacturing is considered in the early stages of product design. Both of these weaknesses are directly related to cost.

First, the axiomatic approach does not provide any means of making judgments between the centrally important tradeoffs posed by possible alternative choices of different materials and processes. Second, at the detail level, guidelines tend to lead designers in an essentially fruitless direction. This is because manufacturing guidelines are invariably intended to make individual processing steps as efficient as possible. Following such guidelines might lead to the design of all of the bend lines of a sheet metal part in a single plane, the avoidance of side holes or depressions in molded parts, the minimization of the number of steps in a part to be made by powder metallurgy, and so on. With this approach, the tendency is to design relatively simple individual components, which will invariably lead to high total fabrication and assembly costs.

A DFM system must therefore be able to predict both assembly costs and component manufacturing costs at the earliest stages of product design. Only in this way will it be possible to design a product that takes maximum advantage of the capabilities of chosen manufacturing processes within the constraints imposed by anticipated production volumes. In many situations, this will simply mean providing designers or design teams with the software tools that will enable them to make sound judgments from a range of

Breakdown of Costs for Two Spindle/Housing Assembly Designs

A. DESIGN USING SHEET METAL HOUSING

	ASSEMBLY	MATERIAL	MANUFACTURING	TOOLING
HOUSING	0.02	1.74	1.56[1]	7,830[3]
BUSH (2)	0.09	0.01	0.06[2]	9,030[4]
SCREW (6)	0.35	0.72	--	--
SPINDLE	0.04	0.26	1.29	--
TOTAL	0.50	2.73	2.91	16,860

1. INCLUDES $1.35 FOR DRILLING AND TAPPING SCREW HOLES
2. MOLDED BUSHINGS HAVE THREE-CORED HOLES FOR SCREW CLEARANCE
3. THREE SEPARATE DIE SETS FOR BLANKING, PUNCHING, AND BENDING
4. TEN-CAVITY MOLD FOR LEAST-COST MANUFACTURE

B. DESIGN USING INJECTION-MOLDED HOUSING

	ASSEMBLY	MATERIAL	MANUFACTURING	TOOLING
HOUSING	0.02	0.14	0.24	10,050[1]
SPINDLE	0.02	0.26	1.29	--
TOTAL	0.04	0.40	1.53	10,050

1. TWO-CAVITY MOLD FOR LEAST-COST MANUFACTURE

FIGURE 3

A comparison of the two spindle/housing assembly designs shows significant cost reductions as a benefit of DFA.

Computer Input Screen for Nylon Housing

ESTIMATION OF INJECTION MOLDING COSTS FOR: HOUSING — THERMOPLASTIC: 6/6 NYLON

DIMENSIONAL DATA

PART VOLUME = 25.51 cm³

PROJECTED AREA = 71.25 cm²

L = 95 mm W = 75 mm D = 57 mm

THICKNESS: MAXIMUM = 2.5 mm AVERAGE = 2 mm

QUALITY AND APPEARANCE

TOLERANCE FACTOR (0-5)?	3
APPEARANCE FACTOR (0-5)?	1
COLORED RESIN?	N
TEXTURED SURFACE?	N

PART COMPLEXITY

OUTER SURFACE OR CAVITY (0-5)?	2
INNER SURFACE OR CORE (0-5)?	0

MOLD COMPLEXITY

STANDARD TWO-PLATE MOLD?	Y
THREE-PLATE MOLD?	N
MULTIPLATE STACKED MOLD?	N
HOT RUNNER SYSTEM?	N
NUMBER OF SIDE CORES OR PULLS?	0
NUMBER OF UNSCREWING DEVICES?	0

AT ANY TIME PRESS: <H>ELP, <V>OLUME, <A>REA, OR <C>OMPLEXITY CALCULATOR

FIGURE 4

Shown are the inputs required to estimate the costs of injection molding the housing.

Results for Injection Molding of Nylon Housing

ESTIMATED INJECTION MOLDING COSTS FOR: HOUSING THERMOPLASTIC: 6/6 NYLON

TOTAL PRODUCTION VOLUME (THOUSANDS)	NUMBER OF CAVITIES	TOTAL MOLD BASE COSTS ($)	CAVITY/CORE MANUFACTURING COSTS ($)	TOTAL MOLD COST ($)	MOLD COST PER PART (CENTS)
100	2	3589	6462	10,051	10.1

MACHINE SIZE (kN)	MACHINE RATE ($/HR)	CYCLE TIME (SECONDS)	MANUFACTURING COST PER PART (CENTS)
1600	72	20.3	23.9

PART VOLUME (cm^3)	PART WEIGHT (GRAMS)	POLYMER COST ($/kg)	POLYMER COST PER PART (CENTS)
26	29	4.69	13.5

TOTAL PART COST (CENTS) = 47.5

SELECT REQUIRED OPTION:

1. SCREEN EDIT
2. SHOW MOLD COST/CYCLE ELEMENTS
3. PRINT RESULTS AND RESPONSES
4. CHANGE BASIC COST DATA
5. CHANGE RESPONSES/POLYMER
6. EXIT

FIGURE 5

The cost breakdown for the proposed injection-molded nylon housing. The total part cost appears in the lower right corner.

choices. These choices may involve designs necessitating increased tooling costs but fewer different parts and reduced assembly costs.

It is anticipated that the product DFM considerations will always start with DFA. To aid designers in implementing these techniques, the authors have developed DFA software[5] that allows a designer to establish an efficient assembly sequence for a proposed new product concept. The software then questions the relationship between the parts and gives an assembly efficiency rating, together with estimated assembly costs.

The DFA process uses the assembly sequence as a vehicle for analyzing the product structure in order to force the design toward more integrated solutions with a reduced part count. This result of DFA is often the most important one in achieving total product cost reductions. Thus, DFA analyses must be supported by techniques that will allow the design team to make early estimates of material, processing, and tooling costs. Only in this way can different designs, with different numbers of parts and perhaps using different materials and processes, be compared before a detailed design commitment is made.

To illustrate this approach, consider the simple spindle/housing assembly shown in *Figure 1*. A DFA software analysis of the proposed sheet metal housing, two nylon bearing inserts, six screws, and a machined steel spindle gives an assembly efficiency rating of only 7%. This low rating arises mainly from the fact that there are 10 separate parts in the assembly, whereas theoretically only two should be necessary.

Application of the DFA minimum part count criteria reveals that the housing and shaft must be separate because of their relative motion. None of the remaining parts satisfy the criteria, however, because they do not move relative to the housing; they can all be of the same material and do not have to be separate in order to assemble the spindle into the housing. These considerations lead the designer to consider a two-part design, and assuming that the bearing surfaces should be of nylon, the alternative design shown in *Figure 2* might be proposed. In this case, the spindle is unchanged, but the remaining parts have been combined into one injection-molded nylon housing with an assembly efficiency of 93%.

Using the authors' techniques for early cost estimating[6], the cost breakdowns for the various parts used in the two designs were obtained (see *Figure 3*). It can be seen that DFA has resulted in significant savings in assembly, material, manufacturing, and tooling costs. Even though this was a DFA study, the largest savings are in material costs, which have been reduced from $2.73 to only $0.40—principally by changing the material for the housing and eliminating the screws. The savings in manufacturing costs can be mainly attributed to the elimination of all the drilling and tapping operations, which cost a total of $1.35. This cost, together with the cost of the screws themselves, amounts to $2.07 and accounts for 34% of the original design's cost. These figures would indicate to the designer the desirability of seeking alternative methods of securing the bushes in the housing if the design with combined housing and bushes was not considered practicable.

The results of this DFA analysis, combined with early cost-estimating methods, illustrate the kind of result that can be obtained by using DFA as the first step in a product DFM study. Of course, it is also possible to achieve savings by considering changes in the product design that are directed at reductions in individual part costs.

For example, the rounded corners in the sheet metal housing (marked X in *Figure 1*) contribute unnecessarily to material and tooling costs. If these corners were squared, the allowance in the strip width of 4 mm per side would not be necessary, thereby resulting in material cost savings of 8%, or $0.14 per part. Additionally, a part-off die would be employed instead of a blanking die, resulting in die cost savings of around $1000. A design change that would reduce manufacturing costs would be the use of self-tapping screws

Worksheet from Machining Software

SPINDLE | **MATERIAL: STAINLESS STEEL—FERRITIC FREE-MACHINING** | **MACHINE: MANUAL TURRET LATHE**

OPERATION NUMBER / OPERATION	<R>OUGH OR <F>INISH	<S>TEEL <D>ISPOSABLE <C>ARBIDE <G>RIND	SETUP TIME (HOURS)	LOAD/UNLOAD TIME, ETC. (SECONDS)	NUMBER OF OPERATIONS	SET TOOL, ENGAGE CUT, ETC. (SECONDS)	VOLUME (in.^{3})	AREA (in.^{2})	MACHINING TIME (SECONDS)	OPERATION COST ($)
1 FACE	F	C	1.41	29	1	9	—	0.15	0.66	0.30
2 CYLINDRICAL TURNING	R	C	0.22	—	1	9	0.03	0.28	2	0.08
3 CYLINDRICAL TURNING	F	C	—	—	1	9	—	0.28	0.50	0.07
4 CYLINDRICAL TURNING	R	C	—	—	1	9	0.28	2.98	16	0.19
5 CYLINDRICAL TURNING	F	C	—	—	1	9	—	2.98	4	0.10
BATCH>10,000		TOTALS	3.48	58		81	0.35	7.50	27	1.29

COST/PART ($): MATERIAL = 0.26 SETUP = 0.01 OPERATIONS = 1.29 TOTAL = 1.56

(MOVE INDICATOR TO REQUIRED ROW/COLUMN/PAGE USING KEYPAD FUNCTIONS)
PRESS <INS>ERT, <DEL>ETE, <C>HANGE, <M>ATERIAL, <H>ELP, OR <O>K

FIGURE 6

Results from entering the machining operations for the spindle appear on the computerized worksheet.

and pierced holes in the sheet metal part to avoid the drilling and tapping operations costing $1.35.

Looking at the costs of the proposed design, *Figure 3(B)*, it is clear that attention should be paid to the design of the spindle in order to reduce material and manufacturing costs. In this case, if the two larger diameter features could be eliminated so that the spindle could be machined from smaller diameter stock with reduced bearing surface diameters at each end, material costs would be reduced from $0.26 to $0.11 and manufacturing costs from $1.29 to $0.85, a total savings of $0.44. Although these considerations are important, they are only the tip of the iceberg compared with the results of DFA analysis.

The early cost-estimating programs that were used to determine the part costs given in *Figure 3* have been developed for use with microcomputers and are based on the results of an extensive research program at the University of Rhode Island. These programs are designed so that they can be easily applied at the sketch stage before detail drawings are available and before precise processing and tooling specifications have been made.

For example, *Figure 4* shows the inputs that are required to estimate the costs of injection molding the housing shown in *Figure 2*. It can be seen that Nylon-6/6 has been selected. This has been obtained from a database that contains material costs and molding parameters.

The footnote in *Figure 4* indicates that support calculators are available for determining the correct responses for the housing's volume, projected area, and geometric complexity. *Figure 5* gives the resulting cost breakdown for the housing. Working together with an injection molding machine database, the program first selects the optimum number of cavities from the tradeoff between cavity manufacturing costs, machine rates, and the number of parts produced for each machine cycle. The program then produces a breakdown of mold, processing, and material costs. These are given in the right-hand column of *Figure 5* and are added to produce the part-cost estimate in the lower right-hand corner.

In a similar manner, the machining program works with two user-editable databases, one for machine tools and one for materials. The user selects the material, specifies the workpiece dimensions, selects the machine tool, and specifies the cutting operation. The program then estimates the machining costs for the operation. As each machining operation is entered, the results appear on a worksheet where allowances are automatically added for nonproductive time and tool replacement costs. *Figure 6* illustrates a portion of the worksheet that results from entering the machining operations for the spindle.

The techniques of DFA and DFM can play a major role in reducing costs and increasing productivity. Recognition of this fact is also increasing the demand for cost-estimating tools that allow design teams to make the necessary tradeoffs at the early concept stages of design. These techniques and tools can play a significant part in helping US industry regain its competitive edge in world markets.

REFERENCES

1. S. Miyakawa and T. Ohashi, "The Hitachi Assemblability Evaluation Method (AEM)," *Proceedings, First International Conference on Product Design for Assembly* (April 1986).

2. G. Boothroyd and P. Dewhurst, *Product Design for Assembly Handbook* (Wakefield, RI: Boothroyd Dewhurst, Inc., 1987).

3. G. Hock, "Designing for Productivity," *Target* (Summer 1987), p. 14.

4. N.P. Suh, A.C. Bell, and D.C. Gossard, "On an Axiomatic Approach to Manufacturing and Manufacturing Systems," *ASME Journal of Engineering for Industry*, vol. 100, no. 2 (May 1978).

5. G. Boothroyd and P. Dewhurst, *The Design for Assembly Software Toolkit* (Wakefield, RI: Boothroyd Dewhurst, Inc., 1987).

6. P. Dewhurst and G. Boothroyd, "Early Cost Estimating in Product Design," *Proceedings, Second International Conference on Product Design for Manufacture and Assembly* (April 1987). ■

Reprinted from *Assembly Engineering*, September 1988

NCR Cashes In on Design for Assembly

For six years now, *Assembly Engineering* has conducted an annual search for the nation's outstanding example of applied assembly technology and thinking. The PAT Award (Productivity through Assembly Technology) is why we do it, and this year's recipient is William R. Sprague, senior advanced manufacturing engineer, NCR Corp., Cambridge, OH.

He was the prime mover in rigorously applying Design-for-Assembly (DFA) methodology during development of a new point-of-sale terminal called the 2760. The product and its associated manufacturing process realized the following benefits vis-a-vis the previous generation terminal:

- 65% fewer suppliers
- 75% less assembly time, with 100% fewer assembly tools
- 85% fewer parts (100% fewer screws)

The new terminal, in fact, has 55% fewer parts than the previous one had screws. To highlight the impact, it was estimated that the cost of just one screw (material, labor, and burden) over the 2760's life was $12,500. Further, total lifetime manufacturing labor cost for the terminal was estimated at $1.1 million less than the prior product, a 44 percent improvement.

Obviously, these numbers are significant; however, even greater savings are expected in areas such as inventory, field service, plant-floor space, production-support costs, and capital equipment. Tooling development, for example, went straight from the design stage to hard (steel) tools rather than the typical intermediate step of using soft (aluminum) tools. Sprague points out that initial tooling only needed minor modification.

Also, the testing and introduction period for the 2760 was remarkably trouble-free. It cleared difficult tests such as EMI emissions on the first pass.

"NCR Cambridge has a major JIT program in progress," Sprague remarks. "The 2760 terminal was designed with that in mind."

Bill Sprague (left), this year's PAT Award recipient, reviews first-off plastic components against a solids model with automation design engineer, John Wallach, and Sandra Smith, a representative of the CAD/CAE system vendor, SDRC.

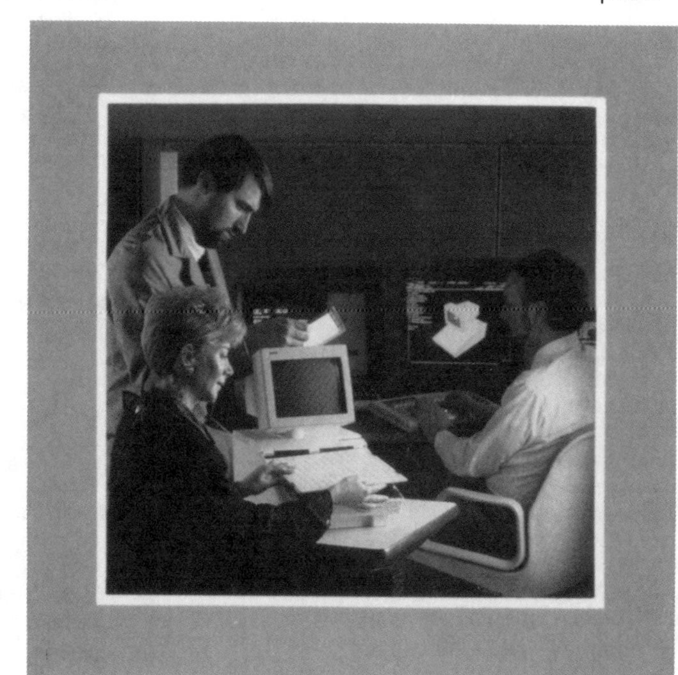

JOHN R. COLEMAN
Editor

Team Work Works

NCR uses multiplant task forces and corporate-assisted pilot projects to explore DFA tools and strategies. The 2760 began as such a project in January, 1987. Mechanical design was completed October, 1987, with commercial roll out expected this month. Sprague and John Wallach (from the Design Automation Group) were the project's champions. Their focus was on the mechanical portion of the design.

"U.S. Industry finally is recognizing DFA as a strategy for leadership," comments Sprague. "But, for it to achieve significant results, you must first overcome two obstacles: Implementing a major cul-

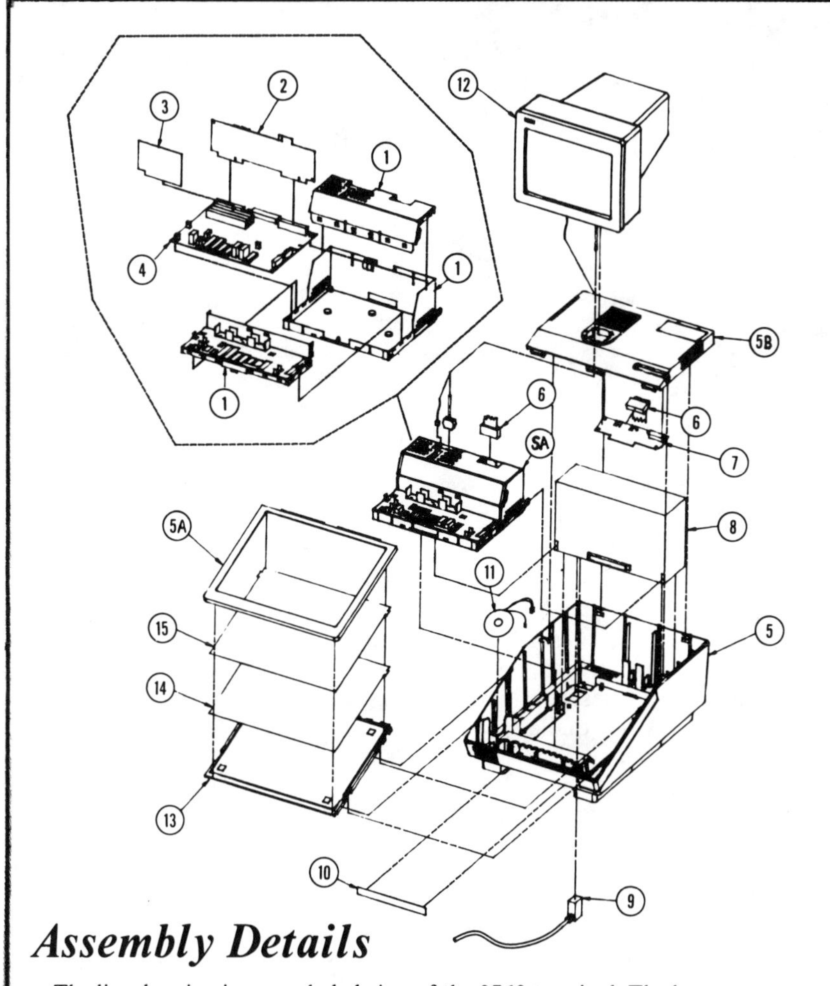

Assembly Details

The line drawing is an exploded view of the 2760 terminal. The lone subassembly is the snap together electronics box (1), which contains three snap and plug-in pc boards (2, 3, 4). The plastic terminal cabinet base (5), like the sheetmetal electronics box, is delivered assembled to the production floor by the vendor as part of NCR Cambridge's JIT program.

The bezel (5A) and top (5B) are removed from the cabinet base to start the assembly process. The lone separate harness (6) is plugged to a pc board (7), which then snaps into the cabinet top (5B).

The power supply (8) is purchased from NCR's Orlando, FL division. It handles either 110V or 220V power, with internal sensing that eliminates the need for switching or multiple power-supply assemblies. It also contains the 2760's cooling fan. The power supply is simply set into the cabinet base and the cabinet top is snapped back on to hold it.

The terminal is turned on its side for plugging the ac power cord (9) into the cabinet base and power supply, and snapping it into a strain-relief tab in the base. The terminal then is turned on its back to apply a single label containing serial number, license, and product information.

The terminal is set upright for sliding in the electronic box assembly (SA), which mates to the power supply for grounding and current, and snaps into the cabinet base to hold it in place. The speaker (11) slides into a holder in the cabinet base for a snap fit and then is plugged into the electronics box. The harness (6) from the pc board (7) in the cabinet top also is plugged into the electronics box.

The CRT (12) cables are fed through a hole in the cabinet top, and then the CRT pivot slides through the hole and is twisted a quarter turn to snap the CRT into place. The CRT connectors then are plugged into the electronics box. Both the CRT and speaker are common to other NCR terminal products.

The keyboard (13) is set into the cabinet front, sliding up to mate to the pc board in the cabinet top and snapping it into place. A keyboard overlay (14) and Mylar cover (15) then slip into tabs in the keyboard (13), and the cabinet bezel snaps back into place to hold them.

The 2760 assembly requires only 15 vendor-sourced parts, and no separate fasteners are used. The terminal cabinet has a number of snap-in escutcheons that can be removed to easily customize it.

tural change in the organization (and gaining acceptance of it); and blending computer-based tools and design process changes in the right mix to optimize development effort."

All astute organizations realize that success comes from the skills and efforts of its personnel. The difficulty is getting the right people, in the right structure, with the right tools, to maximize results.

"We formed a multidiscipline team to simultaneously design the product and manufacturing process for the 2760 terminal," reflects Sprague. "John and I drafted the project plan, then sold it to the director of manufacturing and the director of engineering, making sure all expectations and schedules were in agreement.

"We then established a high-level goal of having quality product and manufacturing process designs in place prior to prototype development and production. Our primary measurements were to meet schedule and cost goals, and minimize assembly time and part count. All other benefits and measurements were derived from these high-level goals and basic measurements."

Team members were invited from product design, customer service, manufacturing engineering, test engineering, product management, industrial design, quality engineering, purchasing, vendor quality, safety engineering, EMI engineering, program management, and human factors. Critical vendors also were brought in as part of a co-destiny supplier relationship. The plastics molder and tool-and-die builder, for instance, were selected early to exploit their knowledge of molding and toolmaking. Suppliers also had to comply with NCR's JIT and SPC programs, and have ability to accept CAD file transfers, which helped reduce tooling quotation and delivery time.

Weekly team meetings were held, with interaction among key members more frequent. Also, presentations were given on how DFA works, and examples of successful applications were distributed.

Each member was expected to develop objectives relative to his area of expertise. Prior to the start of hard design, these objectives were shared with the group to foster an understanding of what manufacturability, testability, and serviceability really meant.

The team pursued design solutions that met or exceeded all objectives, adamantly resisting compromises. A parallel approach permitted simultaneously developing the product design and manufacturing process, which allowed trade-offs between the two as changes were made. This was the key to the pilot project's success.

"Prevention versus Contingency" planning helped isolate problems and point to solutions while the design was still in the CAD system rather than at the testing/prototype stage. Team members openly shared progress and problem reports, eventually developing a consensus before releasing the design. This created multiorganizational design ownership, with all members responsible for meeting the group's objectives. For example, the manufacturing engineer shared with the product engineer (and the rest of the team) in accepting the 2760's manufacturability. They worked up front to define what was expected from the design rather than waiting for a review.

"An important task during the project was upgrading the skills of the team members," recalls Sprague. "At the start of the project, courses were held on using SDRC's GEOMOD solids modeler and the Boothroyd Dewhurst DFA Toolkit software. John and I provided follow-up training and assistance."

CAE: Computer-Aided Everything

The mechanical portion of the design was done from concept to completion on GEOMOD, remaining as a solids model until details were approved by the team. Then the electronic information was transferred to vendors for tool manufacturing.

"Solids modeling was an excellent means for designers and team members to visualize concepts," says Sprague. "The complete assembly, or any element of it, could be viewed just as it would physically appear. This stimulated input from group members because they could see and understand how the 2760 was defined early in the process. In addition, they spent more time working on design solutions rather than just figuring out what parts fit where.

"Subsequently, the designers were forced into greater discipline as details had to be defined at each stage. By using program files, they could make changes without recreating the entire part. The software performed interference checks among the parts to assure the team that form and fit problems were being addressed. The electronically captured design data also automatically yielded volume and mass information to assist in structural analysis and for quotation of molded and cast parts."

NCR's 2760 point-of-sale terminal. Part count and assembly time were tracked during the development effort; using DFA techniques permitted reducing them 66% and 23% (respectively) from original conceptual estimates.

VAX-based CAD stations worked well for design, but were slow when the team wanted to review the completed parts. High-performance Hewlett-Packard 350SRX workstations were trial leased to expedite the on-screen review sessions. Component and assembly details then could be real-time analyzed by group members. This was especially important to the manufacturing engineer because he could quickly interpret product changes and make appropriate manufacturing process modifications.

"CAE analysis was used to leverage overall quality improvements," Sprague continues. "For example, the terminal's cabinet consists of complex injection-molded plastic parts that were analyzed with Moldflow and Polycool II to verify tool design. There also is potential for structural and thermal analysis, which hold promise for further reducing tool changes and improving component quality."

The completed part designs were electronically released to the tooling vendor; no paper drawings were issued. Tool-path information was added and then directly fed to CNC machine tools.

Part geometry was created only once to reduce engineering effort, separate documentation from design, and reduce interpretation error. Documentation for inspection and assembly drawings was generated on an as needed basis. A critical element of this paperless CIM environment involved securing the design's electronic file by developing an archiving system.

Tool of the Trade

Throughout the 2760's development Boothroyd Dewhurst's DFA Toolkit software provided an early analysis of assembly time and sequence, not just a finished-design autopsy. By using a Pareto analysis of the assembly time data, opportunities for improvements were identified to challenge the team. Parts and times were categorized by functional and nonfunctional purposes to highlight the cost of decisions and problem areas. According to Sprague, this enhanced design-solution synergism among group members. The data also was used to keep top management up to date on the team's progress, and was essential to maintaining support of the project.

"The DFA Toolkit focused on key guidelines for improvement," Sprague notes. "We concentrated on reducing assembly time and part count, which significantly impacted the entire organization and our suppliers. In particular, fasteners were eliminated through snap-fit assembly and cables/harnesses were reduced through direct interfacing, which also improved reliability and quality. The remaining parts were designed for Z-axis assembly to minimize handling and insertion time."

Last June, Sprague illustrated the 2760's ease of assembly by manually putting it together in less than two minutes while blindfolded at the Third International Conference on Product Design For Manufacture & Assembly (Newport, RI). Of course, customer acceptance will be the ultimate judge of the project's success, and initial commercial interest for the 2760 terminal is strong. Moreover, several companies that are pursuing their own DFA projects want to replace the formerly used IBM Proprinter with a 2760 for internal demonstrations.

"We have initiated several other projects with greater potential impact, including pc board DFA," boasts Sprague. "Assembly time and part-count reductions are forecast in the 90% range. Design for Assembly is truly a cornerstone of manufacturing excellence."

If you are interested in throwing your hat in the ring for next year's PAT Award competition, see the official rules overleaf.

Reprinted from *Assembly Engineering*, January 1987

Design For Assembly *In Action*

Editor's Note

Database procedures have been developed by the authors as tools to help engineers design products that can be assembled efficiently. These procedures are available in the form of handbooks and IBM compatible software, and address manual assembly, high-speed automatic assembly and the use of general-purpose robots for product assembly. The procedures are being used in a wide range of industries and have resulted in substantial reductions in assembly time and cost. This article is intended to illustrate the effectiveness of Design For Assembly (DFA) by comparing the analysis results of the most popular Japanese dot matrix printer, the Epson MX80, with the recently introduced IBM Proprinter.

Professors P. Dewhurst & G. Boothroyd,
University of Rhode Island

A striking comparison of two different product designs, the Epson MX80 and IBM Proprinter, highlight the importance of the Design for Assembly approach.

The Epson MX80

An Epson MX80 dot matrix printer was disassembled and analyzed for manual assembly using DFA procedures. As a first step, the final assembly only was considered as shown in the exploded view, Fig. 1. A summary of this analysis is shown in Fig. 2. The ***final assembly*** involves 57 separate assembly operations and takes an estimated total assembly time of 552 seconds. Moreover, of the 49 different parts or subassemblies added in the final assembly process only 10 satisfy the criteria for separate parts[1]. This gives rise to a very low efficiency of only 5% and indicates great potential for part count reduction. The low assembly efficiency is also due to the fact that many of the parts are difficult to align and locate and require repositioning for insertion of fasteners. A breakdown of the total assembly time of 552 seconds is as follows:

- Insertion of 21 parts of subassemblies = 211 seconds
- Insertion and securing of 34 separate fasteners = 305 seconds
- Adjustments and reorientations = 36 seconds

As is common in most mechanical product designs, fasteners outnumber the functional parts in the product and represent a major portion of the assembly time. Clearly, eliminating fasteners in favor of an integral fastening design constitutes the greatest potential for reducing part count and assembly time. However, the three separate printed circuit boards, spacers, insulators and connectors offer further significant possibilities for design simplification.

The final assembly of the Epson MX80 involves positioning and securing the main printer subassembly which includes the paper and print-head drive units. The second step in the design analysis involves the ***main printer subassembly***, shown in the exploded view, Fig. 3. The essential design features of this subassembly are a stamped steel baseplate and two steel lower side brackets that are assembled into slots in the base. A cover is secured across the top of the side brackets, and this in turn supports two upper side brackets. Steel rods also stiffen the entire assembly. The main difficulty with this design is that almost all of the functional parts either hang between or on the outside of the side brackets. Most of the parts remain unsecured until the upper front and rear rods are in place and the hexagon nuts are secured against the upper and lower bracket sides.

Good Design For Assembly practice requires the exact opposite of this approach. That is, parts should be added in layer fashion from the base plate and should either be self-locating or secured immediately. In the Epson MX80 design, the base plate supports only a printed circuit board and the printhead belt drive mechanism. The side plates, which are only loosely held in place are required to support all other parts until final securing takes place with hex nuts.

A summary of the DFA analysis for the printer subassembly is shown in Fig. 4. It can be seen that 128 separate operations are involved taking an estimated time of 1314 seconds and giving an assembly efficiency of only 7%. As with the final assembly, a central problem with the design is the large number of fasteners. A breakdown of assembly time is as follows:

- Inserting 65 parts or subassemblies = 708 seconds
- Inserting and securing 40 separate fasteners = 454 seconds
- Adjustments, reorientations and hand soldering operations = 152 seconds

As was the case in the final assembly process, many operations in the main subassembly involve difficult alignments, adjustments, and obstructed access for insertion of both parts and fasteners. The MX80 has a reputation for high quality and reliability, but it is clear that it does not stem directly from the design but must result from skilled and careful assembly work.

---

[1] To qualify as "separate part," part must meet certain criteria involving assembly and disassembly, motion, and material.

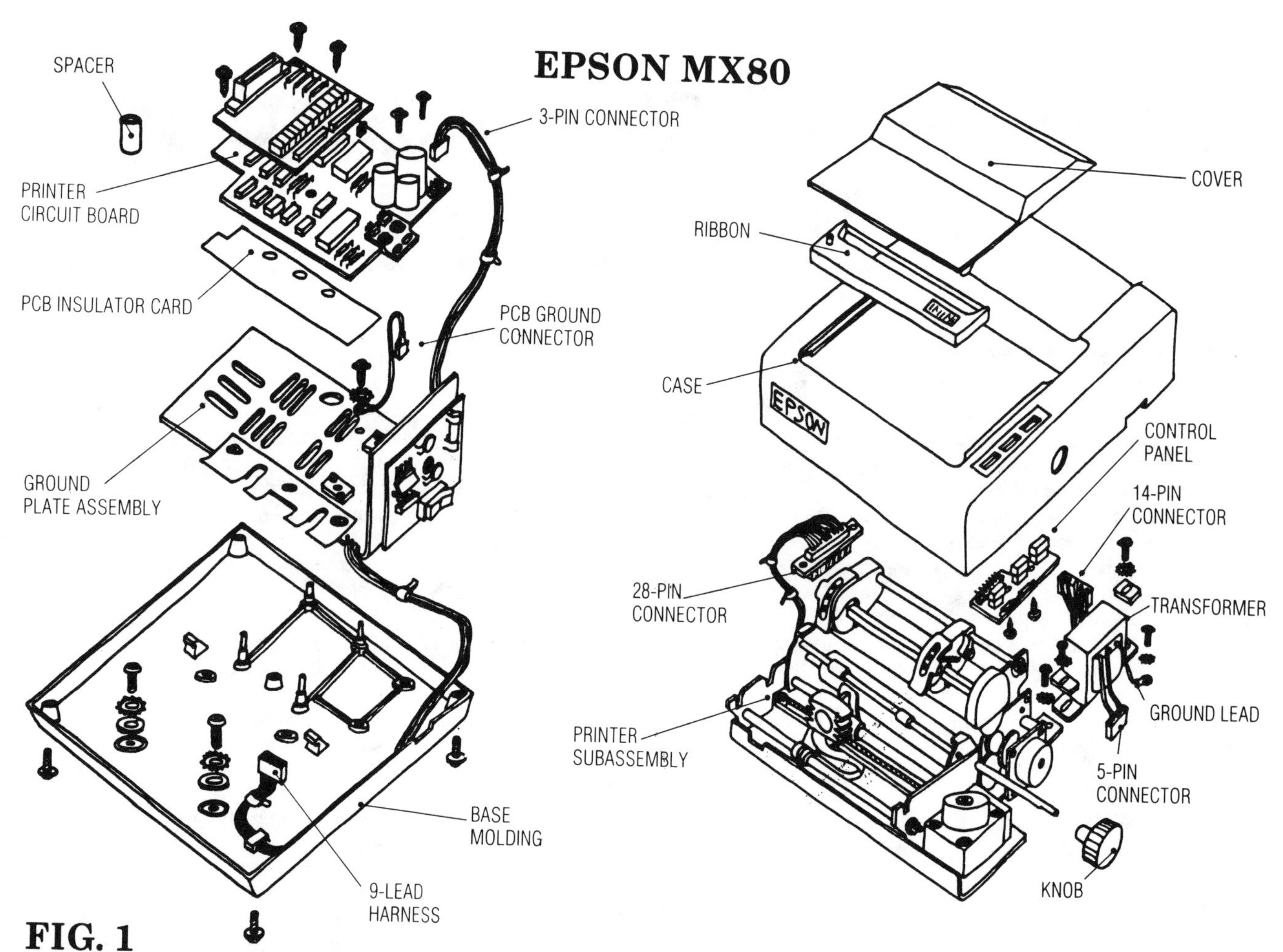

FIG. 1

The EPSON MX80 dot matrix printer final assembly consists of 49 parts or subassemblies as shown in this exploded view.

The IBM Proprinter

Recently the new dot matrix printer called the Proprinter was announced by IBM. It is a replacement for the Epson MX80 printer which has been marketed by IBM under their name and supplied with the PC and XT range of computers. The authors understand that this new printer was designed specifically with robotic assembly in mind, and is a design that involves several revolutionary concepts and avoids the use of screws or other separate fasteners. The final assembly of the Proprinter is being done at IBM's Information Products Division, Charlotte, N.C.

It should be made clear that the authors had no specific knowledge of the labor rates, machinery costs or production parameters appropriate to this particular assembly plant location. Also, it should be understood that any conclusions drawn in the study reported here are subject to the assumptions made regarding these values.

Figure 5 shows an exploded view of the Proprinter with reasonably descriptive names given by the authors to the various parts and subassemblies. In all, there are 32 parts if the four connectors that must be inserted into the printed circuit board are considered separately. The Proprinter is a remarkable design in that all parts or subassemblies snap together during final assembly without the use of fasteners. Even the two motors and the heavy power supply snap into place with fairly simple double motions. It is evident from even a cursory inspection that this is an excellent design for manual assembly. However, its suitability for robot assembly requires more careful analysis.

Before looking at the results of the assembly analysis, the validity of the assumptions regarding labor and equipment costs and production parameters should be considered. ***Cost and Parameter Assumptions:*** Figure 6 presents the values of the important basic costs and parameters used throughout the analysis. These values were deliberately biased in favor of robot assembly. For example, the cost of $50,000 per year, including overhead, for one assembly operator is quite high—a figure of $30,000 is more usual. The cost of one Scara-type robot arm with controls, gripper and sensors of $60,000 is relatively low. The assumption of two shifts working also helps justify automation equipment. ***Manual Assembly Analysis:*** In any Design For Assembly study it probably is worthwhile to perform an analysis for manual assembly first in order to provide a benchmark for subsequent comparisons. Figure 7 presents a summary of the analysis for manual assembly. The total assembly time is 170 seconds and with a labor rate of $50,000 per year ($25/hr), this gives a total manual assembly cost of 118 cents per unit.

As each part is considered in the design analysis, an evaluation must be

EPSON MX80 FINAL ASSEMBLY

Assembly Efficiency (percent)	5
Total Assembly Time (seconds)	552
Total Labor Cost (cents)	383
Total Number of Operations	57
Number of Parts or Subassemblies	49
Theoretical Minimum Number of Parts or Subassemblies	10
Labor Rate (dollars/hour)	25

FIG. 2

Summary of manual assembly for EPSON MX80 shows factors involved in DFA analysis.

EPSON MX80 SUBASSEMBLY

Assembly Efficiency (percent)	7
Total Assembly Time (seconds)	1314
Total Labor Cost (cents)	912
Total Number of Operations	128
Number of Parts or Subassemblies	103
Theoretical Minimum Number of Parts or Subassemblies	31
Labor Rate (dollars/hour)	25

FIG. 4

Summary of DFA analysis for EPSON MX80 printer subassembly points to low assembly efficiency of 7 percent.

made as to whether or not it qualifies as a separate part. It appears that only three parts are candidates for being eliminated through product redesign; namely, the three connectors. Theoretically, these connectors could be integrated into the assembly to which they are connected by the respective leads. This not only would eliminate the leads but would also allow the electrical connections to be made when the parent assembly is inserted. With this design about 12 seconds of manual assembly time could be saved. For a part that is easy to grasp, manipulate and insert with one simple motion, a total assembly time of about three seconds is appropriate. If the 29 parts were all this easy to assemble, then the total assembly time would be 87 seconds. Comparing this figure with the estimated assembly time gives an assembly efficiency of 51%, meaning that the possibility exists for further improvement in assemblability. However, this figure takes no account of practical limitations and, in fact, represents an outstanding achievement. Assembly efficiency figures of less than 10% are more common for this type of product—recall the 5% and 7% values for the Epson printer.

Robot Assembly Analysis

Turning now to robot assembly analysis it should be realized at the outset that this is carried out for three different types of assembly systems: single-station with one robot arm, single-station with two robot arms and the multi-station robot system. After entering data into the *Design for Robot Assembly Program* the summary of results as shown in Fig. 8 was obtained for the multi-station system.

Figure 9 shows how total assembly costs vary for each robotic system and with different production volumes. For a single-station one-arm system the assembly time was estimated to be 211 seconds, representing an annual production volume of 68,000 units. The corresponding assembly cost is 190 cents (point 3 on the graph). For higher production volumes multiple systems result in the same assembly cost. However, lower production volumes obtained by under utilization of one system result in rapidly increasing costs due to the special-purpose equipment that cannot be used for other products. Similarly one single-station two-arm system requires 101 seconds to com-

EPSON PRINTER SUBASSEMBLY

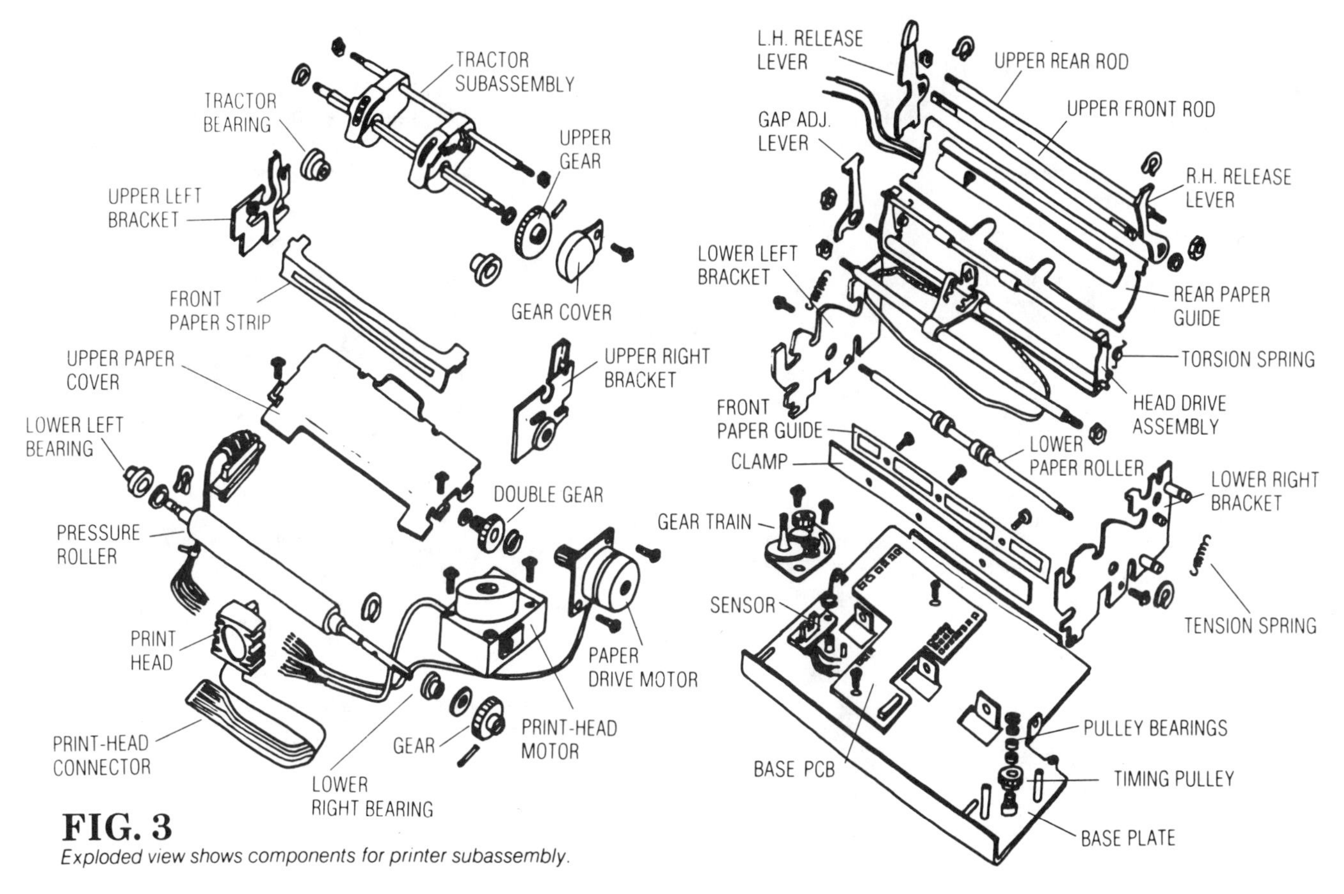

FIG. 3

Exploded view shows components for printer subassembly.

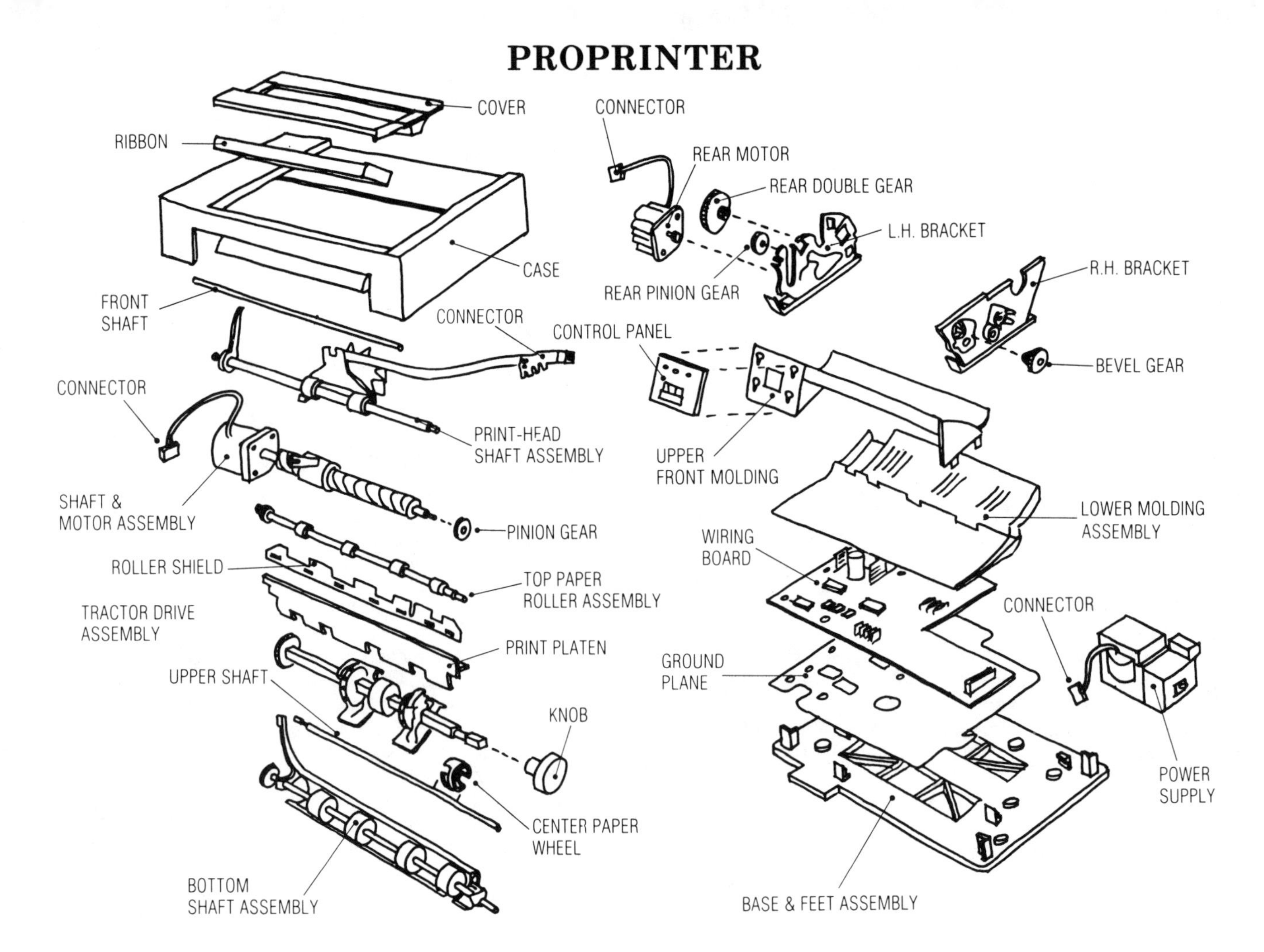

FIG. 5
Exploded view of IBM Proprinter highlights design simplification in this product.

plete the assembly (143,000 per year) at a cost of 150 cents. This condition is point 2 on the graph.

Because multi-station machines can be configured in a variety of ways, they can be designed to match the required production volume. The condition marked point 1 represents the smallest practical multi-station machine fully utilized. This situation results in production of about one assembly per minute, or 240,000 units per year. However, for the Proprinter, a system with only one more station results in the production of two per minute and 480,000 units per year at a significantly lower cost of 136 cents. It was this latter condition that was selected as being the most favorable for robot assembly.

A major problem when considering assembly of the Proprinter using single-station systems is the number of special tools and grippers required. This not only increases the cost of the equipment but lengthens the assembly cycle time due to frequent gripper changes. With the one-arm system about one third of the cost of the total system is attributable to special robot tools. Many of the parts are difficult to grasp and could have been provided with design features that facilitate gripping. However, with the multi-station machine this is a relatively minor problem because the various robots can be provided with suitable grippers for the work at that particular station and the need for gripper changing is avoided. In the case of a one-arm single-station system approximately one half of the cycle time in assembling the Proprinter would be spent on gripper changing.

PROPRINTER ASSUMPTIONS

Cost of Assembly Operator Per Year Including Overhead	$50,000
Cost of Workfixture for Single-Station System	$5,000
Cost of External Manual Workstation for 1 or 2 Arm m/c	$5,000
Cost Per Station of Free-Transfer Assembly Machine	$10,000
Average Cost of Feeder for Robot Assembly System	$5,000
Average Cost of Batch Change Per Feeder	$2,000
Average Time for One Manual Assembly Operation (seconds)	8
Time to Handle and Insert One Part in Magazine (seconds)	4
Cost of One 4-Degree-of-Freedom Robot Arm Including Controls and Gripper	$60,000
Cost of Two 4-Degree-of-Freedom Robot Arms Including Controls and Gripper	$102,000
Cost of One Standard Robot Gripper	$3,000
Cost of Each Workcarrier for Multi-Station System	$1,000
Cost of Magazine or Pallet for Each Part Type	$1,000
Basic Time for Robot Arm Pick and Place Operation (seconds)	3
Additional Time for Final Orientation by Robot (seconds)	2
Number of Shifts Per Day	2
Equipment Payback Period (months)	36

FIG. 6
Assumptions made for DFA analysis of Proprinter are presented here.

FIG. 7
Results of manual assembly analysis for Proprinter point to assembly efficiency rating of 51 percent.

PROPRINTER SUMMARY	
Assembly Efficiency (percent)	51
Total Assembly Time (seconds)	170
Total Labor Cost (cents)	118
Total Number of Operations	32
Number of parts or subassemblies	32
Theoretical Minimum Number of Parts or Subassemblies	29
Labor Rate (dollars/hour)	25

Another problem highlighted in the analysis is that the majority of parts involve multiple insertion motions requiring a more expensive robot than the simple four-degree-of-freedom Scara type. Of course in a single-station system, even if only one part requires a multiple motion (insertion path), the more expensive robot will be required. On a multi-station machine, different robots can be employed at each station. However, all of the five robot stations proposed in this analysis require the more expensive six-degree-of-freedom robot arms. Further consideration probably could have been given to assembly from one direction in the Proprinter.

It should be noted that the design for assembly software assigns a relative cost factor for the robot arm required for each assembly operation depending upon the difficulty of the task. This cost factor is then multiplied by the cost of a four-degree-of-freedom robot arm which is one of the basic costs in the user interactive data base, see Fig. 6. In this way the appropriate arm cost is identified for a single station system or for each station on a robot assembly line.

MULTI-STATION ROBOT ASSEMBLY	
Machine Cycle Time (neglecting down time) (seconds)	28.5
Total Cost of All Assembly Operations for One Unit (cents)	129.3
Total Time for Operator Handling of Parts During Loading of Magazines, and Assembly Time for Parts Manually Assembled (seconds)	112
Size of Batch to be Assembled on One Assembly System or Number of Products to be Assembled During Equipment Payback Period on One System—Whichever is Least (thousands)	600
Number of Assembly Systems to Give Required Annual Production Volume	1
Total Number of Parts or Subassemblies	31
Minimum Number of Parts	26
Number of Stations	10
Total Cost of Robot Arms, Controls and Transfer Devices ($ thousands)	780
Total Cost of Special Workheads and Grippers ($ thousands)	70.0
General-Purpose Portion of Parts Presentation Equipment Cost ($ thousands)	28.0
Special-Purpose Portion of Parts Presentation Equipment Cost ($ thousands)	34.0

FIG. 8
Multi-station robot assembly analysis for IBM Proprinter provides objective view of costs involved.

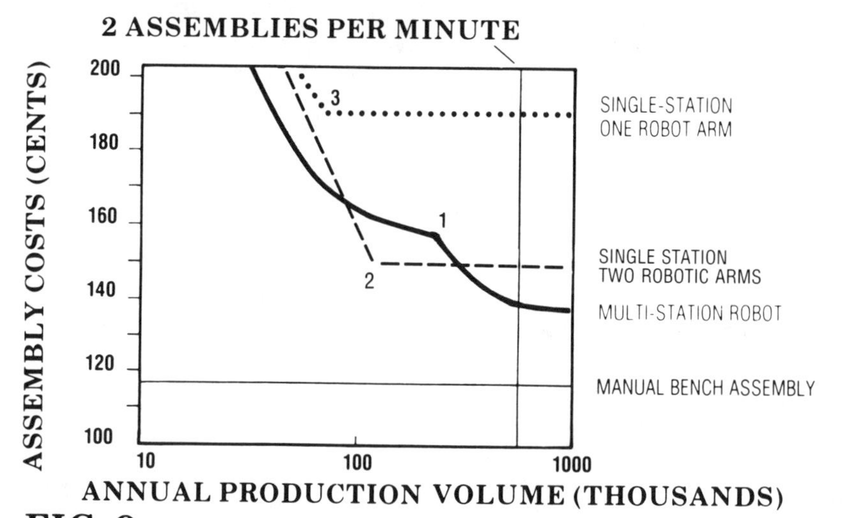

FIG. 9
Cost comparison of several different robotic assembly systems is used in optimizing efficiency and cost.

A significant aspect of this analysis is the time estimated for manually loading the various parts or subassemblies into pallets or magazines for presentation to the robot. Very few parts can be fed and oriented automatically, and an allowance has been made for manual loading time and pallet cost. Regardless of the assembly system used, the total manual time is estimated at 119 seconds; this figure includes 28 seconds for manual assembly of the flexible roller shield and the four flexible connectors. The individual times for manual loading were obtained from the results of the manual assembly analysis by adding the manual handling time to the time of 1.5 seconds for an easy insertion operation. This gives a total time of 91 seconds, representing a cost of 63 cents per assembly. These figures can be compared with a total manual assembly time of 170 seconds and a cost of 118 cents.

The conclusion is unavoidable that if parts are to be manually loaded into pallets, then with a little extra effort these same parts could be inserted into a well-designed assembly such as the Proprinter. It is sometimes argued that the manual loading time should not be included in these analyses because the parts can be obtained from the manufacturers already palletized. Whether or not this can be done without additional expense is highly questionable. It should be realized that if the manual assembly analysis were also performed assuming palletized parts, then the total manual assembly time could be reduced accordingly, and robot assembly would again be difficult to justify in terms of cost.

Conclusion

In analyzing the Epson MX80 printer, care was taken to start with the same level of subassemblies as those used for the final assembly of the Proprinter. Thus the tractor, head drive, gear train and motor subassemblies match almost identical ones in the Proprinter. Comparison of the two product analyses, therefore, is valid as shown in the table, Fig. 10.

It is felt that the Proprinter represents an outstanding design achievement and is highly suitable for manual assembly.

FIG. 10
EPSON MX80 vs. PROPRINTER

	No. Assy Operations	Manual Assy Time
EPSON MX80	185	1866
PROPRINTER	32	170

AE

Reprinted from the ASME *Manufacturing Review*, March 1989

RESEARCH

Design for Assembly of Electrical Products

JEFFREY L. FUNK
Westinghouse R&D Center, 1310 Beulah Rd., Pittsburgh, PA 15235

This paper describes the application of design for assembly and early cost analysis models to the design of electrical products. Four example subassemblies are considered which include printed circuit board (PCB), wire and mechanical assembly operations. The alternative designs are found to have (on average) 32% lower printed circuit board assembly costs, 48% lower wire assembly costs, 73% lower mechanical assembly costs and 76% lower mechanical part costs.

INTRODUCTION

A product's manufacturing costs have an important effect on a product's profitability. However, many firms including the author's firm, Westinghouse, have found that up to 85% of a product's manufacturing costs are typically determined before the manufacturing department becomes involved with the design of a new product. And since design engineers have little manufacturing training or experience and few tools to analyze manufacturing costs, manufacturing costs are not fully considered until the manufacturing department becomes involved when it is too late to make any significant design changes. Therefore, many U.S. companies may be missing a major opportunity for increased profitability.

Manufacturing costs must be considered during the conceptual design phase when less than 50% of a product's cost have been determined. During the conceptual design phase, the product's "functional" design is transformed into a physical design. Mechanical and electrical assemblies (for example, PCBs) are assigned to various functional blocks and wires, or fasteners are used to connect these functional blocks into a working product. At this point, the principles of *design for assembly* can be applied. Numerous design teams have applied these principles and have achieved reductions in manufacturing costs of up to 40% [1].

Design for assembly is a two-step process. First reduce the number of mechanical or electrical parts in a product and second, simplify the remaining assembly operations. Part reduction provides the greatest opportunity for savings in manufacturing costs since a reduction in the number of parts can reduce direct labor, material and overhead costs. Fewer parts means fewer parts to assemble, fabricate, purchase, inspect, store, receive, draw (i.e., drafting), control (i.e., production, planning and control) and count (e.g., accounting). Researchers have found that parts can be combined if:

1. they do not move relative to each other during the product's operation or service;
2. they can use the same materials; and
3. they do not require disassembly during service [2].

Implicit in any analysis of manufacturing costs, however, are tradeoffs between product quality and manufacturing costs and between various categories of manufacturing costs. Designers make tradeoffs between a product's cost and its size, appearance, reliability and serviceability. Further, alternative designs may affect assembly, fabrication, purchasing, inventory and other overhead cost categories in conflicting ways. For example, a new injection molded part may reduce assembly costs but it may also increase purchasing and inventory costs because it is a non-standard part. Therefore, design engineers need a simple method to estimate, analyze and compare the manufacturing costs of the various design alternatives and compare these costs to the differences in product quality of each alternative.

Cost models and data bases of various manufacturing processes have been developed to assist the product designer in making these tradeoffs. The basic labor cost model is summarized in Appendix 1. Data bases have been developed for mechanical, printed circuit board and wire assembly, injection molding and machining. Also, data bases are being developed for sheet metal fabrication and manufacturing overhead activities such as material control (e.g., purchasing, inventory, production control) and quality control (e.g., inspection). These data bases include the time to perform various assembly, fabrication and overhead operations, the cost of materials such as sheet metal blanks and injection molding resins and the mold cost due to adding specific features to an injection molded part. A major objective of the process time and mold cost data base is to relate manufacturing process times/costs to parts and assembly characteristics so that the designer does not need to know or understand the elements of the manufacturing processes.

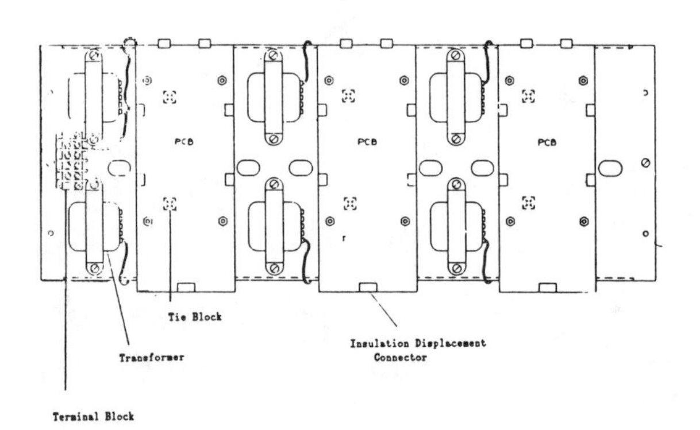

FIG. 1. Base drive assembly

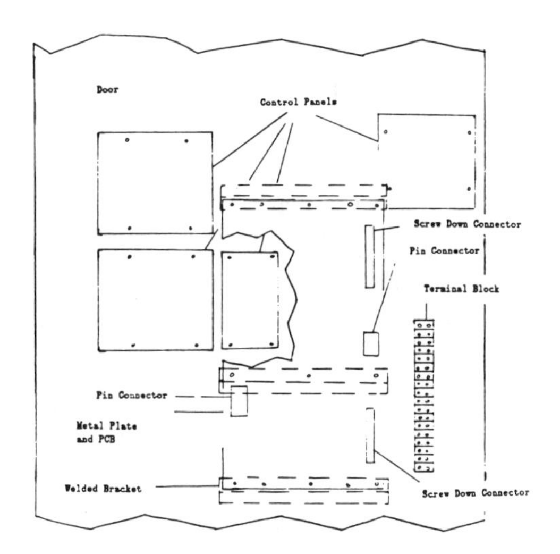

FIG. 2. Door assembly

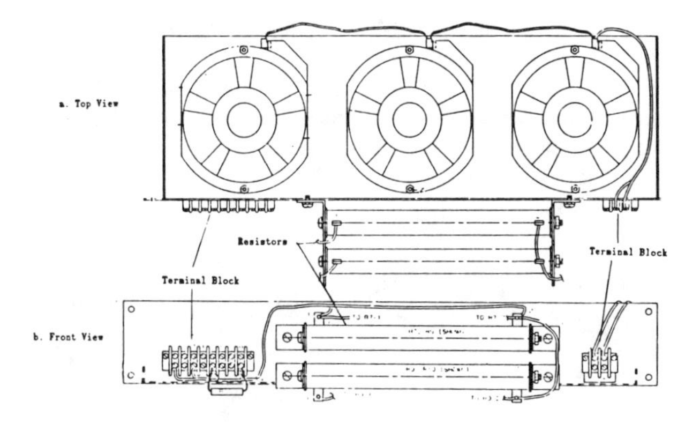

FIG. 3. Fan assembly

ELECTRICAL PRODUCTS

There are many applications of the design for assembly principles to electrical products primarily because, unlike mechanical products, there are almost no moving parts. For example, consider the electrical assemblies shown in Figs. 1, 2, 3 and 4 which do not have any moving parts (the first constraint to combining parts). The base drive assembly has three PCBs, six transformers and one terminal block mounted on a sheet metal panel and connected with wires as shown in Fig. 1. The door assembly has one PCB and one terminal block mounted on two sheet metal plates, three control panels attached directly to the door and the PCBs and terminal blocks are connected with wires as shown in Fig. 2. The fan assembly has three fans, four large resistors and two terminal blocks mounted on a sheet metal plate and connected with wires as shown in Fig. 3. The capacitor chassis assembly has eight large round capacitors attached to a copper plate with clamps and connected with metal busbars as shown in Fig. 4.

With respect to the second constraint to part reduction, although the electrical components use different materials than the mounting (i.e., fasteners) or connecting (e.g., wires) parts, the fasteners and panels and the wires and termination devices do not usually have different material requirements so they can be combined. Therefore, many of the fasteners, brackets and wire terminals are potential candidates for elimination. In addition, the functions of individual electrical components can often be combined into a single component. However, this option is limited by the type of components offered by component suppliers.

The third constraint, disassembly for service, may be the greatest limiting factor for combining parts. Components often fail during the product's operation so they must be removable for repair or replacement. Therefore, mechanical fasteners are often used to mount large components to sheet metal panels since the disassembly of a plastic snap or force fit might damage a plastic panel. However, components can still be snapped to a plastic base if the savings in initial manufacturing costs are greater than the potential increase in assembly costs during the service of the product. The designer must consider a number of factors in order to choose the lowest cost alternative such as: the frequency of required service for each component, the relative manufacturing costs of snapping components onto a plastic plate vs. attaching them to a sheet metal plate with mechanical fasteners and the cost of discarding the plastic plate when disassemblying the parts. this article, however, will only analyze manufacturing costs and leave the analysis of service costs for future research.

The second design for assembly principle, simplify the remaining assembly operations, can also be applied to this type of product. Electronic components that have the correct body type (e.g., DIPs, axials) and lead spans for the available automatic equipment can be automatically inserted. Some components are also easier to manually insert in PCBs than other components and some wiring and mechanical operations are easier to manually perform than other operations. For example, spade terminals are easier to connect to large electrical components than ring terminals. Mechanical parts are less ex-

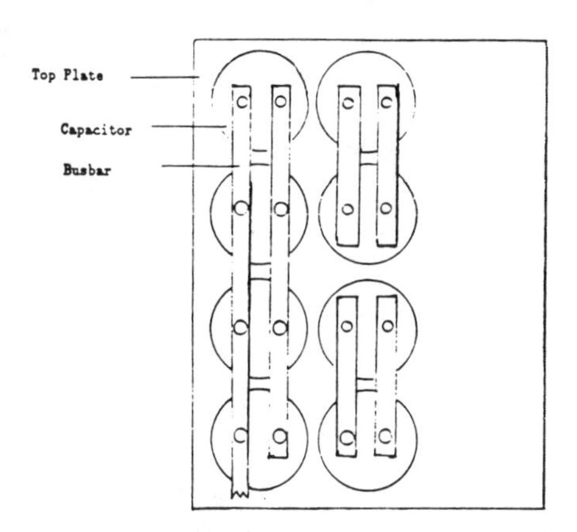

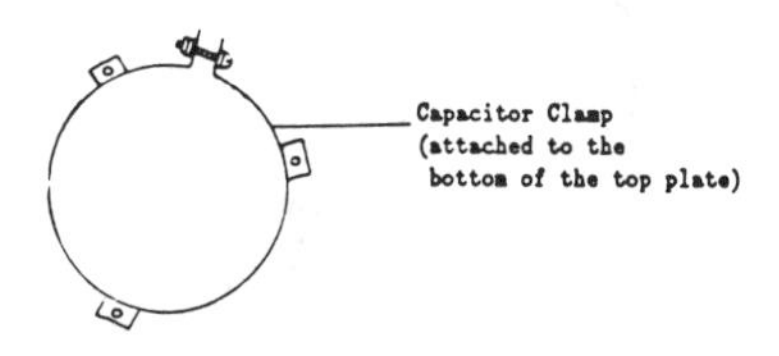

FIG. 4. Capacitor chassis assembly

pensive to assemble when they are easy to handle and they can be inserted from one direction.

PRINTED CIRCUIT BOARD ASSEMBLY

The number of components/component type is shown for two PCBs in Table 1. These PCBs are part of the door and base drive assemblies. The Westinghouse division which assembles these PCBs has automatic DIP and axial insertion machines and several semi-automatic insertion machines which are used to insert the radial and non-standard components. Therefore, in the first PCB, 12 DIPs and 56 axials are automatically inserted and in the second PCB, 48 DIPs and 168 axials are automatically inserted. The remaining components are semi-automatically inserted except, for a few components which are manually inserted and hand soldered after the wave solder operation.

Part reduction is one way to reduce the assembly costs for these PCBs. For example, first, the seven connectors on the base drive PCB and the six connectors on the door PCB can be combined into one connector on each board. Second, the spacers and tubes used underneath nine of the transistors and one varistor can be eliminated if sufficient space is allowed between the transistor cans and the traces on the PCB.

The remaining assembly operations can also be simplified by choosing components which can be automatically inserted and/or wave soldered. For example, first, since radial capacitors and diodes are usually available as axial components and the existing factory only has automatic DIP and axial insertion machines. Therefore, the use of axial leaded capacitors and diodes on the example PCBs will allow the automatic insertion of 32 more components in the base drive PCB and 89 more components in the door PCB although the effect of these design changes on material costs is unknown. Second, the four components inserted on the back sides of both PCBs can be placed on the front sides thus eliminating the hand solder operations and allowing the automatic insertion of one more axial component.

A comparison between the existing and new PC assembly costs can be made using the cost model described in Appendix 1 and the PCB assembly process times which are shown in Tables 2 and 3. These times are a partial list of a more complete data base for PCB assembly that also includes the insertion time for 6, 8, 10 and 12 leaded radials and non-DIP components whose leads do not require cutting (e.g., switches). The data base was developed from time studies done by the University of Rhode Island [3] and Westinghouse.

The cost comparison is shown in Table 4 using the models and data base of process times. These cost estimates assume setup times of 15 minutes and 10 minutes respectively for automatic and semi-automatic insertion, a lot size of 10 and a costing rate of $40/hour. The new base drive PCB is 23% cheaper to assemble than the old design and the new door PCB is 36% cheaper than the old design. Retrieval and kitting costs are reduced because a greater number of components are now kanbanned between the storage area and the automatic insertion equipment. Insertion costs have been re-

Table 1. Number of components/component type on two PCBs

Component type	Function	Base driven PCB	Door PCB
DIP	ICs	12	48
DIP	Other functions[1]	0	3
SIP	Resistor Networks	8	20
Axial	Resistors	29	83
Axial	Diodes	26	68[2]
Axial	Capacitors	2	20
Radial	Capacitors	30	80
Radial	Resistors (variable)	2	17
Radial	Diodes	2	9
Radial	LEDs	2	25[7]
Large Radial	Capacitors	4	11
Large Radial	Transformer	0	2[8]
Large Round Radials	Transformer	6	8
3-Legged Radial Cans	Transistors	8[3]	5[4]
3-Legged In-line Radials	Transistors	6[5]	3[9]
3-Legged In-line Radials	Varistors	4[6]	1[10]
Connector	External Connection	7	6
Pin		<u>0</u>	<u>32</u>
Total		148	441

1: e.g., switch
2: 1 backside, 1 too large for automatic insertion and 1 hairpin
3: each transistor requires spacers between it and the PCB
4: 1 transistor has a lead that requires tubing
5: 4 transistors require heat sinks
6: 2 varistors require heat sinks
7: 3 LEDs are placed on the reverse side
8: 1 transformer requires a heat sink with 2 bolts and 2 nuts and 1 transformer requires 4 bolts, 4 spacers and 4 nuts
9: heat sinks
10: 1 spacer w/lead holes

Table 2. Processes and times for PCB assembly

Process	*Typical time*
Retrieval	33 seconds/component retrieval
Kitting	0.7 seconds/component
Automatic Insertion/ Placement	1.8 seconds
Semi-Automatic Insertion	(See Table 3)
Manual Insertion	(See Table 3)
Wave Solder	3.5 minutes/board
Final Assembly	(See Table 3)
Rework	360 seconds/incorrectly placed components

Source: Reference [2] and Westinghouse

Table 3. Process times for manual, semi automatic and final assembly of through-hole components

Component Type	*Manual*	*Semi-automatic*[1]	*Final assembly*
DIPs	18.0	20.0	93.0
Axials	18.0	12.6	30.0
w/1 sleeve	33.0	27.6	45.0
w/1 sleeve	48.0	42.6	60.0
2-Leaded Radials	12.0	5.6	29.0
w/ring type spacer	17.5	13.1	34.5
w/spacer w/lead holes	19.0	14.6	36.0
2 sleeves	42.0	35.6	59.0
3-Leaded Radials	11.5	8.8	34.5
w/ring type spacer	16.0	14.8	40.0
w/spacer w/lead holes	18.0	16.3	42.0
w/heat sink	37.5	35.8	61.5

1 Light directed assembly
Source: Reference [5] and Westinghouse

Table 4. Assembly cost ($) for existing and new PCB designs

Process	Existing designs		New Designs	
	Base Drive	Door	Base Drive	Door
Retrieval	1.20	3.60	1.10	1.00
Kitting	.67	1.90	.40	1.00
Automatic Insertion	3.30	5.80	3.80	7.10
Semi-Automatic Insert.	5.00	17.62	2.10	8.92
Wave Solder	2.64	2.64	2.64	2.64
Final Assembly	.00	1.00	.00	.00
Rework	.20	.67	.10	.50
Total Assembly Costs	13.20	33.32	10.14	21.32

duced due to the greater use of automatic insertion and the simpler manual insertion operations. Final assembly costs have been eliminated by placing all of the components on the front of the PCB. Rework costs have also been reduced due to the greater use of automatic insertion.

WIRE PREPARATION AND ASSEMBLY

The number of connections/connection type and assembly costs are shown for the four existing assembly designs in Table 5. The existing design uses a variety of methods to connect the various electrical components to each other and to terminal blocks both in and between the example assemblies.

Part reduction is one way to reduce the wire preparation and assembly costs in these example assemblies as shown in Table 6. First, directly connect the various electrical components as opposed to using terminal blocks as intermediate steps. This strategy will eliminate two terminal blocks and markers and four spade lugs on the base drive assembly, one terminal block and 29 spade lugs on the door assembly and two terminal blocks and markers and five spade lugs on the fan assembly.

These terminal blocks are included in each subassembly to simplify the wiring operations at final assembly since the correct terminal (each terminal is numbered on the marker) on the terminal block is easier to locate than the correct component. However, a direct connection is probably a cheaper method for three reasons. First, the extra assembly time to connect the components at final assembly is probably much less than the time to attach and wire a terminal block during subassembly. Second, components that are to be electrically connected can be placed close to each other thus reducing the number of connections between subassemblies.

Third, there are other ways to make the connections at final assembly appear clear to operators. For example, if an injection molded base is used instead of a sheet metal base, numbers and tie offs can be made part of the injection molded base as shown in Fig. 5 so that the wires can be "tied off" in a marked position during the subassembly stage and thus the final assembly operation is still simple and straight forward. The cost of such an injection molded part is estimated using Reference [5] and other features of the part are described in the next sub-section. The extra features required to tie the wires off will only slightly increase the mold cost.

A second way to reduce the number of parts is to use wiring methods (insulation displacement and screw down bare wires) which do not use terminals or only have one terminal for multiple wires (flat cable) and thus have the lowest wire preparation and assembly time. However, flat cable, insulation displacement and screw down wires are typically used for making a connection to a PCB and a flat cable requires multiple connections between points. Therefore, a flat cable cannot be based on any of the example assemblies, but an insulation displacement connector can be used instead of the 23 pins (one eight pin and one 15 pin connector) that are presently attached to the PCB on the door assembly.

A third way to reduce the number of parts in a wire assembly is to replace the 20 tie wraps in the base drive assembly and the seven tie wraps in the fan assembly with "tie offs" that are integrated with an injection molded base such as the one described above. The wires can easily be supported through such a tie off and the mold cost will only increase slightly due to these tie offs.

The assembly costs of the existing and new designs can be compared by using the cost model described in Appendix 1 and the wire preparation and assembly process times shown in Table 7. These times include the time to cut, strip (not re-

Table 5. Number of connections/type of connection and assembly costs for the existing wire assemblies

Assembly	Type of Connection: Insulation displacement	Solder wires	Spade lugs	Ring lugs	Tab Recept.	Pins	Tie wraps	Assembly cost ($)
Base-Drive	72	12	4				20	30.39
Door			29			23		42.82
Fan		13	7				7	17.10

Table 6. Number of connections/type of connection and assembly costs for the new wire assemblies

Assembly	Type of Connection: Insulation displacement	Solder wires	Spade lugs	Ring lugs	Tab Recept.	Pins	Tie wraps	Assembly cost ($)
Base-Drive	72	12						23.30
Door	52							10.97
Fan		13	2					12.54

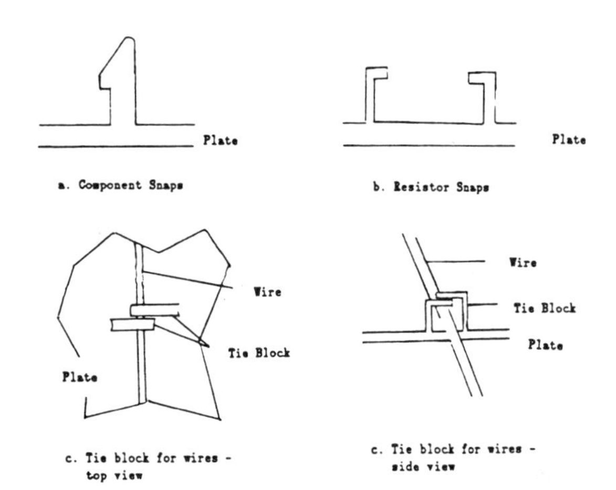

FIG. 5. Various geometrical attributes of the injection molded parts

quired for insulation displacement and flat cable), terminate (only required for flat cable, terminal lugs, tab receptacles and pins), apply protective plastic tubes (terminal lugs and tab receptacles), attach the wires and run one end of a two foot wire. Bundling time is not included in these times. Table 7 is a partial list of a more complete data base that includes the time to manually prepare and assemble multi-conductor cables, automatically prepare and assemble single wires and bundle wires using a variety of methods. The data base is from a time study done by the Westinghouse R&D Center at the Westinghouse Automation Division and from a study done by the University of Colorado [4].

Assuming a setup time of 5 minutes, a lot size of 10 and a costing rate of $40/hour, these design changes reduce the total assembly costs significantly as shown in Tables 5 and 6. The wire assembly costs on the base drive assembly are reduced by 23%, on the door assembly by 74% and on the fan assembly by 27%.

MECHANICAL ASSEMBLY

The number of components/component type and the mechanical assembly and material costs are shown in Tables 8 and 9 for the existing and new designs of each subassembly. The existing designs use a large number of fasteners to mount the various electrical components on sheet metal and other plates.

Table 7. Preparation and attachment time for various connectors₁

Type of connection	*Time (seconds)/connection*
Flat Cable₂	2.4
Insulation Displacement₃	10.4
Screw Down Wires₃	19.9
Solder Wires	57.2
Spade Lugs	73.8
Ring Lugs	86.2
Tab Receptacles	80.0
Pins	42.3

1: Includes the time to cut, strip, terminate, attach and run a two foot wire
2: Assumes there are 26 wires in the flat cable
3: Assumes there are 5 wires in the connector and they are run between points simultaneously
Source: Westinghouse and Reference [4]

Three ways to reduce the number of parts in the existing subassemblies are considered. First, parts that only provide structural support for the assembly can be combined into a more complex part. Second, simple, standard off-the shelf parts such as brackets and plates can often be combined into a single custom part. Third, separate fasteners can be eliminated by using integral fasteners which are designed into the parts that provide structural support.

The sheet metal plates in the existing designs of the base drive, fan and capacitor chassis assemblies can be replaced by injection molded plates where most of the components are snapped to the plastic plates and the plastic plates are snapped to the product's frame using the snaps shown in Fig. 5a. These snaps are designed to fit the size of the mounts on each component. In addition, large projections are designed into the plastic plates to support the heavier components.

In the base drive assembly, the existing design uses a large number of separate fasteners, brackets and spacers to mount three PCBs and six transformers on a sheet metal plate and to attach the sheet metal plate to the product's frame. In the new design, the PCBs and the transformer clamps are snapped to a plastic base and the plastic base is snapped to the frame thus eliminating all of the fasteners and spacers.

Table 8. Number of components/component type and costs ($) for the existing designs

Part type	*Base*	*Door*	*Fan*	*Capacitor chassis*
Electrical Components₁	9	1	7	8
Screws and Bolts	30	19	20	56
Washers	18	13	24	40
Nuts	24	6	10	14
Plates	1	2	1	3
Spacers	12	13	0	0
Component Bracket/Clamps	6	0	4	8
Terminal Blocks/Markers	4	1	4	0
Tie Blocks	6	0	3	0
Connectors	18	6	0	0
Other	0	4	0	8
Total Assembly Costs	14.91	11.46	8.44	19.13
Material Costs of Plates	21.00	21.00	21.00	24.70
Other Material Costs₂	2.62	4.46	2.98	5.79

1: Electrical components include PCBs, resistors, fans, transformers, fuses, contactors, busbars and capacitors
2: Excluding electrical components and the welded brackets on the door assembly

Table 9. Number and type of components and costs ($) for the new designs

Part type	*Base*	*Door*	*Fan*	*Capacitor chassis*
Electrical Components₁	10	1	7	8
Screws and Bolts	0	9	0	10
Washers	0	9	0	20
Nuts	0	9	0	0
Plates	1	0	1	1
Spacers	0	9	0	0
Component Bracket/Clamps	6	0	0	0
Connectors	3	1	0	0
Other	0	4	0	8
Total Assembly Costs	3.33	4.83	1.70	4.36
Injection Molded Bases	7.62	.00	7.51	8.31
Other Material Costs₂	0.00	.96	0.00	0.00

1: Electrical components include PCBs, resistors, fans, transformers, fuses, contactors, busbars and capacitors
2: Excluding electrical components and the welded brackets on the door assembly

The tie blocks are eliminated by making them part of the plastic plate as shown in Fig. 5c and 5d. The terminal blocks and markers were eliminated earlier by directly connecting the electrical components.

The transformers could probably be snapped directly to the plate, thus also eliminating the clamps in the base drive assembly. However, for purposes of example and since the clamps are purchased as part of the transformers, the new design does not try to eliminate the clamps although their elimination may reduce total manufacturing costs even further.

In the fan assembly, the existing design also uses a large number of separate fasteners and brackets to mount three fans, four large resistors, two terminal blocks and to attach the base to the product's frame. The new design eliminates these fasteners and brackets by snapping the fans and resistors to a plastic base and snapping the plastic base to the frame. The snaps used for the fans are similar to those used for the PCBs and transformer clamps although they are little longer. The snaps used for the resistors depend upon plastic bending for fastening as shown in Fig. 5b. The resistors can be mounted in each corner of the plate since the terminal blocks have been eliminated. The tie blocks and terminal blocks are eliminated for the same reasons as in the base drive assembly.

In the capacitor chassis assembly, the existing design attaches 8 capacitors to a top plate with 8 clamps, 32 screws, 16 washers and 8 nuts. The capacitors are electrically connected with 8 metal busbars, 16 screws and 32 washers. The top plate is attached to the product's frame with two brackets, 8 screws and 8 nuts.

The new design eliminates these fasteners, clamps and brackets by placing each capacitor in a cup contained in a plastic base and snapping the base to the product's frame. The capacitors are still electrically connected using the same method as used in the existing design. Although the plastic base provides the structural support for the capacitors, the busbars hold the capacitors in a fixed position.

In the door assembly, a sheet metal plate is not replaced by a plastic plate. The existing sheet metal plates include four brackets which are welded to the door and one flat plate which is attached to three of the brackets with fasteners. A PCB is attached to the flat plate, a terminal block is attached to one of the welded brackets with fasteners and four plastic control panels are attached directly to the door with fasteners. The new design eliminates all of these sheet metal plates by mounting the PCB directly to the door and integrating the terminal block with the PCB. The new design also adds snaps to the 4 control panels thus eliminating their separate fasteners.

The mechanical assembly and material costs for the existing and new designs are compared in Tables 8 and 9. The assembly costs are estimated using the cost model described in Appendix 1 and a data base of process times developed by the University of Rhode Island [2]. The material costs in the existing design are for fasteners, sheet metal plates, spacers, brackets and clamps, terminal blocks, tie blocks and connectors. They are based on the purchase price of the parts. The cost of the electrical components or the welded brackets on the door assembly are not included.

The material costs in the new designs are for the fasteners in the door assembly and the injection molded bases in the other assemblies. The cost of the injection molded plates includes the material, injection molding operation and mold base and cavity cost. These costs are estimated using a methodology developed by the University of Rhode Island [5].

These cost estimates are described in [6] and summarized in Table 10. The estimates are for polypropylene parts that have the same width and length as the existing sheet metal parts. The costs shown in Table 9 include a 20% markup over the costs estimated in [6] to account for the supplier's profit. The estimates assume a costing rate of $40/hour for mold making, a part thickness of .375" and a total volume of 20,000 parts for the capacitor base and a thickness of .25" and a total volume of 10,000 parts for the remaining assemblies. Different assumptions are made since the capacitor base holds more weight than the other assemblies and it is used twice in the example product as opposed to once for the other assemblies.

The new designs reduce the assembly and material costs significantly as shown in Table 9. The assembly costs assume a setup time of 5 minutes, a lot size of 10 and a costing rate of $40/hour. The new design for the base drive assembly reduces the assembly costs by 78% and the material costs by 64%. The new design for the fan assembly reduces the assembly costs by 80% and the material costs by 65%. The new design for the capacitor chassis assembly reduces the assembly costs by 77% and the material costs by 73%. In this assembly, the total material costs of the top plate and the top plate and the top brackets are much more expensive than the single injection molded part.

The new door assembly reduces assembly costs by 58% and material costs by 96%. The material costs are primarily reduced by eliminating the sheet metal plate which is not re-

Table 10. Material, operation and mold costs for the injection molded plates

	Base drive plate	*Fan plate*	*Capacitor Base*
Dimensions			
Projected Area (cm^2)	1625	1084	993
Length (cm)	65	61	25
Width (cm)	25	18	14
Depth (cm)	1.00	5.1	18
Thickness (cm)	.63	.63	.95
Machine Characteristics			
Clamping Force (MN)	10	10	10
Clamping Stroke (cm)	10	10	10
Maximum Power (kw)	120	120	120
Dry Cycle Time (sec)	12	12	12
Material Cost			
Material Type		Polypropylene	
Part Volume (cm^3)	1024	1081	2370
Cost/Part ($)	1.42	1.50	3.28
Operation Cost			
Fill Time (sec)	1.4	1.4	3.2
Cool Time (sec)	86	89	196
Reset Time (sec)	15	17	20
Cost/Part ($)	1.21	1.26	2.50
Mold Cavity Hours due to:			
Cavity Area	431	304	282
Geometric Complexity	80	170	170
Screw Threads	0	0	0
Side Pulls	260	260	0
Tolerance Level	20	20	20
Surface finish	40	41	50
Texture	0	0	0
Mold and Total Unit Cost			
Mold Base Cost ($)	3970	323	1950
Mold Cavity Cost ($)	33240	31800	20880
Total Volume	10000	10000	20000
Total Cost/Part ($)	6.35	6.26	6.93

placed by an injection molded plate in this assembly. If the cost of the three sheet metal brackets and their welding operations are considered, the assembly and material savings would be greater. In the case of the control panels, this analysis assumes that the panels with snaps will not cost more than the panels without snaps. This would be true if they are available as standard, off the shelf parts.

SUMMARY AND CONCLUSIONS

This report demonstrates how assembly and material costs can be reduced by applying the principles of design for assembly and by using early cost analysis tools. These principles and tools are applied to several subassemblies in a typical Westinghouse electrical product. The alternative designs have on average 32% lower PCB assembly costs, 48% lower wire assembly costs, 73% lower mechanical assembly costs and 76% lower mechanical part costs. In addition, since most of the cost savings come from reducing the number of parts in these assemblies, there should be a similar reductions in purchasing, receiving, inspection, storeroom, drafting, production, planning and control, inventory, inventory control and accounting cost.

ACKNOWLEDGEMENTS

This research was partially supported by the National Science Foundation, Grant #DMC-8513930. MR

REFERENCES

1. Boothroyd G and Dewhurst P (Nov. 10, 1983). Design for Assembly: Selecting the Right Method. *Machine Design*.
2. Boothroyd G and Dewhurst P (1987). *Product Design for Assembly*, Boothroyd and Dewhurst, Inc., Wakefield, RI.
3. Boothroyd G and Shinohara T (1986). Component Insertion Times for Electronics Assembly. *Technical Report*, Department of Industrial and Manufacturing Engineering, University of Rhode Island.
4. Ostwald P (1985). *American Machinist Cost Estimator*, McGraw-Hill, Inc.
5. Dewhurst P and Archer D (1987). Cost Estimating for Injection Molded Components. *Technical Report*, Department of Industrial and Manufacturing Engineering, University of Rhode Island.
6. Funk J (1987). Design for Assembly. *Technical Report*, Westinghouse R&D Center, Pittsburgh, PA.

APPENDIX 1

Basic Labor Cost Model

Unit assembly, or part cost (UC), can be modeled using Equation 1 where MC is the material cost, SUT is the setup time, LS is the lot size, PT is the process time, NO is the number of operations, CR is the costing rate, MDC is the mold cost and VOL is the volume. Material cost depends on the cost/volume of the material and the part's volume. Setup time can range from seconds to hours depending on the operation and the amount of emphasis on reducing setup times. Since it varies so much, however, this project is not developing a data base of these times and the user of the models must input a value. Lot size can range from one to thousands depending on the assembly volume and the amount of emphasis on JIT manufacturing. It is also input by the user. A data base of process times are being developed and examples of this data base are shown in Tables 1, 2 and 5. The costing rate depends on employee wages, equipment costs and other overhead costs. Mold cost depends on various attributes of the part and a method of estimating the mold cost of injection molding parts is described in [5]. The costing rate, the number of operations and the part volume are input by the user of the models.

$$UC = MC + (SUT/LS + PT * NO) * CR + MDC/VOL \quad (1)$$

Jeffrey L. Funk is a Senior Engineer at the Westinghouse Research and Development Center. His research interests include the economics of manufacturing, design for manufacturing, economics and management of design for manufacturing, automation, and CIM. Presently, he is studying Japanese design and manufacturing techniques in Fukuoka Japan.

Designing for Productivity Saves Millions

Reducing the number of a product's components and its manufacturing steps starts in the design phase.

RUSS GAGER
Technical Editor

Designing for productivity is a process which combines both design and manufacturing disciplines. A screw may be a quick and easy solution to hold two assemblies together for a designer, but for the manufacturing department, driving that screw for hundreds of thousands of parts instead of making the assembly snap-fit can cost far more than the design time saved.

Simplifying an assembly by reducing the number of parts can also provide astronomical savings for a company mass producing products. Many times, even the costs of additional design time, retooling, specialty fasteners, and new materials do not outweigh the savings generated by the more efficient design.

Whether manual or automatic assembly is used is many times less important than that the analysis to simplify the assembly was done. Some companies have simplified an assembly, done a cost analysis on automating it, and decided to stay with manual assembly. But those cost savings from simplification of the assembly remain, and sometimes are even larger because there is no additional capital expense from the automated equipment to pay off.

The basic principles of Design for Productivity were outlined by Geoffrey Boothroyd and Peter Dewhurst of the University of Rhode Island (formerly of the University of Massachusetts). Boothroyd and Dewhurst have developed a software system for IBM PC, XT, AT and compatible personal computers which encapsulates their Design for Assembly Handbook. A 45-minute instructional videotape is also available.

The handbook and software detail a system which uses numbers to evaluate whether a product is being assembled most efficiently. The seven software programs operate by entering the names of a product's parts on the screen and answering basic questions about each part.

As each question is answered, a worksheet is automatically created on the computer screen. Complete analyses for the basic assembly systems are printed out after the questions are answered.

Boothroyd and Dewhurst are also working on a Design for Electronic Assembly handbook to analyze the most efficient methods of assembling components on a circuit board. Another way of speeding the design for assembly analysis is also being investigated. This method would include the software in CAD/CAM systems so that ultimately the manufacturing cost of a product or assembly could be determined by pressing a button with a minimum of inputs.

DFA analysis used extensively

Many companies are using design for assembly analysis regularly. Hitachi has been teaching a method for years. Xerox and IBM teach its employees the entire Boothroyd/Dewhurst system.

IBM offers a two-day seminar on Design for Assembly methodology at regular intervals at its IBM corporate Quality Institute.

"We stress the idea that successful assembly automation hinges upon purposeful design for automatic assembly," says Morris Krakinowski, a senior engineer at IBM.

A major project at IBM involved design for assembly evaluation of 1,300 drawing that make up 50 subassemblies. A functional prototype was developed and then analyzed for manufacturability.

"We run the design analysis software on an IBM PC, which requires no programming skills," says Krakinowski. "But if users are not carefully introduced to the system, the results can be somewhat misleading.

"The terminology is simple, but the meaning of the terms are rigorously defined. For example, the meaning of a term such as 'part is easy to handle' may appear self-evident, but it has a very specific meaning in the context of the analysis system. That is why we use the two-day workshop to introduce users to the subject and give them enough of a working knowledge of the system."

General Electric has been conducting Design for Assembly workshops throughout the company. They have resulted in reduction of parts requirements by an average of 20 percent and cut assembly labor by an average of 40 percent. Handling and inventory costs have also been reduced due to the improved designs.

The 2 1/2-day sessions are led by two DFA consultants from GE. The first day explains DFA evaluation and breaks the group of 20 attendees from design engineering, manufacturing and quality personnel into small teams which each analyze two practice assemblies.

On the second day, the small teams analyze their own products. Then on the third day, the teams brainstorm ideas to improve each product's assembly.

At Black & Decker, the Boothroyd Dewhurst design for assembly technique has been used extensively. One product which has used it is a specially designed, high-strength flashlight for firefighters called the Sun Lance.

The number of parts on the flashlight were reduced from 30 to 10 and the number of manufacturing processes dropped from 16 to five. Assembly time was reduced by 60 percent. Among the techniques used besides design for assembly techniques were eliminating adhesive bonding and soldering.

Hamilton Standard Controls, a unit of United Technologies, has also designed many of its products for productivity using the Boothroyd Dewhurst approach. Besides their electronic controls, the company has also used design for productivity in the manufacture of two relays.

Direct labor on the Type 91 relay was substantially reduced by designing it for automation. There was no net reduction in parts or assembly operations,

Designing automated manufacturing *for the Type 91 relay at Hamilton Standard Controls in Logansport, Ind. resulted in a substantial reduction of direct labor. In the photo at the upper right is shown robotic assembly of the magnetic subassembly and terminal block subassembly. In the photo at bottom right is robotic assembly of the movable contacts and actuator to the terminal block assembly. Shown in the large photo at left is final assembly riveting of the terminal block and magnetic subassembly. Behind the feeder bowls, terminals are soldered.*

Sequential stacking *motions are used for this Whirlpool vacuum cleaner motor assembly (the enclosure of which is shown being inserted on the line). The assembly consists of the electronic controls, four screws, a mounting bracket, the seal and then the motor.*

but some of the metal fabrication processes were changed so they were more consistent. The use of feeder bowls, pick and place machines, and automated testing resulted in the labor reduction. Within four months, the line was running at 70 to 80 percent of its rated output.

The Type 134 relay was designed using the knowledge the company had gained by redesigning the Type 91 production line. The cover and base of the 134 relay snaps together instead of using screws, and fly winding equipment, which is new to the company, will be used to wind wire on the relay's bobbin.

The standard winding method which had been in use involved putting the relay's bobbin on an arbor and rotating it at high speed. But since the 134 relay has terminals which are long relative to most relay bobbins, there was the possibility that the centrifugal force of rotating the bobbin would throw the terminals out of place. But with the fly winding equipment, this mishap won't be possible because the bobbin will be stationary and the flyer will rotate the wire around it.

The company also has a proprietary way of maintaining the critical dimension between the magnet core and the pivot point of the armature. A patent is being applied for it.

Designing vacuum cleaners

When designing its new vacuum cleaner line, Whirlpool used several design for productivity methods including the Boothroyd Dewhurst method. One method used was classifying parts by their material properties and designing them to be multifunctional.

An example of this is the wraparound bumper on the canister vacuum cleaner. It serves not only as a bumper, but as a seal for the dust bag cover.

The motor mount, which is made of an injection moldable thermoplastic rubber, also serves as a vibrational dampening mount and as a seal to separate the two vacuum chambers of the machine.

Another productivity method was to classify parts by production tooling processes by incorporating more features into one mold. The number of parts and molds were reduced in several locations of the canister vacuum cleaner.

A new swivel caster design reduced the number of parts from 20 to four. The base of the canister includes the support for the rear wheels and is retained by snap catches. The front handle on the base also serves as a bumper support, and the base includes the cord reel support.

The net benefit of these sophisticated tooling ideas was that the molds are pieces of fixed automation--they reduce the number of assembly operations.

The third productivity method used was to design the product for assembly efficiency. Linear motion is used extensively in the assembly of the vacuum cleaner. For example, the motor was designed to use straight line assembly motions.

Included in the motor assembly is one of the electronic controls, four screws, a mounting bracket, the seal and then the motor. These components are all assembled in a sequential stacking operation where the net benefit results from a simplified and more efficient assembly method.

DFA rules

The basic principles of Design for Assembly relate to two areas: the ease of component handling and assembly of a product or subassembly; and whether the minimum number of parts have been used for a product or subassembly.

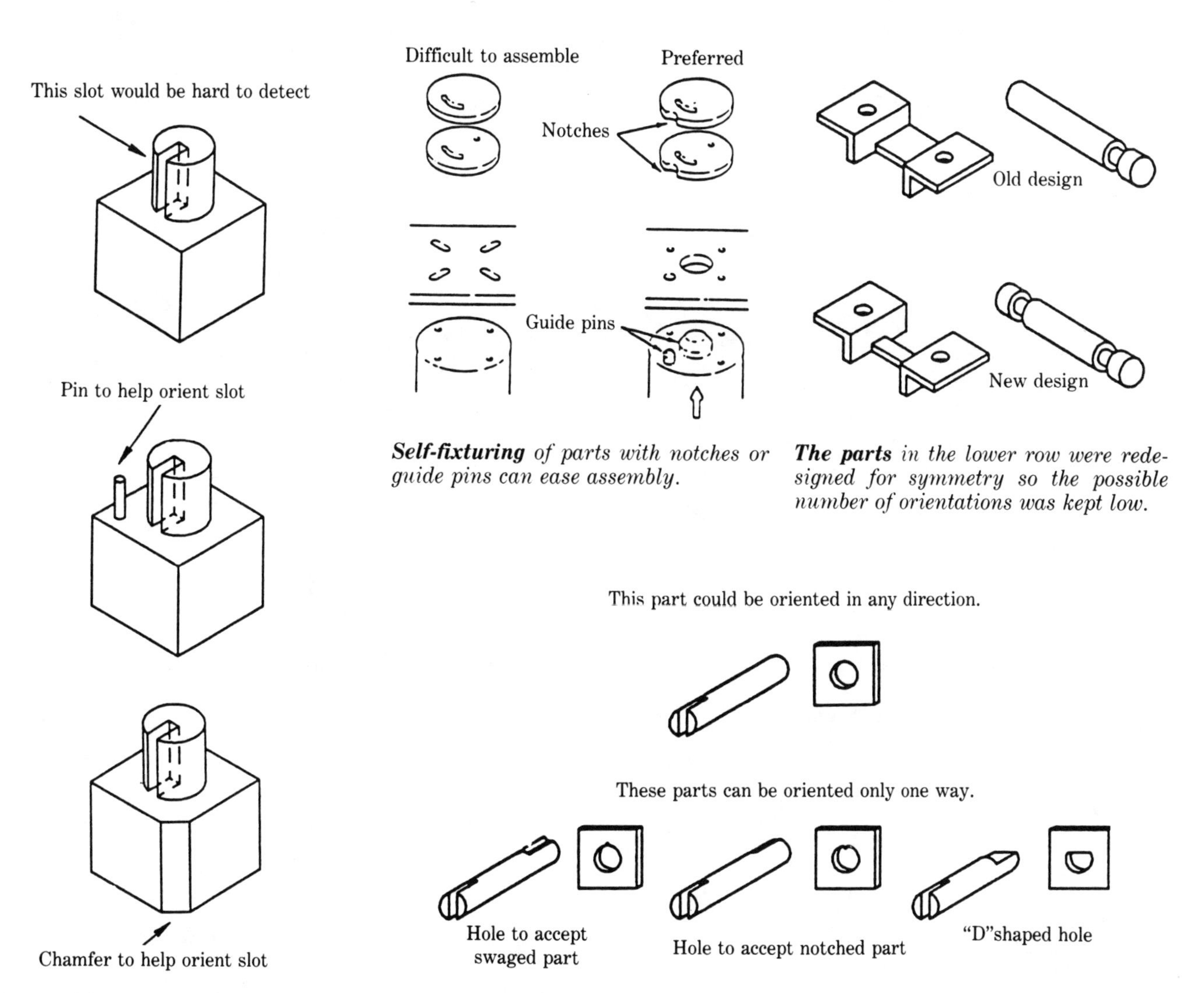

***External features** can be added to orient parts.*

***Self-fixturing** of parts with notches or guide pins can ease assembly.*

***The parts** in the lower row were redesigned for symmetry so the possible number of orientations was kept low.*

***If parts** can't be designed symmetrically, their asymmetry should be increased so they can be oriented only one way.*

The criteria Boothroyd and Dewhurst established for the minimum number of parts test was:

- o Does the part move relative to its mating part during operation or service?
- o Do the parts have to be of different materials?
- o Do the parts have to be disassembled during assembly or service?

The assembly efficiency is then calculated by dividing the theoretical minimum assembly time by the estimated assembly time. The theoretical minimum assembly time is assigned by the Boothroyd Dewhurst method according to the product. The time value is for ideal parts and is standard for all parts.

The estimated assembly time is a theoretical minimum assembly time plus time penalties for problems in handling and inserting assembly parts. If a part is eliminated, the assembly time for that part is zero.

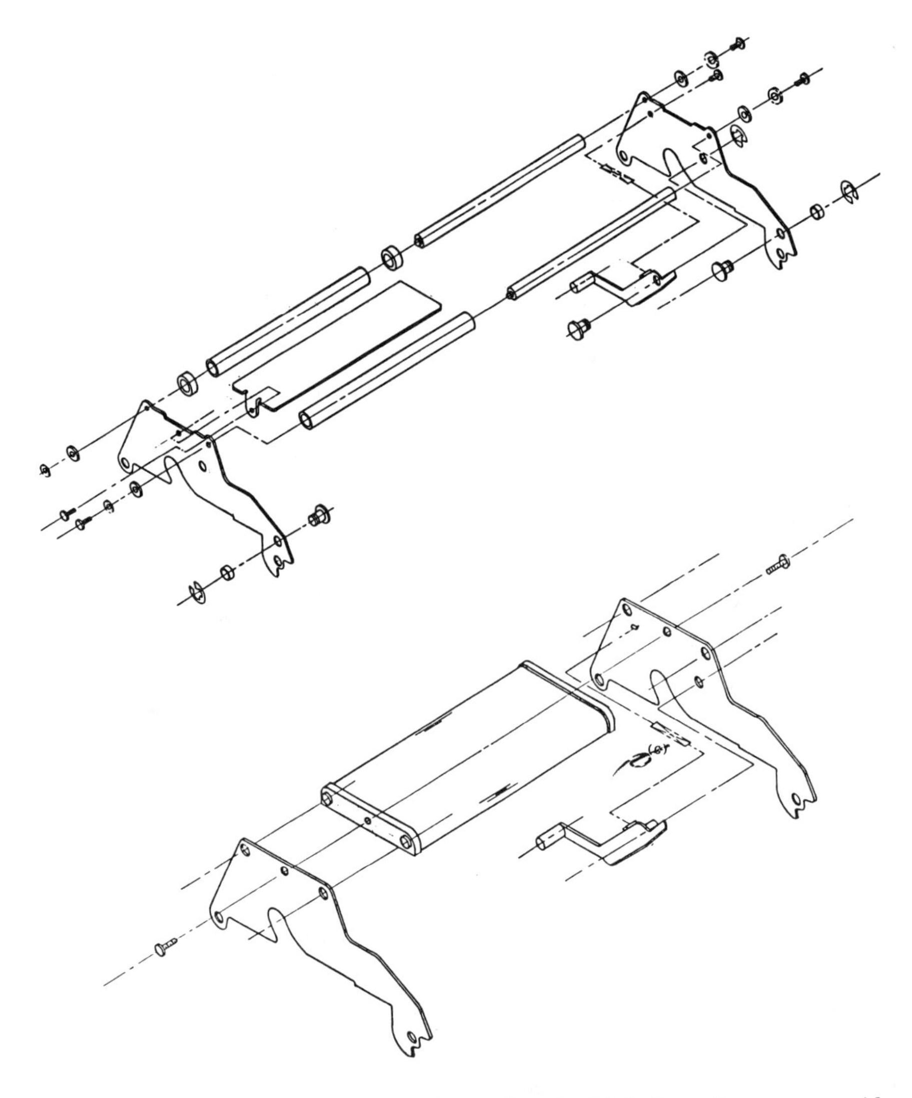

***The redesign** on the bottom was one of three done by NCR Corp. for a paper guide assembly in a teller machine. It reduces the total number of parts for the assembly from the 33 shown on top to seven.*

A general rule of designing for assembly is to avoid fasteners, especially non-standard ones. But when they are necessary, the same type of fastener should be used. If washers are necessary, they should be attached to a bolt or screw ("captured") rather than separate. Also, tapped holes should be avoided and self-threading fasteners substituted.

In manufacturing, "pancake" assembly, in which parts are layered on top of each other along the Z axis, can be used not only to simplify assembly motion, but also to reduce the number of fasteners. To provide compliance or "give" in part assembly, so accuracy errors can be forgiven, chamfered or beveled part edges are helpful.

Simple mechanical processes should be used in manufacturing when possible. Difficult ones such as welding or brazing, soldering, and certain adhesives should be avoided.

Nesting a part in an indentation in the work surface which matches its contour can eliminate clamping the part down during assembly. Cables which must be connected, as on a loudspeaker, should have connectors attached to them so the cable end can be located. Electronic components should not be connected to a circuit board with cables, but rather mounted on a slave circuit board and the slave board inserted into the main circuit board.

Other design for assembly ideas are shown in the accompanying diagrams.

Reprinted from *Mechanical Engineering*, April 1989

Integrating Product And Process Design

In a global quality improvement effort, Perkin-Elmer is striving for designs that require the fewest possible levels of manufacturing. Existing products are being redesigned to eliminate process problems, and new products are planned so that such problems do not have a chance to arise when they reach production.

Gina Goldstein
Associate Editor

Not long ago, analytical instruments such as spectrophotometers, liquid chromatographs, and thermal analyzers could be expected to last as long as 10 years. But today, with new competitors continually entering a market that is itself in constant change as a result of technological progress, the useful lifetime of these instruments rarely exceeds three to five years. In such an environment, a company that can come up with new products quickly and at a comparatively low cost has a clear competitive edge.

Global Improvement Plan

In 1984, Perkin-Elmer Corp., the world's largest supplier of products and services for chemical analysis, instituted a global quality-improvement plan extending across both product and process lines at its Connecticut Operations Sector in Norwalk. Inventory turnovers were to be increased by reducing setups, lot sizes, and parts, and by improving scheduling policies and communications with vendors; manufacturing operations were to be consolidated to improve work flow and in-house lead times. Internal and external product failures were to be reduced by a factor of 10 over the course of five years. In product development, the period from completion of the first prototype to market was to be reduced to 6 months for new models or accessories and 12 months for new products. Teams of representatives from manufacturing, marketing, quality assurance, service, and engineering would work together to achieve this goal; design for assembly and design for manufacturability techniques would be used to create more integrated products, each of which was to contain 40 percent fewer parts than its predecessor. Finally, CAD and manufacturing systems would be integrated and design rules standardized so that products could be designed at one site and built at another or manufactured at several sites concurrently.

Consolidated Manufacturing. Since electronics are used in all of Perkin-Elmer's instruments, a pilot project in just-in-time manufacturing was established for the assembly of printed circuits. As Joseph F. Malandrakis, the company's vice president of operations, recalls: "When we first looked at the master plan, we had two totally different manufacturing operations—an instrument group and a semiconductor group. The manufacturing manager of the semiconductor side and the instrument manager sat down and agreed that it really didn't make sense for us to have a facility here and a duplicate facility a mile and a half up the road. It's a common process; let's share the technology.' " Accordingly, the entire printed circuit board operation for instrument and semiconductor products was consolidated with the instrument manufacturing plant.

The specific goal in printed circuit assembly was to improve inventory turns by reducing setups, increasing productivity, and standardizing parts. New automatic insertion equipment has reduced setup times by 84 percent, allowing the department to reduce lot sizes and coordinate them with the number of instruments actually being built on the other side of the floor.

"Before we had this equipment," Malandrakis says, "it probably took 30 minutes to an hour to set up for a job. Now it takes 10 minutes. With the old equipment, we might have built 100 boards or so, and 80 of them would have gone into stock to be used up over a few weeks, or even months. Now we're building boards in lot sizes of about 20.

"We did an analysis to determine which parts are used on 80 percent of the boards," Malandrakis continues, "and they're loaded on that equipment full time. The other 20 percent are on or off depending on the setup of the job. But our goal is to load up just once and never change it for any product." Since large buffer, work-in-process, and raw material inventories tend to obscure problems, these reductions have greatly improved efficiency.

The new equipment also enabled an entire initial inspection operation to be eliminated, resulting in significantly fewer failures from handling. In just three years, Perkin-Elmer has seen a 100 percent improvement in reducing circuit board failures the first time through the test equipment. The goal this year is to realize another 100 percent improvement.

With the initial inspection operation no longer necessary, the next step is to expedite assembly even further by means of vendor cooperation. Malandrakis explains: "Through statistical process controls, I want to be able to go straight from my vendors, right on to the equipment. And that's probably from six months to a year down the road." Previously nonstandardized parts like the power supply are being standardized, and a new vendor quality rating system is helping the company to evaluate its suppliers as to their quality and delivery. Other goals are to greatly reduce the number of vendors from the current level of over 1000 and to work with them more closely.

All together, these improvements have greatly increased yields. "In just two years," Malandrakis says, "the lead time for printed circuit boards has been reduced by 60 percent."

When it comes to manufacturing the instruments themselves, many of the same goals and strategies were implemented. All the instrument division's products are built in a single assembly area. "It's hard to envision," Malandrakis observes, "but last year we had 40,000 more square feet devoted to building instrumentation. What we have now is all the process groups circling the final assembly and test area, feeding it parts. We have it set up so the instruments flow from assembly to the test cells to shipping. You don't need cluttered workstations, which actually tend to accumulate inventory and cause inventory and technical problems. Now it's all out in the open."

As the various instruments queue up, workers in assembly and test pace the line themselves. "They can see what's coming in," Malandrakis says, "and they have the flexibility to move to another instrument or accessory as needed." Another benefit of consolidation is that communications have improved and led to greater efficiency. "Since the test people are so close to the assembly people, they can walk over and say, 'Hey look; you forgot a lockwasher here.' And the assembly person can fix it and check the rest." Moreover, if rework is required, only five or six instruments tend to be involved because the setups are kept down and only a few units are on the line.

Previously, it would take two or three days to assemble one of the company's typical IR products, the FT-IR-1600. Today, the FT-IR-1600 can be assembled and tested in roughly eight hours. One reason for this is self-diagnostics. "We do some manual diagnostics," Malandrakis explains, "but there are computer programs that allow these units to test and diagnose themselves. And the software was written by a manufacturing engineering software group as well as development engineers." This high level of automation, coupled with a highly manufacturable design, facilitates training. One technician in each of four test cells can test four instruments at the same time.

Another change is in part numbers. Six years ago, the IR spectrophotometer would have had four times more components than it does today. As a result of this drastic reduction, according to Malandrakis, the instrument "is a pleasure to build, test, and ship out. Now we're finding out that in the vast majority of our installations, the customer has this product up and running even before the service engineer gets there. And that's setting the standard for new products."

Teamwork

In its new products and models, Perkin-Elmer is striving to design in-

Manufacturing line for Perkin-Elmer's IR spectrophotometers. A machine can now be assembled and tested in eight hours instead of three days.

struments not only with fewer parts, but also with fewer configurations. To meet these goals, product development teams composed of representatives from quality control, engineering, manufacturing, purchasing, marketing, sales, and service participate from the earliest stages. The FT-IR-1600 was the first instrument built with the help of one of these product development teams under a new set of development guidelines.

The team concept has also been applied to quality improvement. "Problem-solving teams formed directly from the labor base are identifying problems in their own areas and fixing them. These are the types of problems I would never have enough time to look into," says Malandrakis, "but they know where the problems are because they work there every day and they can fix them just like that."

So far, eight teams have been formed to devise comprehensive strategies for improving products and processes, such as the infrared and atomic absorption spectrophotometers, and processes, such as PC board assembly. Another process area that has participated in the improvement plan is the sheet-metal operation, where the use of a computer-aided process planning system (CAPP) is now standard in determining fabrication requirements. Brigham Young University's group technology software program is used to calculate material uses, operation sequences, finish requirements, and standard labor hours. This information is then used to automatically create and route parts, create bills of material, and create and route NC tapes.

Consistent Plans

The manufacturing engineer, prompted by the program, supplies data directly from the part drawing to establish a part data base. The process plan generated includes all the paperwork required to introduce the part order into the MRP system for subsequent fabrication. At the moment these forms are processed through the keypunch group for implementation into the manufacturing system, but eventually that step will be eliminated and the CAPP data will be electronically transferred directly from the group technology software to the MRP system.

"Before the CAPP system was brought in, if you asked three different manufacturing engineers to provide a process plan for a particular part, chances are you'd be given three different process plans," Jack Herman, manager of the fabrication manufacturing engineering group, recalls. Now, the decision-making expertise captured within the group technology software enables engineers to create a single, consistent process plan. The two primary benefits of the system are, first, that the paperwork moves through the system much quicker, and second, that consistency in the way parts standards are set is ensured. "If we find we're not quite right," says Herman, "we have a common data base and can go back and fix it.

"When we did an evaluation of this system," Herman continues, "we knew there was going to be a cost saving, but it turned out to be really fantastic." A typical sheet-metal part that in the past took an hour to complete can now be processed in 10 minutes. And a weldment, which used to take four hours, today requires only 40 minutes. The CAPP system has also proven cost-effective and dependable in identifying the labor and material constituents of total part costs. The data base of labor content and rates permits fabrication costs to be estimated at an early stage in the design process. Finally, the drastic reduction in the labor involved in generating part documentation has given engineers the opportunity to spend more time on the shop floor. "Before, 30 percent of the engineer's time was devoted to clerical work," Herman says. "Now that time can be devoted to thinking up new ways of improving processes on the floor."

Perkin-Elmer is now developing another CAPP system for its machine shop parts fabrication plans. The mill/turn machining centers were chosen as the first module for developing CAPP in this area. The effort will be carried into the various milling machining areas until all the required CAPP modules have been completed within the machine shop operations. The tooling classification for perishable cutting tools will initially be used to identify optimum machine, feed/speed, and type of cutting tool based on part configuration and material. This will pave the way for standardization that will minimize redundant tooling and inventory investments.

In a related program, the fabrication manufacturing engineering department at Perkin-Elmer has applied paperless factory software to the procedure for obtaining documentation for work order releases. Enhancements to the existing part routing program now allow a combined route/op sheet to be created directly on the MRP system's CPU; printouts of part lists are then obtained directly from the MRP system.

Before the paperless enhancements were added, Herman says, five days would pass before all the paperwork was in place on a work order. The process entailed many time-consuming steps: obtaining a route sheet printout from the MRP system, determining if an op sheet was required, pulling a copy of the op sheet from a file, making a list of all the required drawings, making copies of op sheets, and refiling them. The on-line routing and operation data base has eliminated keypunch operations; hard-copy op sheets; the need to obtain, reproduce, and refile op sheets; and the need to review documentation and manually prepare a list of required drawings. As a result, the process now takes three days.

Role of CAD/CAM

No improvement plan would be complete without CAD/CAM. Perkin-Elmer is now running an Intergraph CAD/CAM program on two converted DEC VAX machines for tool design and for NC programming of sheet-metal parts. Drafting boards have been completely eliminated in tool design, except when old fixtures have to be updated. As a result, blueprints can now be sent to the tool room for fabrication in about a quarter of the time it took in the past. The library of components—tooling plates, pins, and clamps—has also helped in the effort to standardize parts.

The CAD/CAM system has had a similar effect on the NC programming. In addition to eliminating errors of calculation, the system has helped to minimize shop setups. The library of standard tooling for sheet-metal parts in the system data base has been reduced from 600 to 16. Eventually, it should be possible to make all sheet-metal designs with those 16 tools. A verification routine enables NC programs to be tested on screen so machines on the shop floor do not have be tied up.

Thus, quality improvement is a principle that is being enforced throughout the company, from design engineering to the shop floor. As Joe Malandrakis says, "That's the way you're going to succeed in this type of marketplace. You can't keep doing the same old things over and over; you've got to continually improve your entire process." ■

A Visit to Motorola's Bandit Plant

How They Brought Home the Prize

The results are in and Motorola has won the first Malcom Baldrige prize for quality. How they did it will be analyzed and dissected for months to come in every magazine you can find. Here is our little glimpse into the thinking, planning and realization that helped them do it as epitomized by their Bandit pager facility in Florida.

Tom Inglesby
Editor

Florida has gotten a bad reputation lately. *Miami Vice* and other entertainment have skewed the public's perception of this sunbelt state. Ever since the movie *Smokey and the Bandit,* a certain lawlessness and disregard for convention seem to have attached themselves to the Florida image. Perhaps that had something to do with the choice of names for a Motorola automation project, one that in many ways threw convention to the winds and developed new laws of integration. The project and the end product both bear the name: Bandit.

But Bandit, the product, and Bandit, the plant, aren't as foolhardy as Burt Reynolds' Bandit in the movie. While Burt burned up the highways, Motorola's Bandit is burning up the competition—all overseas built—in a market that is growing quickly throughout the world. They are doing it without "smoke" and mirrors; they're using US know-how, managerial techniques that bring people fully into the project, and a design for manufacturability scheme that has tightened the product while improving its assembly. We aren't privy to the deliberations that went into the selection of Motorola for the Malcom Baldrige Quality Award, but having toured the Bandit works, we think this project may have had something to do with their winning.

So fasten your seatbelt, put on your Stetson and let's take a quick trip to Gatorland where Motorola is making—out like—a Bandit.

Sun, sand and success

Northerners coming into the Ft. Lauderdale airport in winter are usually surprised at how fast they feel hot and sticky. The humidity in December can be a shock when combined with 70 plus degrees of temperature.

The area boasts a high quality of life rating due partially to its warm weather and maximum sunshine

All photos courtesy of Hewlett-Packard.

Special jigs were designed to allow orientation for machine vision inspection.

and partially to the relaxed, somewhat laid-back style of the Deep South. Getting people to move here isn't usually a problem, getting the right people can be. In the case of Motorola's Pager Division in Boynton Beach, the bandits first struck by "highjacking" quality people from throughout the company.

The Motorola bandits "stole" more than just the best and the brightest from the organization, they stole ideas. As Russ Strobel, Bandit's engineering manager, recalls, they were after something more than warm bodies.

"We wanted to eliminate a major cause of manufacturing headaches, the *not invented here* syndrome. NIH would kill us, given the chance. In the past, we rewarded the designer who came up with the newest, the greatest, the latest thing. We should have been looking for the designer who took something that basically worked and optimized it.

"So we wanted to foster the idea of building on past successes and previous designs," he continues. "We have a sign outside the lab that says it all: *Please don't leave the area without leaving us a good idea.* We need ideas from anywhere we can get them to prevent reinventing the product. That's why we called it Bandit."

Strobel and Director of Manufacturing Operations Scott Shamlin might admit they also "stole" a few people they wanted, although they won't use the term. "We had to staff the program," Strobel agrees, "but we didn't steal anybody. Motorola has a program called IOS—internal opportunity system—where you post job openings for anyone in the company to apply. We didn't go up to managers and tell them we were taking someone. We didn't have to, the project was so popular that we got applications from all over. It took us about three months to staff it with 20 people."

And while the staffing was going on, the work was, too. Strobel, Shamlin, "a tooling engineer and a couple of Double-Es" were evaluating pagers and manufacturing approaches. Three attacks were considered: make the new receiver (pager) capable of automated assembly by using surface-mount technology (SMT); adapt an existing receiver that was already close to SMT; or develop something new in both product and process.

The Bandit facility, located within a larger manufacturing plant, is as spotless as a cleanroom. The common color scheme of white and blue is carried out in uniforms as well as equipment. Not visible in the picture but highly visible in person is the pride of the workers at all levels of "Team Bandit."

"We looked at it and looked at it," remembers Strobel. "When we put it down on paper—in a matrix of cost, performance, space and risk—it was clear that the only way to meet our schedule was to adapt something that was pretty close to SMT already. We just didn't have the time to develop all the SMT components we needed for the other approaches."

The receiver they decided to adapt is called a *five-spot* design. It has both oscillator and preselector in one package—there are four spots on the crystal blank for a four-pole preselector filter and one spot for the oscillator, therefore the name. The design was more efficient in part count, eliminating about 15 components from the standard unit, and all the required integrated circuits (ICs) and support chips were already available. "We hope to fool around with those components some in the next generation," Strobel hints.

As it is, the new design, a hybrid if you will, is so compact that the battery—an AA cell—is the largest component. "We could make more progress in size reduction if we could get a better, smaller battery," comments Strobel, "but a lot of our customers cry that they don't want the pager any smaller."

Because of the tremendous number of variations of the Bandit—millions of combinations are possible—each unit is essentially a one-of-a-kind, lot-size-of-one manufacturing marvel. By the way, although we use the term "Bandit" throughout to identify the pager made in this facility, it is actually a model within the Bravo series. Other Bravo units with slightly different component counts and assembly requirements are made in the same building, but without the advantages of the automation used with Bandit.

Stealing space

The Boynton Beach facility is a large, practically windowless building. The Bandit team needed space, and management found some—in a storage area. "They had a bunch of stuff—scaffolds, stored things—in this area and the space was being used but not utilized," recalls Strobel. "We had a dinky lab back in the bowels of the factory, next to the sprinkler system and fire alarms. We were in pretty close quarters at first."

Buildings don't build product, people and equipment do. The approach the Bandit bunch took was to codevelop the product and the processes, back and forth, back and forth. Again, Strobel reminisces.

"We knew we were going to have to populate the bottom of the board with about seven components. That meant we'd need three robots. We knew how fast the machines were and that dictated how many we'd need to maintain the production schedule. We also knew the dimensions of the equipment so an industrial engineer used AutoCAD on a Compaq computer to chunk out little designs of how the process would go and how it would look. As we decided on things, the physical layout, he'd redraw the facility.

"We got lucky, I guess. We knew enough about what we wanted that we got pretty close to what we actually ended up with. It would have been nice to use simulation, to be able to make decisions based on running 'what-if' programs, but we didn't simulate—we just went out in the shop and measured it."

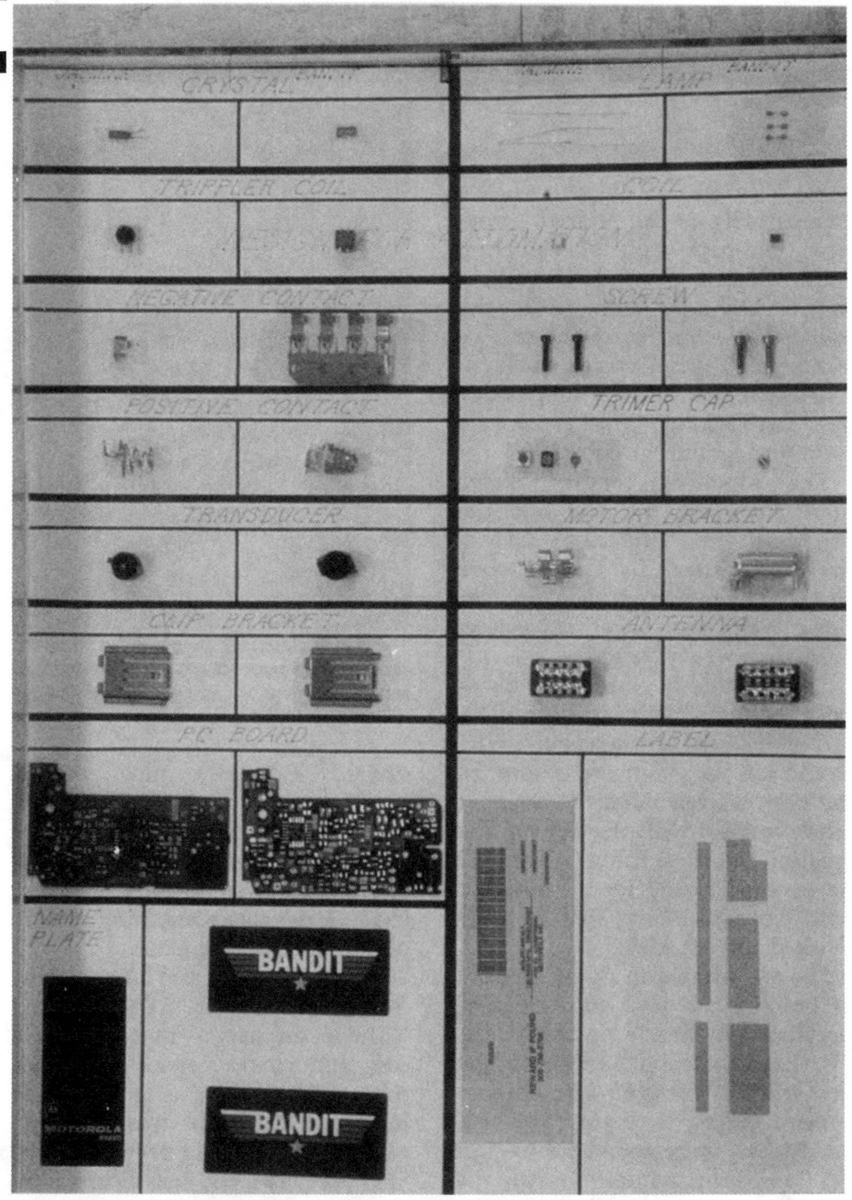

Design for manufacturability played an important part. Without such exercises in simplification—as represented by the change in parts from previous pagers (left) and the Bandit—and an eye on how things can be assembled easier, automation would be wasted.

Follow the road map

Strobel explained that taking measurements was easy because they have a detailed "road map" of the Bandit factory-in-a-factory posted on the wall outside the actual manufacturing facility. This is up for several reasons, not the least as a motivator to the workers. Bandit is only part of the manufacturing that goes on at this plant, and it has its own climate conditioned room separate from the other production. People not involved in Bandit come by, look through the windows and can track the operations on the "map." It also makes it easier for guides to explain the advanced systems to visitors, especially customers. Good "PR" internally and externally.

What they found was that a customer's order, placed through the Chicago office and stored in an IBM 3090 computer, is downloaded to the division's IBM 4381 where a bill of materials and shop order are generated. Everything is based on the customer input—color of housing, label, with or without a vibrating signal, frequency and other variations—and the options are translated into sequences for the eventual machine operations.

The order goes to a Stratus computer where serial numbers are scheduled and matched against the customer's order. Then to a Model 825 Hewlett-Packard RISC (reduced instruction set computing) computer that supervises five HP1000 computers used as controllers for the Bandit factory. It's here that Strobel confesses to not getting exactly what the team wanted.

"Our computer guys wanted the latest Unix box on the floor, not an older architecture like RTE (Real Time Executive). But we couldn't find a Unix system that would respond to an interrupt. Again, we went to minimizing the risk factor and to building on what we had seen work.

"We prepared a variety of benchmarks—three months from project start we'd have this, six months into it we'd have something more—and then a six month period to make everything work together. It was a phased approach," he explains.

"The very first thing we did," Strobel continues, "was to hook up a couple of robots to the computer to demonstrate they'd respond to commands. The HP1000 had to control Seiko robots in this prototyping exercise. Then we tried it with the HP1000 and a Panacert machine, then the computer and other equipment. Once we got one cell working, we'd move on to the next."

There were cells that had been "prototyped" in actual production with previous pager models. There were also new ideas and new production that the team had had no exposure to before. According to

Strobel, some areas were the toughest. "The housing assembly, putting the chassis into the housing, putting on the back cover and driving the screws, putting on the belt clip, all the mechanical assembly was pretty new to us from the standpoint of automation.

"We hadn't ever used a computer to send instructions to a robot before. That was a real reach-out in our opinion—a high-risk endeavor—but something we thought was critical to the success of a factory building in lot sizes of one."

Throughout the development of the processes, the battle cry was "Make sure you can get English instructions here and machine actions there." The major emphasis was to drive customer orders into robotic actions. But even with this practical approach, the issue of quality was never far from the surface.

"Hewlett-Packard kept trying to focus on SPC (statistical process control) systems, and coming from an engineering background, I was sympathetic," Strobel confides. "Scott (Shamlin) had been disappointed with a computer project and said, 'Don't mess with all that stuff, we can get it later.' Well, we've got it all running now.

"In retrospect, it was probably a good idea that we didn't spend a lot of time trying to figure out how to collect SPC data. Even though, in our black little hearts as computer designers, we knew we were going to have to have it—and even put in a line controller with the knowledge we'd use it to run the SPC software at the line level."

Planning for the future, while installing what was necessary for the present, is a hallmark of the Bandit facility. They found that HP's 1000 series cell control computers, the A400 models, needed more "horsepower." Instead of scrapping the controllers and starting over, they considered changing the configuration to make six cells out of five and add another computer. Strobel says, "It was easy because our software backplane allows us to add and delete cells, move functionality around, talk from computer to computer without paying any attention to which process we're trying to get to. The backplane routes it where it needs to go."

That backplane was codeveloped by Hewlett-Packard's Atlanta Project Center and allows integration of various software developments on multiple vendors' computers, transparently to the user. Using computer aided software engineering (CASE) tools, Motorola can cut its software development from years to months.

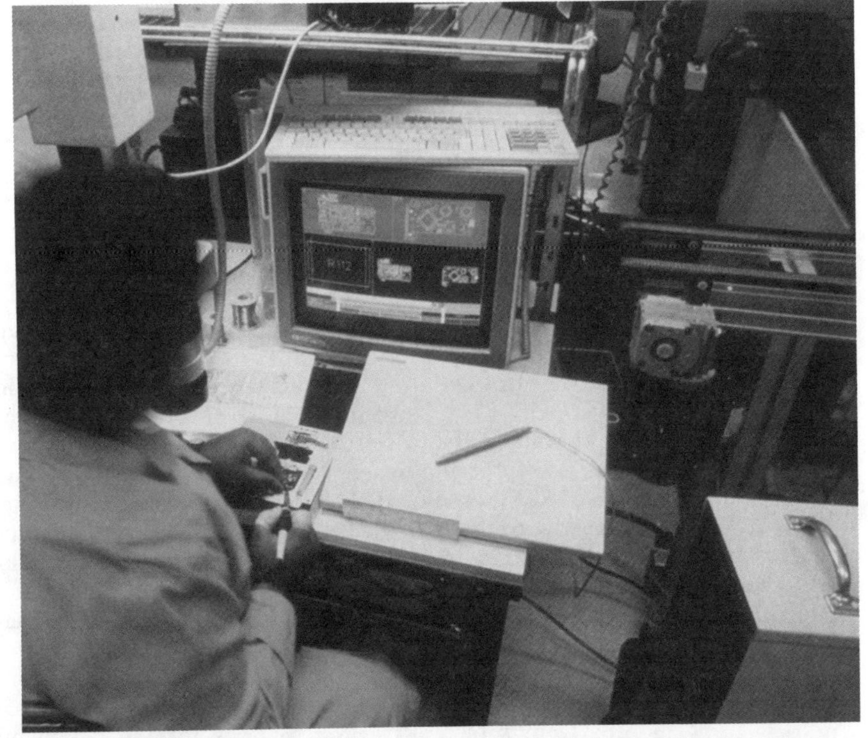

Assembly diagrams and troubleshooting information can be displayed at the workstation to eliminate paper documents.

Through a window brightly

The Bandit facility is a glass-enclosed structure within the larger pager factory. This means there is no place to hide mistakes. And to make sure, bright red and yellow lights along the production line warn management and technicians alike when something goes wrong.

At the head end of the line, slightly off the actual shop floor, a monitor and control computer system operates under the careful watch of an engineer. Equipped with HP touchscreen terminals, the operator can move through a network of menus with the tip of a finger to access immediate capacity, status, operation, maintenance and other information on any or all of the machines.

Charts for SPC and simulations of where product is in the build cycle can be displayed on color monitors at this and other locations around the line. Workers use the workstation displays for assembly diagrams, troubleshooting information, test and conformance data as well as serial number tracking.

The first three cells, where the boards are stuffed with components, are replicated in Puerto Rico—but the work there is done manually. According to Shamlin, "We run these cells in parallel with our offshore manufacturing. They were the first we set up—starting at the front and working around the line seemed logical. To exercise the new line, we produced some boards for use in the Japanese market, in the standard Bravo pager. It allowed us to tweak the machinery while we were developing and installing the rest of the system."

But there were some problems sharing production facilities with actual products while developing systems for a new line. Shamlin recalls, "In the daytime, Bandit engineers worked on the line, hooking up hoses, plumbing and wiring. At night, the 280 team came in to run production (280 refers to the 280 MHz frequency band of the Japanese pagers). It was somewhat aggravating to come back the next morning and find all of the switches in different positions, reels loaded with different components. But we got some miles on the machinery to prove it out, to get some confidence that we were doing the right things."

The differences between the Bravo—a two-board receiver that is at least partially assembled manually—and the Bandit are mostly skin deep. In designing the Bandit for automated assembly, many components were changed in size or shape, SMT systems were applied and a double-sided single board was used.

Strobel explains, "Bandit is actually ultraconservative in design because we knew we had to achieve a Six-Sigma* process at the end. That's why we went to components on both sides of the board. When I joined the project, the design had all the components crammed onto one side—two mil lines, two mil spaces, custom packages that didn't even exist—and the first thing I told Shamlin was that it wasn't going to work. You can't get Six Sigma out of this state-of-the-art spacing.

"So we ended up with some parts on the bottom of the board and had to add a whole cell to populate the underside. It was a conservative design strategy because we took things that we knew would work."

Shamlin adds, "We tried to minimize any invention so we could save our resources to concentrate on things we knew we had to develop. We had to create processes where robots would insert components, solder leads, add switches, things that hadn't been done by robots. We even have a robot that tunes the receiver—that's never been done before."

"That's all part of the idea of overcoming the NIH problem," chimes in Strobel. "We concentrated on using good ideas wherever we could find them so we didn't have to reinvent things." Indeed, they go so far as displaying a sign with NIH surrounded by a red circle with the familiar red slash through the center.

Making Bandits pay

Automation is nice; it can make things no one even wants faster than ever before. But that's only part of the Motorola Bandit story. Using computers to drive cells, control machines and produce unique products one at a time is interesting, but becoming a familiar story. Let's dig a little deeper into the lessons learned from Bandit. Let's find out what the payback has been.

"I tell people we learned four things from Bandit," Shamlin starts. "First, create interdisciplinary teams; then reward risk taking; third is to enforce outrageous goals; and fourth, which might sound strange, is to avoid Japanese management techniques. Oh, yes, there is a fifth: we learned it can be done."

Expanding on some of those points, Shamlin continues, "The part about avoiding Japanese management techniques is important because we are two decidely different groups of folks. Team building here is different than in Japan. In Japan, they are constantly reinforcing the team as a single entity; in the US, the team is composed of individuals and their roles are emphasized. Both are teams, but the thought of using Japanese management techniques with a team of US individuals doesn't work very well.

"I think we haven't spent enough time and effort considering the strengths we have as a nation, as a

**Six Sigma is defined by Shamlin as a rate of 3.4 defects per million parts. This is an extremely high level of quality. He uses this analogy: The odds of a person getting on an airplane and reaching his or her destination alive is about 6.2 Sigma; the odds of the same person and his or her bags arriving at the same place at the same time is about 4.1 Sigma.*

At the time of our visit at the end of 1988, Bandit was approaching Five Sigma, "with fairly traditional technology and approaches," according to Shamlin. He adds, "Five Sigma can be reached with hard work and smart decisions; you can't work hard enough to get to Six Sigma. You have to work at the process level to predict the next defect and then head it off with some preemptive action. That's the way you get to Six Sigma. You have to have a smarter toolbox and that's what we are developing now.

"You have to have the experience to practically smell a defect condition starting. We have to be able to inject a machine controlled or human controlled modification to prevent the condition from developing while the process is still within an acceptable range. We aren't there yet but we're further down that road than when we started. It takes a lot of compute power to predict rather than detect and we're doing some neat things in this area."

Machine vision systems attached to Seiko robot manipulators allow the determination of pass/fail for many component placements.

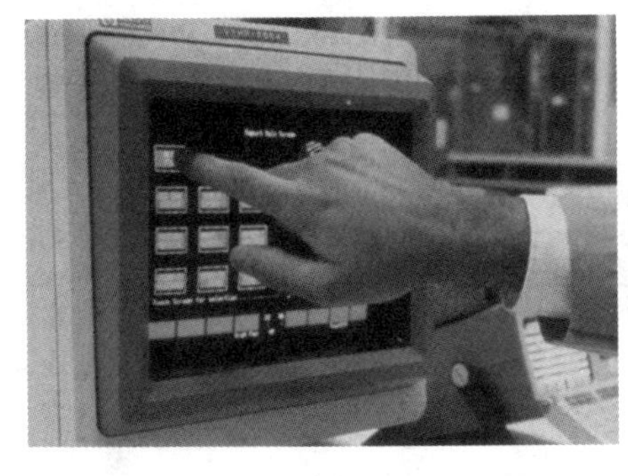

The control and monitoring facility for the Bandit line is located within the environmentally conditioned room. Controls feature Hewlett-Packard touch screen terminals to allow the operator to access information with the tip of a finger.

culture. In fact, maybe too much time is spent downgrading some cultural aspects of our society and not enough on how we can use them as competitive weapons."

Strobel adds to the discussion on team building by saying, "In the beginning, we didn't know what we couldn't do.

"What we tried was clearly a generation ahead of anybody else—to drive the factory by a bill of materials and produce in lots of one. These things were talked about, but no one had implemented them.

"We were always up-front with our supplier team. We told them what we were up to and what we needed—or thought we needed—to push the envelope a little further. We admitted we didn't know exactly how to do it or whether anyone else had ever done it before. We asked for some unusual accommodations from them and told them that when it was successful, they would share in the success," Strobel goes on.

"We had always heard that it took a young, entrepreneurial company to react quickly, to move back and forth easily. What we found was we got as much entrepreneurial spirit and flexibility from large, well-established companies as from the smaller ones. It was clearly a matter of picking the right companies."

Strobel then relates a story, which sounds vaguely familiar to every manager implementing a new system. "There was one company that, up-front, had us spending more time with their attorneys than with their engineers. That was a signal to us that we had picked the wrong people to work with."

Shamlin picks up on the story, "We were sharing information with our suppliers that was proprietary and they in turn were doing the same with us. We were moving so fast, we were forced into it. We began to lose vendors who weren't willing to participate at our speed, by our rules."

Speed trap

Shamlin, who worked for many years in the NASA space program, puts it in that perspective. "When we were trying to put vehicles on the moon, the old guys would say, 'If you don't like our schedules, talk to God. He's the one who put the planets in their rotation. We have no choice.'

"With Bandit, too, the schedule was our religion. Any supplier candidate who couldn't operate in that environment dropped out."

Meeting schedules is perhaps the hardest part of managing a manufacturing—or any other—operation. At Motorola, the schedule took on a life of its own. Strobel: "We weren't going to miss any of those major prototype dates. When those cells were to be on-line, they would be on-line. When the product had to be in a certain position, it would be. The equipment had to be, the computers had to be, everything had to be. The date was the date and we never missed one."

"Well, we did miss one or two," Shamlin interjects. "But we made them up within a day or two and by the next date we were on schedule."

Throughout the project, quality of the product and the process was important. With a strong team of suppliers on the outside, a strong team of engineers and designers on the inside, Bandit moved along at scheduled speed. At no time was Shamlin or Strobel willing to compromise either quality or schedule.

"Think about it realistically. There is no such thing as slipping a little," Shamlin argues. "There's a mentality that is absolute poison. It says a little bit more won't hurt. Of course it hurts.

"The way you attack a problem really makes a difference," he goes on, getting spirited. "When you're willing to accept less than what you said you were going to accept, that's a matter of integrity. If you build a team where you authorize a disruption in what you expected, that's a much bigger issue than just letting the schedule slip for a day or two. You just won't be able to keep the competition at bay with that type of attitude."

Dealing with suppliers and making them stick to your schedules can be frustrating. In one case, Shamlin got a taste of his own medicine from one of his suppliers. Nancy Ewing of Hewlett-Packard: "Shamlin and Strobel were always saying they wanted partners to provide expertise in various areas. So we approached the project as partners. That meant a two-way street."

"I hate to admit it," Shamlin admits, hating it all the way, "but Nancy would come in and give me performance reviews. She would actually chew me out because we'd make decisions contrary to the stated plan. She was right and we modified our behavior."

Moving on and moving up

Bandit is fact. The Bandit facility is turning out pagers at the planned rate, with the quality expected at this stage of the project. More effort is around the corner, as the program heads for that goal of Six Sigma.

The Bandit team has become something of a cause celebre within Motorola. And that *wasn't* one of the goals. As Shamlin puts it, "We didn't want to build a group of superstars who wanted to get all the hot projects. We built the team to re-establish the level of expectation, put it together to develop a strategy and then dispersed them like seeds that will hopefully have a multiplying effect.

"We're doing that now," he concludes. "Some guys will be involved in the next generation programs, some in propagating the technology we have now, some will continue operating Bandit. We hope that by the end of 1989 we'll be able to point at some who have broken off on their own. Breaking up a team has both positive and negative connotations. Hopefully, this time it's all positive."

Strobel wraps it up by saying, "If we had had 100 percent success at every stage, it would mean we didn't define the project the way our charter was written—to go where no man has gone before. If you go where you think no one has been before and you find beer cans, then you didn't exactly set your criteria properly." (MS)

Presented at CASA/SME AUTOFACT '89, October 1989

A Case Study of Simultaneous Engineering

By Claus Madsen

ABB Robotics, Incorporated

Presentation Outline:

- Simultaneous engineering.
- Fixed budget approach.
- Selection of partner.
- Chronology of a simultaneous engineering project.
- Simultaneous engineering team.
- What is needed to create a successful "simultaneous engineering environment.
- A case study.
- Improvements for future projects.

Simultaneous Engineering

The objective behind simultaneous engineering is to optimize the value received for the invested capital. Value here means better quality, more automation, more flexibility, and less money spent to cover the unknown. The result will be a world-class system using proven technology with low piece cost, high reliability, and best possible product quality.

A simultaneous engineering project works with budgeted funds that are disclosed when the vendor is selected and when the system is designed within budget. The purchase order is awarded with no competitive bidding.

Fixed Budget Approach

The fixed budget approach is one of many ways to approach simultaneous engineering. The main idea is to work backwards from a market price, ex. - a car manufacturer has to decide whether he can build a certain component price competitively. The market price is known, now the manufacturer has to figure out his cost of producing the component. He makes a budget based on his experience, selects a vendor, and basically asks him to supply a piece of machinery for that cost. The vendor knows that he will get the job if he can create a feasible solution within the budget.

The customer establishes a team consisting of people with the needed information. The vendor does the same. The two teams now work together to develop a feasible solution. All information needed will be revealed during this process, and the two teams normally develop a team spirit that will carry on into the actual project. This allows for maximum utilization of information.

Selecting a Partner

When selecting a partner, the customer creates a team of representatives from the customers organization to conduct the selection. If a partner is to be selected for an engine program, the team most likely would look like this:

- Purchasing Agent
- Corporate Manager of Manufacturing Engineering
- Plant Manufacturing Engineer
- Assistant to General Manager, Powertrain Division
- Chief Tool Engineer
- Various Staff Members Reporting to the Above

Once the team is established, the selection process can start. The criteria for selecting the partner is obviously different from customer to customer and from job to job, but will basically look like this:

1. Must be a design and build house.
2. Best in the business, world-wide, for the specific application based on practical experience and creativity. But, avoid yesterday's technology.
3. Vendor should be able to take known solutions for each of the individual processes involved and create the most cost-effective total solution.
4. Prime vendor should be the one that controls the quality of the final product, not just based on which type of vendor would have the largest cost basis in the project, eg; conveyor manufacturer.
5. Before selecting a partner, determine which is more important - a cheap solution for today's problem, or a flexible solution that can be adapted to future products and processes.
6. Vendor must be developed into a partner, which means they should have the financial strength to be around not only for the life of the project, but also to support you in the years to come.

7. Vendor should have a history of providing successful systems and properly supporting them during and after the launch.
8. Vendor has sufficient level of sophistication such as CAD and simulation capabilities as well as project management.

The selection phase could easily take two to four weeks, but this is an investment in time that is well taken. The evaluation involves documentation of the above eight points, as well as interview of the partner's management.

Chronology

After the selection of a partner, a simultaneous engineering project can take place. The Trenton project followed these steps:

- Team Established (Vendor and Partner)
- Initial Line Up
- Concepts Formulated
- Budgtary Costs Estimated
- Selection of the Best Concepts
- Detailed Description of the Solution
- Solicitation of Firm Quotes
- Final Selection of System Configuration
- Computerized Simulation of the System
- Final Pricing
- Presentation of the Solution to the Customer by the Teams
- Purchase Order Released

The above listed steps took six weeks from beginning to purchase order release. During each step, there was a consensus between the team members, and every decision was a team decision.

Simultaneous Engineering Team

The team that was built had, obviously, representatives from the customer and the vendor. We normally support the team with the following persons:

- Application Specialist, that has an in depth knowledge regarding the actual application.
- Project Manager, that has experience in running projects.
- Proposal Engineer, responsible for cost analysis and cost estimates.

- Sales Engineer, who has an already-established, good relationship with the customer.

The customers part of the team will be different from job to job. In this case, the team had this configuration:

- Tool Engineering Supervisor
- Tool Engineer
- Plant Engineer
- Industrial Engineer

Team meetings were as frequent as two to three times per week, depending on the need. At these meetings, it's usually not necessary that the entire team attend, but only the team members that need to meet to discuss the agenda.

What is Needed to Create a Successful "Simultaneous Engineering Environment"?

- Appoint team members with authority to make decisions, and with the authority to deal directly with product design.
- Create a reasonable time schedule and overall goal for the simultaneous effort.
- Utilize creative solutions from sub-suppliers and award them.
- Start with a reasonable budget.
- Eliminate every unnecessary cost that does not improve quality or performance of the system.
- Genuine commitment from management to support the team.
- Total openness and full cooperation between the two organizations.

Only when the above-mentioned commitments are present can the team accomplish new and innovative solutions without re-inventing the wheel.

Advantages of Simultaneous Engineering

From the first team meeting until final production rate is met, there are several benefits to both organizations; benefits that are obvious and some side benefits that are less obvious but just as important.

- A very big advantage using simultaneous engineering is the fact that the customer will get the best solution for the available money. There are several reasons behind that fact. First, the partner knows the facts that are normally unknown, therefore, the job includes less risk and pricing does not include a big "safety net". Also, all creative solutions are being evaluated and understood by the team, so solutions that would normally not be evaluated are now investigated and, if feasible, the solution will be utilized in the final system.

- Allowing the time needed to investigate different solutions normally means that the final system incorporates more flexibility to accommodate future changes than the traditional way of designing and buying systems.
- A benefit to the partner is the instant feed-back from the customer when questions are asked and information is needed.
- Another very important benefit is that the customer's team members experience a high degree of satisfaction from having their expertise utilized. They have tremendously valuable information that is normally not used, but that when utilized enhance any solution. Also, because they know what is expected from the organization that is normally running the line.
- If new technology has to be used, simultaneous engineering allows the time to properly test new techniques before integrating the solutions into the system.
- As the system is being developed, the plant takes ownership of the solution at each step and at each level. The feeling of ownership is carried over into the installation and start-up phase, and all the small problems that will always occur during launch are normally solved in a friendly manner since the team feeling exists on a very personal level. We have had incidents where we gave the customer services and parts free to solve a problem during launch, and since it is a fixed budget situation, no money is available for N.O.R.'s. Later, during the same launch, the reverse situation happened and was solved within the team. Since most problems can be solved by the team, upper management involvement at both the vendor and customer is minimized.
- Elimination of N.O.R.'s. Only major changes will be covered, ex. - product changes.

A Case Study

Recently ABB was selected as a simultaneous partner in an engine program. Our task was, together with the customer, to develop a reliable, flexible, and cost-effective cylinder head assembly system. The system data was given up front, timing, budget, volume, product design, floor space, but not a final process. After a kick-off meeting, approximately a month was spent intensively by the teams to develop the process and a corresponding layout.

During that time, some unique solutions were developed which would encompass our cost/function goals. To mention a few unique approaches:

- Feeding Equipment
- Washers
- Valve Loading

As each solution was developed, all information was disclosed by both teams, and based on SPC-values and gained experience, we could foresee problems and design around those.

Finally, as the process was agreed upon, the layout was generated. To determine buffer sizes, a simulation was performed. The result was helpful in finalizing the layout, and in determining bottlenecks.

Improvements for Future Projects

The projects ABB has been involved in could be improved for future projects with regards to:

- A more detailed plan with more milestones and objectives has to be established up front, and management should be involved when the milestones are reached.
- When to start simultaneous engineering is another issue that could be improved upon. We have found that the earlier the start, the better. The amount of involvement will increase as the product design gets more and more firm.
- The customer's ability to estimate the system cost up front. The budget we are faced with is often too low, and that is obvious to the entire team as costs are being established and all cost sheets are available. A low budget does not necessarily give the customer what he is asking for.

Reprinted from *Manufacturing Engineering*, August 1989

Under the pressure of a one-year countdown, Xerox's New Build Operations turned to a team concept to design, manufacture, and assemble a new copy cartridge. Here's the story of

A Team Approach to Success

By John M. Martin
Consulting Editor, Endicott, NY

With one year to go before a rollout, your company's decision makers realize that it's no longer strategic to do business with the supplier that they had selected to manufacture and assemble a certain product. They ignore the fact that the product design team has been working with that vendor for two years on the project, forget the fact that prototypes have already been built.

Instead, your division—a low-volume, high-complexity assembly house that thinks in terms of lot sizes of less than 100 for finished product—will receive the assignment to design, manufacture, and final-assemble a subassembly in lot sizes that may go into the millions.

What do you do? If you're the New Build Operations (NBO) organization within Xerox, you roll up your sleeves, put together a team, meet with design and development to see what they need, and establish a working facility for the product in 12 months flat.

The keys to NBO's success? A cooperative team approach, solid relationships with outside vendors, a strong just-in-time (JIT) orientation, a formal quality program, Leadership Through Quality (LTQ), that is now entering its sixth year, and the active participation of union management and line workers—the latter called "industrial staff" at the company.

Clean, roomy, and well-lit stations add to worker satisfaction.

Xerox's NBO (Webster, NY) is an assembly house for the corporation. Project Manager Alan Makovsky recalls that he first got the build assignment for the company's "customer replaceable unit"—a disposable copy cartridge for the 5018 and 5028 copiers—in January 1987.

"The first thing we did was assemble a small team of three or four people. We added some key industrial staff to the group as well, and then dedicated the first six weeks of the project to understanding the copy cartridge: what the design intent was, what the problems might be, how quality should be controlled, and so forth."

After that phase, Makovsky says, "We went to the HRDC, our human resources development committee, and described our needs to them. They match people with specific requirements throughout the organization, taking into account individual career development aspirations as well as who best fits what job."

The up-front analysis work was done in conjunction with the product design team (PDT). Tom Lynch, Xerographics section manager for the 5028 program, recalls: "We had started this project five years ago. Two years back we got into development work and the prototype phase."

Then, suddenly, the PDT had to work with a new manufacturing team. Given the abruptness of the decision, the pressure to put together a manufacturing facility in only a year so that the product rollout wouldn't be delayed, and the historical walls between design and manufacturing organizations in general, the degree of cooperation between PDT and NBO was truly impressive.

"Manufacturing's input was critical," says Lynch, "particularly its input relative to the processes for the manufacturing and assembly work. For example, until we sat down together and saw how manufacturing was going to approach the wire stringing operation (electrified wire is used to charge the key component in the cartridge, a photoreceptor) with a totally automated station, we really didn't have a good handle on the design."

Jerome M. Pfundtner, head of quality assurance and new business development at the NBO site, reports that "when we reviewed the design, we were looking for opportunities to make things more manufacturable. We wanted to eliminate things like adhesives and tapes, foam gaskets, springs that would tangle. We wanted snap fits, orientation features that would facilitate assembly, things that would fail-safe the process."

Out of 30 recommendations, he says, 16 were accepted. "Give the product design team credit for accommodating us. They were on the verge of introducing a product, far enough along that these changes could have been risky. That shows you how our culture is really changing at Xerox."

One of the ways in which culture change has been accomplished has been by concentrating on "the number one objective: customer satisfaction," says Lynch. "That's the only way we can work together to drive costs down and quality up."

Our Leadership Through Quality program broke down the barriers, grouped us in this problem-solving mode

Where possible the PDT/NBO team opted for simplicity of design, so that the same parts and processes could be used over and over again. "We achieved that by brainstorming, by creativity. There was synergism during our sessions, where design, manufacturing, and field engineering got together to cross each other's functionalities and exchange the ideas that would eventually provide us with the solutions we needed."

Lynch and others involved in the project credit Xerox's LTQ program as "the process we used to bring us together across the various functionalities, the tool that grouped us in this problem-solving mode, broke down the barriers, and let us concentrate on the common objective: delivering quality to the customer," he says.

The quality program was launched at Xerox in 1984. "LTQ is an umbrella—it's how we manage the company now, using better communication, rewards, and recognition, and lots of other tools of quality," says James T. Horn, vice president of quality for Xerox's business products and systems group. "Quality means meeting the requirements of the customer," he concludes.

That Xerox defines "customer" in two senses—the traditional outside purchaser of its products and the internal individual or department dependent on another individual or department to get a job done—played a big role in facilitating the spirited cooperation between design and manufacturing on the copy cartridge assignment.

The line they eventually chose to build the product is an amalgam of automated and manual assembly. "Our automation philosophy was not to go full tilt, but rather to do it gradually, waiting until the line matured a little, until we understood the product and processes better," says Makovsky, adding that "down the road we expect to add vision sensing to enhance existing automation equipment with location and checking capabilities, as well as more automated cells."

Pfundtner states that "some design features required automation as a technological solution, such as the ultrasonic welding station. In other areas, where operator repeatability would be difficult to achieve or where direct labor costs would be prohibitive, we chose automation."

Further, he adds, the facility was designed for flexibility, so that it could accept automation, accommodate additional segments of automation, or accommodate traditional techniques, depending upon future design changes or process refinements. This philosophy led, for example, to a conveyor line built in eight-foot increments—each with its own air, power, and communication lines—which can be easily plugged together and modified as required.

The team was also committed to optimizing the company's JIT strategy. In fact, NBO hopes that its facility will become a benchmark for JIT within Xerox. "We designed the line for no banks, no floats, one in and one out," Pfundtner says. "We opted for small stations with small capacity. We went to our vendors and said, 'We must have JIT.'"

The cartridge subassembly is comprised of some 60 parts. NBO and its materials management support people selected virtually all local vendors. "Only one part comes from outside the US, and probably only two or three from beyond an eight-hour drive from Rochester," Makovsky states. "And 60% of the vendors are within a one-hour drive."

Fortunately, Xerox's materials management organization had fine-tuned its vendor network prior to the copy cartridge project. Its centralized commodity management program has winnowed the company's original production supplier base from 5000 to about 300.

"We picked the best of the best, benchmark suppliers in special components, and established long-term relationships with them. We trained them in statistical process control [SPC], JIT, even our LTQ program. Some of these vendors had been selected prior to our startup in early 1987, others had to be sourced."

The company's thrust into JIT revolved around getting its vendors to supply components in a continuous flow. The aim, naturally, was to keep inventory low and quality high, and materials management has been able to get supplier cooperation by offering higher volumes and long-term contracts.

JIT depends on quality. With a limited amount of components in the pipeline, out-of-specification parts can't be tolerated. And since the copy cartridge, once it's put together, cannot be repaired, the quality issue is paramount.

Quality was also tackled by "designing out" opportunities for failure, by ensuring, for example, that two parts simply could not be put together backward. Other techniques employed included embedding SPC within the automated equipment controllers, designating key stations for monitoring and sampling, alarm procedures, and bar code labeling and tracking for date/belt lot/time of day in case a specific lot has to be found.

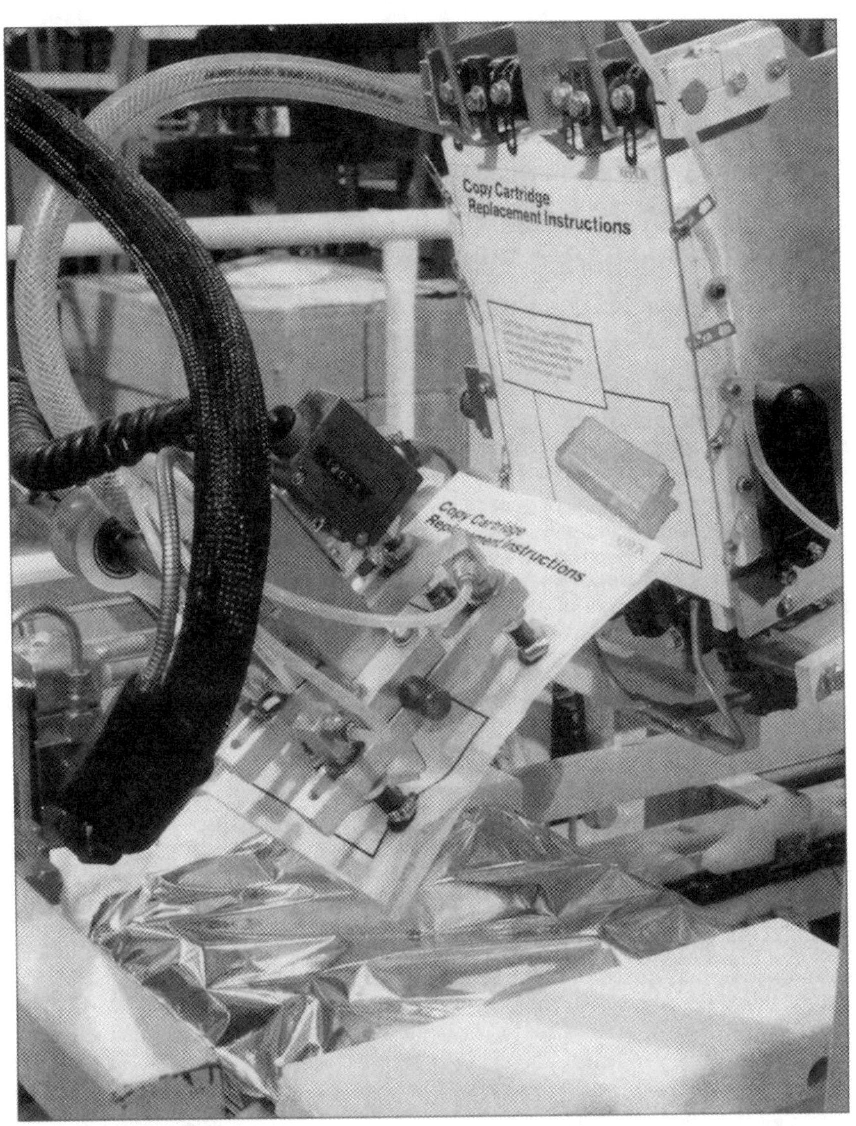

The NBO team broke down traditional manufacturing barriers. Here, customer instructions are automatically inserted during packaging.

Perhaps the biggest contributor to quality, though, and certainly one of the most interesting aspects of the project, was the participation of the assembly workers at every phase of the venture. "We got their involvement as early as possible: at the stage when we were meeting with designers, and when we went out to our equipment vendor sites," says Pfundtner.

Barcy Smith, an industrial staff member who was on the project right from the start, describes the process. "I went to the team meetings once a week, beginning in January 1987. They asked for my input to get things going. They even sent me to Illinois to check out a company and a line they were considering. It's a very different feeling from just being an operator. Now, I help design workstations, help make design changes on the parts."

As Bob Levitsky, shop-floor chair for the Amalgamated Clothing and Textile Workers Union at the Xerox facility, explains it: "One of the reasons we went along with the company on this was because we both feel that we have to balance human needs with business needs in order to accomplish our tasks. It has worked out well here because the people have seen that they're being treated like individuals, they're getting the same treatment that management gets, and nobody is trying to hurt anybody. Plus, we were able to remove these barriers to cooperation while still staying within the terms of the union contract."

Cindy Paxson, a robot tender on the line with responsibilities that include maintaining the automated packaging stations, says, "I love my work now. It makes me think a lot more. It's not the same thing every day. And quality is the result. Everyone on the team has a say. We're all encouraged to express our opinions every morning."

Paxson is referring to the daily 10-minute meetings before each shift. There, the staff review quality data collected from the day before, talk about what happened yesterday and what to watch for today. Every Friday there's a two-hour meeting to examine the entire week's operation. Problem-solving teams often break off from the general meeting to concentrate on particular line segments.

"We call it 'quality time,'" says Pfundtner. "We know we're losing production time, but we feel it's time well spent. We want to put as much responsibility in the hands of the industrial staff as possible. We feel that it's the key to being successful. If they are not successful, the business will not be successful."

Makovsky agrees: "We've got very highly participative people here. They've really helped us. For example, they identified areas and parts that would be susceptible to contamination from light, dust, or handling. They also set up unusual work practices and policies for the line, such as no drinks at the workstations and

The biggest contributor to quality was the participation of assembly workers

covering up the parts at night with cloth to support the stringent quality requirements of the product.

"We've even got an MVP-of-the-month award. The team nominates and votes for the winner. He or she then gets Jerry Pfundtner's parking space for the month," says Makovsky. And, he adds, if you've ever seen a Rochester winter, as it howls down from Canada across Lake Ontario, you'll know a parking space close to the plant door is no prize to sneeze at.

"Our industrial staff is special," Pfundtner says. "They chose to be here. This is a high-quality, tight-tolerance, high-volume project. There are easier assignments they could have gotten. Their union leadership was very proactive in supporting these participative work practices. And our management team made a point of trying to remove barriers for them."

The workers themselves are very happy. "I think it's great," says Paxson. "I hope Xerox will always be this way, and that we will become the benchmark for product delivery within the corporation."

Adds Smith: "As long as we can have input, as long as they ask for our help, I feel better and work harder. I think both Xerox and I are better off for it."

The NBO management team had the same kind of experience as the industrial staff. "The philosophy was virtually unconstrained thinking," says Pfundtner. "Get out of the box, give everyone freedom, and consider no idea dumb. Management has to be comfortable with empowering people to do things, has to attempt to remove all the barriers. We all must be willing to change.

"I saw my role as taking away any obstacles that prevented us from being successful," he concludes. "As a result, some very dynamic things happened, ideas got implemented, and we saved a lot of money." ■

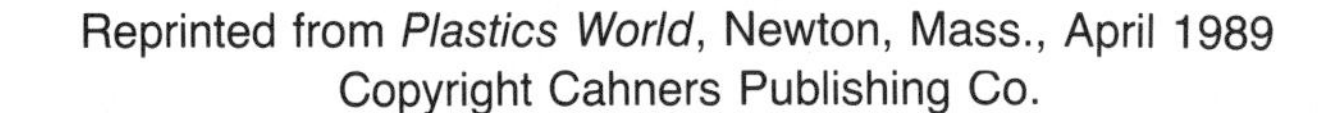
Reprinted from *Plastics World*, Newton, Mass., April 1989
Copyright Cahners Publishing Co.

Design-for-assembly slashes cost of Hoover's all-plastic vacuum cleaner

Clever design created for automated plant cuts labor and floor space by half, saves 119 parts, and eliminates all internal wiring

BY BERNIE MILLER, EXECUTIVE EDITOR

In an industry given to incremental change, the Elite series of upright vacuum cleaners stands out for its range of innovative ideas. Developed by Hoover Co., North Canton, Ohio, which marketed the first "electric suction cleaner" 80 years ago, the all-plastic Elite is a testimonial to the success of an arranged marriage.

All the related engineering specialties were wedded into an integrated project team to develop a cleaner that competes in quality and cost with units made anywhere in the world. The team's assignment was to develop a production-ready design and a customized new facility to produce the vacuum cleaner.

The result: a vacuum cleaner that was not only functionally sophisticated, but also intensively engineered for economical manufacturing and assembly. To obtain these economies, Hoover built what is probably the industry's most highly automated production plant.

The Elite project could hardly be called a low-risk decision for Hoover, which was recently acquired by Maytag Corp. It involved replacing the Convertible, a well established line that had accounted for more than 70% of Hoover's upright vacuum-cleaner sales. And it required a commitment of $30 million for new equipment, plant rearrangement, and related development costs—a swallow-hard investment even by today's standards. But after a year in production and with well over 1 million units sold, Hoover's gamble has clearly paid off in the marketplace.

The all-plastic Elite is also a manufacturing milestone. The project launched an interactive system for design and production planning that will greatly affect Hoover's approach to developing future models. A project team was set up having the responsibility not only to design the Elite, but also to ensure its producibility by planning the processes and working with suppliers to select materials and develop the production equipment and tooling. Besides design engineers and plastics specialists, the team included process, manufacturing, and tooling engineers and a representative from the corporate financial staff.

The systems approach and the opportunity for simultaneous engineering permitted continuous, two-way feedback that led to a functionally efficient and, equally important, an economically producible design. Start-up problems and lead-time delays were minimized because the process and tooling engineers on the team could review changes in part design and check them out with suppliers before part designs got too far downstream.

"While we always had communications between design and manufacturing, we had not integrated them as completely as we did for the Elite," states Keith Minton, vice president of manufacturing. As one of the benefits of the interactive design and manufacturing planning, Minton notes that the development lead time was cut by 25% compared to earlier programs.

Phillip Zepp, manager of manufacturing engineering, points out that, despite the high level of automation, the Elite went into production on schedule. "There's no way this could have happened if each of the engineering groups had worked on their parts of the project separately," Zepp says. "As another argument for simultaneous engineering, we've had essentially no quality problems in the plant or in the field," he adds.

Designed for producibility

Except for a few metal fasteners, shafts, and springs, and motor parts, the Elite is a totally injection-molded plastic design. Four major parts are in ABS: the base, hood, and upper and lower handles. Other polymers used include thermoplastic polyester, polycarbonate, nylon, and acetal, polyvinyl chloride, and polypropylene.

The real Elite story, however, is not its total plastics construction. The few metal remaining subassemblies in the Convertible were converted to plastics—the agitator and handle-release assemblies, for example—resulting in large reductions in the number of parts and assembly operations. Nevertheless these metal conversions were part of Hoover's much larger objective of designing the Elite to streamline and automate its production to the greatest possible extent.

Among the basic goals were eliminating mechanical fasteners by designing for snap-in assembly; avoiding repeated turnovers of the base by enabling all parts except the agitator and wheels to be installed from the top; eliminating manual assembly by insert molding; and reducing part handling by integrating part functions and designing for automatic assembly.

How well these goals were reached is revealed by Minton. Compared to the Convertible, its predecessor (which, incidentally, had been steadily refined over the years), the Elite:

- Cuts direct labor in half;
- Saves about half the manufacturing floor space;
- Reduces the number of components from 215 to 96;
- Requires 12 fasteners instead of 56 because almost all components snap into place;
- Weighs less than 12 lb, a 25% reduction;
- Reduces seven electrical interconnects to zero; all internal wiring is eliminated;
- Uses no painted parts; all color is molded-in; and
- Requires only one turnover of the base to install components; many were needed previously.

Hoover's Keith Minton (left) and Phillip Zepp check agitator-brush unit at discharge end of automatic-assembly machine.

Simplification by design

Three components, shown on p. 38, illustrate some of the elegant design-for-production thinking that went into the Elite.

The agitator, the motor-driven ribbed brush that sweeps the carpet, was redesigned from a fabricated metal assembly that required 52 manufacturing operations. For the Convertible line the unit was built up from a painted metal tube (the body). The formed metal "beater" bars were riveted in place; the bristles were held in molded bases that slipped into spiral slots in the body. The bearing holders and end caps were machined castings and stampings.

The Elite's agitator assembly, in contrast, is all plastic except for the center shaft and a pair of ball bearings. The benefit from converting to plastic was not so much from the parts count—just four fewer parts—but from the major reduction in manufacturing operations. The plastic design requires only 13 manufacturing steps, a 75% reduction. The splined bearing holders, bearing seals, and end caps, previously machined or stamped components, are injection molded. The body, made in a four-cavity tool, is molded with integral ribs and bristle-mounting surfaces that previously required separate stampings. A phenolic drive-belt pulley, robotically loaded into each cavity, is insert-molded onto the body during the molding cycle. The same robot demolds the bodies and places them in tote pans.

The agitator is automatically assembled on two custom machines. The first drills holes into the body for the bristles and inserts them. The second assembles the rest of the components to the body. The completed agitators are automatically tested and placed in tote boxes for transfer to the assembly line.

The handle-release lever, which unlocks the cleaner's handle from its vertical position, is now a one-piece acetal molding that includes an integral spring. In the Convertible the same function was performed by a seven-piece assembly of stamped and machined parts that required a preassembly riveting operation and four manual operations as part of the cleaner assembly sequence.

The motor field assembly ranks as one of the most imaginative design features in the Elite and "an excellent example of design for manufacturing," Zepp comments. Besides its basic function as a stator, the unit also contains plug-in terminals for all electrical circuits in the cleaner, including the headlamp, motor brushes, line plug, and, most unusual, the on-off power switch (which is activated by a plastic rod inside the handle).

"By centralizing the connection points and using plug-in terminals, we eliminated all internal wiring and splicing operations," says Larry Mancini, process and tool engineering manager. He notes that the Convertible cleaner required nearly 6 ft of internal wiring.

The field assembly is built up around two very complicated moldings, made from glass-reinforced polybutylene terephthalate (PBT). The mating PBT parts create a frame that anchors the stator laminations, provides slots to seat the electrical terminals and a snap-in pocket for the on-off rocker switch, and furnishes coil forms for the stator windings. The field is assembled by automatic equipment that sandwiches the prestacked laminations in the PET forms, inserts the insulation strips, six terminals, and the rocker switch, winds the coils, and interconnects the terminals. The winding operation would be impossible to do manually.

A second automatic line assembles, insulates, balances, and trims the armature assembly. The two subassemblies each receive about 20 automatic quality checks as they move through processing.

Of the various components that Hoover designed for the Elite, Mancini says development of the field assembly benefitted most from the use of the firm's McAuto computer-aided design (CAD) system. The CAD system not only detailed the PET moldings, but analyzed the geometry of the entire assembly for component interferences and dimensional tolerances.

Just-in-time support

To produce the Elite, Hoover essentially set up a plant within a plant organized for just-in-time operation. In addition to the assembly line, this dedicated area produces, mostly on automatic assembly and test equipment, the subassemblies to support the assembly line. These include the

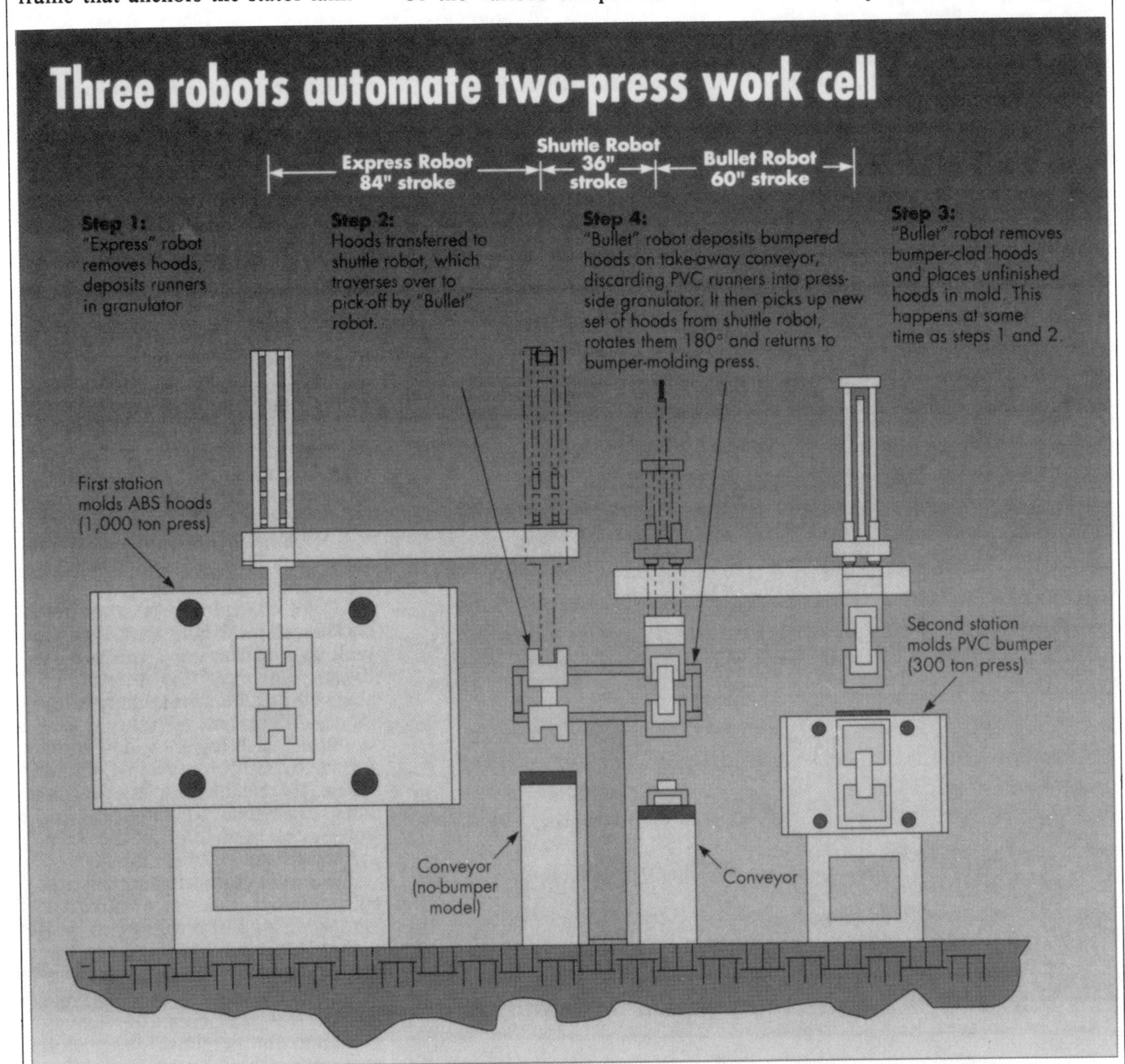

Three robots form double-play team that transfers ABS hood to second press for molding polyvinyl chloride bumper in place, then deposits finished parts on conveyor. System was engineered by Conair Martin.

wheel assembly, agitator, and the motor field, armature, and motor assembly in which they are used.

"The basic idea is to dovetail the subassembly and assembly operations, reduce inventories of high-value subassemblies, and save the handling costs that would be incurred if these components were produced elsewhere in the plant," explains Minton.

The sub-assembly area is served by a just-in-time parts-storage and delivery system controlled by an MRP-II computer program. As the subcomponents come off their machines—the molded agitator body, for example—they are robotically packed in bar-coded trays of known capacity, which then move into an overhead rack system. When an assembly machine needs more parts, it signals the system computer to release a loaded tray to the conveyor system, which delivers it to the robotic parts feeder at the machine.

In addition, the assembly area's operations will soon be monitored by a Hoover-developed computer system to collect production statistics, quality-test results, and performance and service data on the equipment.

Handling automation also extends to the molding department that supplies parts for all of Hoover's products. Virtually all the molding machines for Elite parts, mainly Reed Prentice presses, have been equipped with Conair-Martin robotic devices that load and orient non-droppable parts in tote trays.

The most unusual setup is the three-robot grouping at the presses that mold the ABS hood (top cover) of the cleaner base and the bump strip of flexible PVC. This is a two-step operation. First, the hood is molded in a Reed Prentice 1,000-ton press; it then is transferred to an adjacent 300-ton Reed press which injection-molds the bumper around three edges of the part. Dovetail grooves molded into the hood help anchor the strip. Hoover has three of these hood-molding cells in operation.

The robotic installation, engineered by Conair Martin, Agawam, Mass., is detailed in the drawing on p. 37. To start the sequence, a robot demolds a pair of hoods from the 1,000-ton press and transfers them to a horizontal shuttle. The shuttle, in turn, hands them off to a third robot on the bumper press, pivoting each cover 180° before the transfer to suit the bumper-mold gating. The bumper-press robot demolds and deposits the two completed hoods on a belt conveyor before taking the waiting pair from the shuttle.

Both press robots also pick off the sprues and runners and drop them into granulators for immediate recycling. The three parts handlers are controlled by a pair of Allen-Bradley programmable controllers—equipped with trouble-shooting diagnostics—that are interfaced with the presses' microprocessor control systems.

For earlier models this operation was entirely manual. Operators unloaded the hoods and put them in bags for transfer to another department where the bumper strip was cut, positioned, and fastened at a riveting machine.

How Hoover's design streamlined assembly

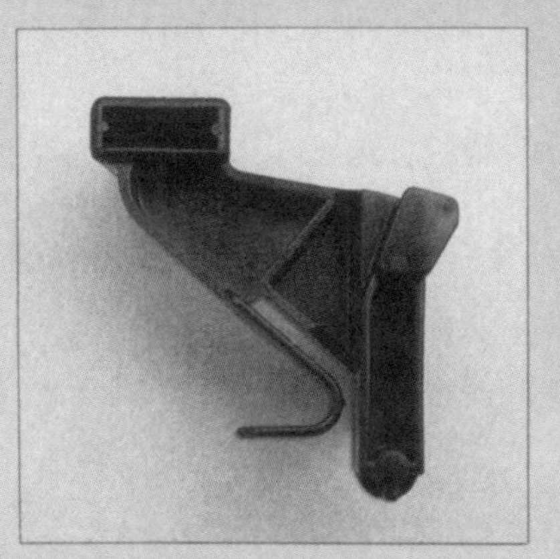

Handle-release lever consolidates seven stamped and machined parts, including return spring, into a single molded acetal part.

Motor field assembly, molded and wound on an automatic line, embodies spectacular design concept. Spring terminals in plastic stator provide plug-in connections for all electrical circuits, eliminating all internal wiring.

Agitator assembly reduces 52 manufacturing operations to 13 steps. Integral ribs and bristle seats and insert-molded drive-belt guide on agitator body are major time-savers.

Twelve models, one line

Unlike the highly automated production of the various subassemblies, Hoover chose to use operators to assemble the Elite on a moving-belt line. The relatively small number of components, most of which simply snap into place, makes manual assembly economical.

Model changeover is simplified because all models in the series are built up from the same base and use the same or physically interchangeable components. Also, many of the model-differentiation features, such as outer-bag styles, trim, and accessories, are simply added to the carton at the packing station. By this time, more than 50 functional and quality checks on 100% of the cleaners has guaranteed high quality.

"The Elite costs substantially less to build than the Convertible, but we didn't automate simply to save money," Minton says. "The new line also is equal or better in performance, quality, user features, and reliability. And long term, our investment and our merging of design and engineering thinking, will pay off in the ability to meet competition from anywhere on the globe." □

Reprinted from *Manufacturing Engineering*, September 1988

Simultaneous engineering, along with transitional and continuous improvement efforts, comprise Ingersoll's ambitious partnership program

Converting Customers to Partners at Ingersoll

By Robert N. Stauffer
Executive Editor

The customer is king! But today, that isn't good enough. Manufacturers intent on maintaining or expanding market share in their field now need a more intimate working relationship with supplier companies who provide the necessary production equipment and processes.

At The Ingersoll Milling Machine Co. (Rockford, IL), that relationship is being established with a growing list of customers who have opted to work with Ingersoll under an arrangement that the company simply calls Partnerships. It's a wide-ranging program that covers a lot of bases. In broad terms, it includes three types of activities—simultaneous engineering, transitional programs, and continuous improvement. The success of the program to date has led to a new corporate culture at Ingersoll, one that gives the company an even more solid footing as a leading builder of transfer lines and special machines.

Ingersoll is achieving significant savings during the development and construction of systems like this through the use of simultaneous engineering.

This isn't really a brand new effort by Ingersoll. It has been doing simultaneous engineering projects for about three years. But today, there's a more formal, more directed approach to the program. And the company has impressive evidence that its efforts are paying off—both for its customer/partners and itself.

For starters, here are some figures that John B. Droy, vice president, simultaneous engineering, uses to get the listener's attention. Back in 1978, the company built a transfer line for machining in-line, four-cylinder engine blocks. That project involved 62 customer-initiated changes at a cost of $1.3 million. This year (1988), Ingersoll completed a vee-type block line for the same customer. Using simultaneous engineering, the new line was completed with only seven changes at a cost of $436,000.

"This time," says Droy, "the type of costly changes that had to be made in 1978 were largely taken care of in the simultaneous engineering phase before the customer ordered the equipment. In addition to those savings, we documented savings of $750,000 in modifications of customer specifications. And, of course, we did this without any adverse affect on the reliability of our product."

Not to be overlooked in this particular situation is a significant reduction in the time required to implement the program. The 1988 program took 23 weeks less time than the 1978 project. In terms of monetary cost, the shorter implementation period provided an additional savings of close to $750,000. Altogether, the bottom-line savings for 1988 over 1978 came to just under $3 million, strong testimony to the effectiveness of simultaneous engineering.

These and other cost savings are impressive, but they are just part of a bigger objective that Ingersoll has set for itself. That goal is to help customers refine their product, make it more manufacturable, and in the process better define the steps required to make it.

"One of the big benefits is a reduction in the time required for the customer to introduce a new product," Droy notes. "I think we'll be able to take out about 30% of the time. That's very important today, when flexibility and fast response to the market are so vital in staying competitive."

The Ingersoll simultaneous engineering group now includes about 20 people, with eight to 10 more expected to be added this year. Five people will soon come on board at offices in Troy, MI. One basic requirement for all personnel in the group is experience. Work of this type is no place for someone right out of school. As new people are added to staff, they're given a special booklet that explains the simultaneous engineering concept and Ingersoll's approach to it, how to start a project, key reports required, scheduling of various steps in a program, and other similar points. The concept is still so new that even experienced people joining the group must be brought up to speed.

Each of the three partnership activities has a clear-cut niche in Ingersoll's new program. Typically, a simultaneous engineering project involves the design, development, and building of new equipment to be used in manufacturing either a new product or an existing product. A continuous improvement project deals with existing equipment. A transitional program is aimed at assuring that all of the advancements and benefits created through simultaneous engineering are indeed transferred to the floor of the customer's plant.

The scope of the simultaneous engineering activity involves all of the following:

- Improve new product designs to make them less costly to manufacture
- Address flexibility issues
- Plan and implement advanced technology—machine tools, cutting tools, software, and procedures
- Integrate new technologies into existing processes.

Expanding on the simultaneous engineering concept, Droy doesn't view it as just a new buzzword for practices that have been around for a long time. "I think it's new and different. One reason it got started and has caught on is that cutbacks and attrition, especially in the auto companies, have left fewer experienced people to handle this kind of work. And then there has been the big impetus from the impact of foreign competition.

"What's also new, or at least getting more emphasis in all of this," he continues, "is the need for trust between the vendor and customer. This includes trusting that information will be handled confidentially and that costs will be fair. The simultaneous engineering approach involves a departure from the competitive bid scenario. That's new. In this new concept, the customer usually works with just one supplier. Some people question whether the customer can count on a competitive price in this arrangement. But that has not been, and should not be, a concern. The customer has a good feel for price, and it has to be realistic. For us, it means we can redirect our resources and put our time and energy into improving both the customer's product and our own."

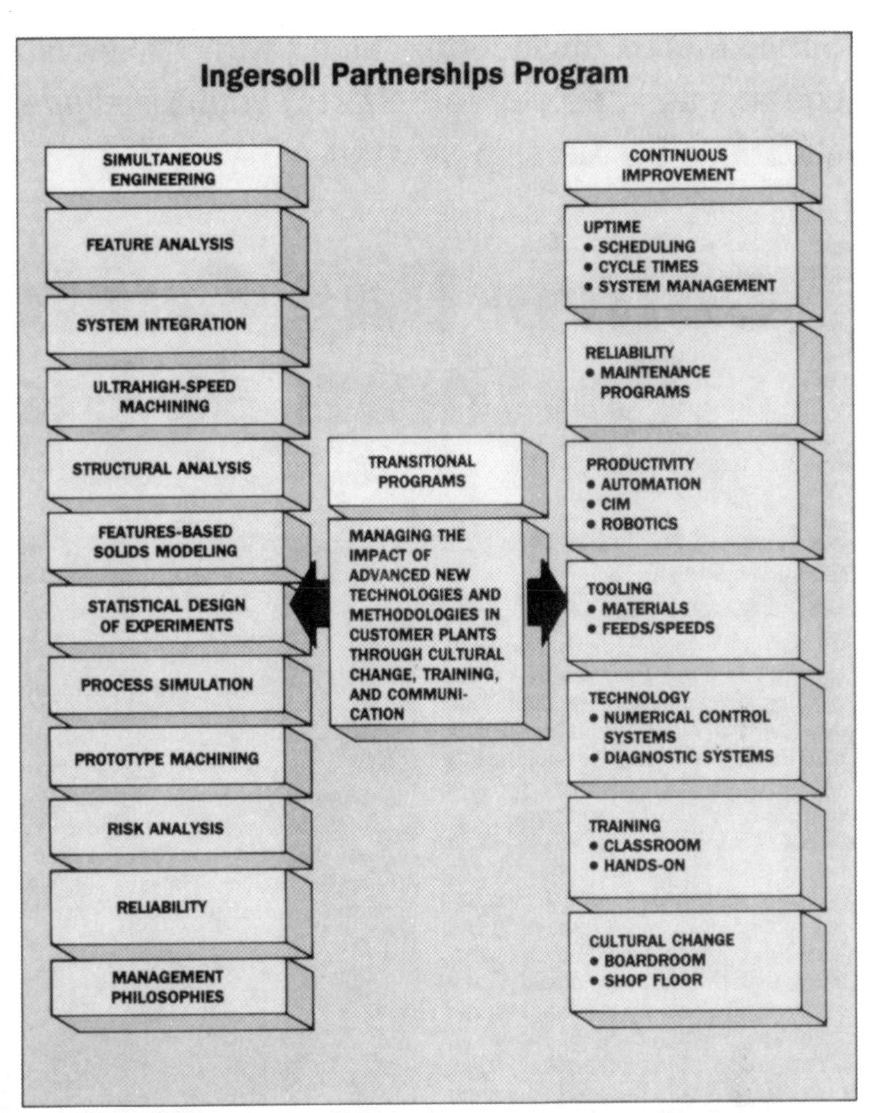

Simultaneous engineering, transitional programs, and continuous improvement comprise the Ingersoll Partnerships program.

One of the main responsibilities for Ingersoll is to better define the process, the tolerances, and all of the other things associated with the process. The company's engineers start with the customer's product drawings and do a feature analysis. They look at all the things that have to be done to add value to the product. Basically, this gets down to all of the holes and surfaces that have to be machined.

"We look at these features in several ways," Droy says. "One thing is cost, not just initial costs to create the features, but the ongoing tooling costs and operating costs associated with those features. Another concern is standardization. This can pay big dividends. For example, we recently reduced the number of different hole sizes in one part from 50 down to 30."

Implicit here is the fact that the simultaneous engineering effort is built around the teamwork approach. Once a project is started, the Ingersoll team includes representatives from such groups as Engineering, Design, Manufacturing, and Purchasing. One of the benefits is instant communications. With everyone together, decisions get made, action comes quickly, and the program moves ahead faster.

The bottom line of simultaneous engineering, according to Droy and others at Ingersoll, includes the following four areas of benefit:

- Capital savings
- Fewer changes during equipment engineering and build
- Shorter overall time to implement a program
- Better overall communications.

One of the members of Ingersoll's simultaneous engineering group, Joe Eisele, program manager, provides some insight into the give and take that has marked the progress on a year-old simultaneous engineering project for a new engine line. "We started with concept drawings before we had any initial hard drawings, just to get some early budget figures. Then as the product designers progressed with the engine design, we updated our machine processes and budget figures. We've spent the last year refining and analyzing the new engine with the customer's designers and manufacturing people."

An early benefit of simultaneous engineering on this project was an agreement by the team regarding the position of locating and clamping surfaces on the engine block that will be used throughout the process. "Without a collective look at these features," Eisele says, "we might have completed the part prints and then had to go back in to add the manufacturing lugs and clamping surfaces. This was a big plus. And if we find that some of these lugs still aren't right, we'll have time to modify them before the customer gets into the casting design and build stage."

For this particular program, Ingersoll engineers also plan to do some tooling tests that will help determine whether a second pass is needed in various holes. Eisele: "Right now we think we have to ream the critical holes, but if we use new, top quality drills, we may be able to hold the diameter and eliminate the usual ream pass. Without simultaneous engineering, we wouldn't have the time to do that kind of testing. The potential savings in equipment are significant because we would normally do reaming at three or four stations in an engine line like this."

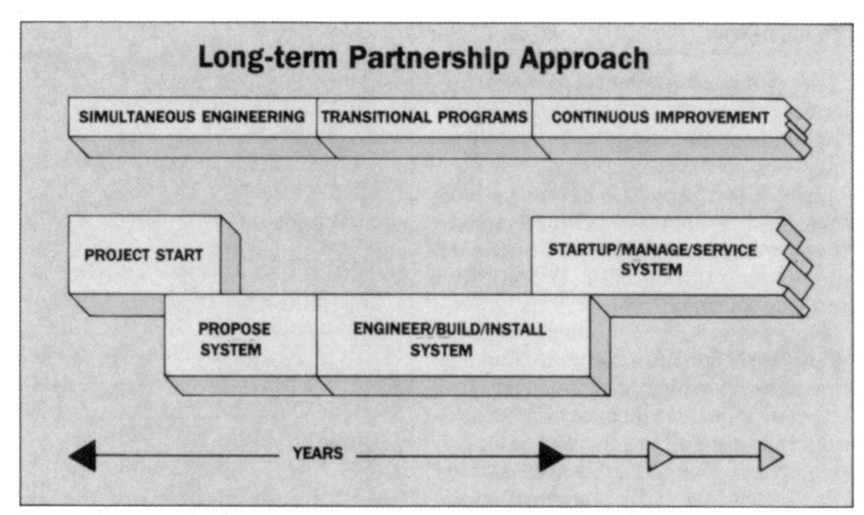

The usual steps involved in the evolution of a new system are blanketed by the three elements of Ingersoll's new program.

On this project, Ingersoll also worked with a number of suppliers in what amounts to a simultaneous engineering relationship. In one case, a supplier of locators proposed a fixture to use on the prototype machines. It will simulate the fixturing to be used in the actual process line. Close contact is also being maintained with a number of various other suppliers of peripheral equipment.

Ingersoll recently held a Supplier's Day and spent much of the time discussing the simultaneous engineering concept and how it can work between Ingersoll and its suppliers.

Simultaneous Engineering: What Is It?

Simultaneous engineering has been defined in many different ways. In the simplest of terms, it is a merging of the efforts of product designers and manufacturing engineers to improve manufacturing processes and products. Here are the views of several practitioners of the concept.

W. David Lee, director, Arthur D. Little Center for Product Development (Cambridge, MA)—"Simultaneous engineering is the process in which key design engineering and manufacturing professionals provide input during the early design phase to reduce the downstream difficulties and build in quality, cost reduction, and reliability at the outset. Another important result is reduced incompatibilities among the development objectives of business strategists and engineering and manufacturing professionals. Manufacturers have found that successful simultaneous engineering requires a 'marriage' among team members and a multidisciplinary culture. To encourage this approach to product development, many manufacturers have sought ways to combine various areas of expertise in one location."

Henry W. Stoll, manager, Design for Manufacture, Industrial Technology Institute (Ann Arbor, MI)—Stoll offers the following Four Cs of Simultaneous Engineering:

- **Concurrence**. Product and process design run in parallel and occur in the same time frame.
- **Constraints**. Process constraints are considered as part of the product design. This helps ensure parts that are easy to fabricate, handle, and assemble and to facilitate use of simple, cost-effective process, tooling, and material handling solutions.
- **Coordination**. Product and process are closely coordinated to achieve optimal matching of needs and requirements for effective cost, quality, and delivery.
- **Consensus**. High-impact product and process decision-making involve full team participation and consensus. ■

The intent is to eventually have the concept work as well in that direction as it does the other way—between Ingersoll and its customers.

Ed Thomas, staff engineer, has been deeply involved with a simultaneous engineering project aimed at developing the optimum process for making automotive pistons. Features-based solids modeling figures importantly in this project. Thomas explains that analyzing the effect of clamping and locating in various ways and the effect of cutting and centrifugal forces reveals what happens to the part before any machining is done.

Thomas and his team worked with a supplier of special gages that can handle the accuracies associated with pistons. "Our customer agreed that this step was necessary. So we looked into the situation early on to resolve the issue before production started on the pistons. That's what simultaneous engineering is all about. This is the third piston machining program we've done under simultaneous engineering, and improvements have been made each time."

Without transitional programs to put the results of simultaneous engineering to effective use, it would be "business as usual," according to Droy. "After we have carried out all of the steps in a detailed simultaneous engineering project, we start to question how both we and the customer will implement the results. That gets us into things like training and all of the cultural changes that have to take place to achieve significant increases in the efficiency of manufacturing equipment in the customer's plant.

"That has been one of the big differences between the US and Japan. They've had that philosophy in place for many years—constant improvement, teamwork, training, quality circles, and the like. We're just getting started here in the US."

Continuous improvement programs at Ingersoll include the following activities:

- Identify productivity limitations
- Increase productivity of existing processes
- Plan and apply new tooling concepts
- Identify improvements in support areas
- Increase reliability
- Participate in new systems launch and in achieving goals.

Activities under the continuous improvement label often involve working with a customer to improve the efficiency or output of an existing line. The goal that most transfer line users are aiming for is an efficiency of 80% or better. Historically, such equipment has operated in the 50-60% range.

"To achieve those efficiencies," says Droy, "the reliability of the equipment not only has to improve, but the management philosophies applied in its use must also change. Our continuous improvement work includes a variety of projects. An important one is our effort to incorporate the advanced tooling that is now available. In that regard, we're building spindles that will operate up around 50,000 rpm."

Ingersoll currently has 14 simultaneous engineering projects under way. Before the end of the year, it expects to start 12 more, plus seven continuous improvement projects.

Droy: "The simultaneous engineering concept is catching on and expanding rapidly. The ideas involved in this concept are so sound and fundamental that we should eventually see them applied even in the smallest job shops. Simultaneous engineering will develop at the speed that people understand it, trust it, and believe in it. ■

Key Elements of Simultaneous Engineering

A strong believer in simultaneous engineering and its impact in manufacturing, Joseph D. Carney, president, Gilman Engineering & Manufacturing Co. (Janesville, WI), offers the following as among key elements and features:

- Simultaneous engineering means total commitment of resources to a project on the part of both the customer and the equipment supplier
- Ideally, the customer and supplier team should be in place prior to the start of product design
- Without the technique of simultaneous engineering, there simply is not sufficient time for adequate customer/supplier interface
- Like any team effort, simultaneous engineering mandates a free exchange of information
- Simultaneous engineering provides both customer and supplier a degree of flexibility previously unattainable in the industrial environment
- Simultaneous engineering provides the supplier engineers with a thorough understanding of the product and process requirements
- From any standpoint, simultaneous engineering is really a win-win situation for product design, manufacturing, and the equipment supplier
- The key benefit of simultaneous engineering is its synergistic expansion of knowledge for the task at hand. ■

Presented at The Third International Conference on Design
For Manufacturability and Concurrent Engineering, December 1989.
Reprinted courtesy of the Author.

DFM Support for Simultaneous Engineering ©
Product Design For Manufacturability and Assembly©

Lada Zajicek
Apple Computer, Inc.

IA. IMPORTANCE OF DFM (DFA) IN PRESENT INDUSTRY

Only in recent years the term DFM (DFA) has appeared in trade magazines, university curriculums, conference agendas and workshops. Despite its present publicity, DFM - Design for Manufacturability (DFA - Design for Automation/Assembly) - is not a new branch of engineering. I myself have successfully been applying the principles of DFM (DFA) throughout my entire engineering career of twenty five years. The theory behind DFM (DFA), is to create the 'perfect' design. In this instance 'perfect' stands for the best design possible in terms of its aesthetics, efficiency, practicality, easy assembling and manufacturing qualities as well as lowest overall cost. Providing the industry with all of these, DFM (DFA) has been hailed by many as the key to domestic and worldwide competitiveness. A well established and well organized DFM (DFA) department is in my opinion the blueprint for a successful product and ultimately a successful enterprise.

We all know that theory and putting that theory into practice are two completely different things. So how then does a DFM (DFA) department implement its theory of a 'perfect' design? Realizing the significant potential of DFM (DFA) many companies have established a DFM (DFA) department in the recent years, implementing it into their overall operating structure. In this rather short period of time, DFA has played a major role in the overall production and release of many new products, a subject we shall expand on later. My goal is and has always been, to put DFA's entire theory into practice. In deed, do I strive to design a product that will not only look great and be state-of-the-art, but will also be easy to assemble, easy to manufacture and finally be economical.

In order to achieve these high goals and standards, I stress three objectives: organization, a highly qualified staff and communication within the DFA department(s). Preferably, the department should be divided into two groups, a DFA for PCB Assembly and a DFA for System Assembly, under one common management. This allows each group to concentrate on the two quite diverse types of assembly on one hand; on other, mutual management provides continual control and cooperation among both groups. Thus inspite their different tasks and fields, both groups are able to continually work together and understand each others issues and problems as well as those of other departments. This environment of 'open communication' and 'specialized task' has proven to be quite efficient and effective. Another possibility is, of course, to create one DFM department with coordinated functions. In either case, DFM (DFA) engineers must be highly qualified and specialized. Not only must they be creative and innovative designers, but they must be practical designers with extended experience as well. This is very important knowing that the core of DFA is design. Indepth knowledge and practical experience in manufacturing, process development, parts fabrication, such as sheet metal, diecasting, injection molding, PCB fabrications, etc., are highly recommended. In addition thorough knowledge in electro-mechanical design and details like requirements for various testing methods such as RFI/EMI, ESD, cooling, etc., are also very helpful. Keeping in mind the demanding objectives of DFM (DFA) and realizing DFM's involvement in so many other aspects of engineering, such an extensive and specialized background becomes essential. Last but not least, the DFM engineer must be an effective communicator and a great teamplayer.

Continuing training and education are also highly important, because they allow the designer to be up to date on recent developments within his or her company and the industry as a whole. DFA seminars, educational meetings and classes as well as competitive analysis sessions are excellent tools to provide this competitive edge. Personally, I find competitive analysis to be a most effective learning tool. Supplying both insight and inspiration, this concept adds yet another dimension, namely that of comparison. Seeing a similar task, objective, solved with a different design strategy - that of our competetor - competitive analysis provides the ultimate advantage.

Now that we have established the criteria for the optimal DFA engineer, we still do not know how the DFA works. Actually the methodology is quite simple considering the high goals: first of all the DFA staff must be an efficient and a well organized team, and furthermore a design(s) must be 'perfect' on the first attempt.
In order to achieve this, DFM (DFA) of course has to work very hard. But hard work alone is not enough, DFM (DFA) must create a fine communication network between the departments involved in a project(s) in order to succeed. Meetings between all the different departments must be held on a regular basis. These meetings have to be very constructive, whereby the current stages of a particular design are tested, whereby design improvements, suggestions and ideas are directly implemented into the design at each and every stage of the design. It is only because of this constant open communication network, that redesigning, meaning having to go back and doing it over again, can practically be eliminated. With other words, as a design is being developed, it is continually being improved upon. Thus the ever so famous last minute change before the deadline has successfully been eliminated.

Some practical examples of my recent successful DFM (DFA) activity are the Apple Personal Computers MAC II, MAC II/CX and MAC SE, MAC SE/30, the SCSI HD 20/40/80, the Tape Back, the AppleCD SC™ and the famous Mouse Program, all of which have been developed during my employment with Apple Computer, along with many other product designs of my long standing career. Applying this theory, the principles of DFA, in many of the designs, I use the properties of plastic to their fullest and thus am able to eliminate screws and fasteners and replace such with design features such as snaps, hooks, bosses, locators, etc. Furthermore, the Macintosh 2 and Mouse 2.0 were designed in an 'assembly oriented product design' manner, which means, they can be assembled either manually or by automation.

IB. ASSEMBLY ORIENTED PRODUCT DESIGN

'Assembly oriented product design' actually means, that a product should be designed in such a manner that assembly expenditures are reduced to a minimum, whereby production costs are reduced to a maximum. Assembly oriented product design can :

- reduce expenditure for automation
- make automation possible in the first place
- decrease expenditure for manual assembly
- use existing production line for a new product(s)
- create a flexible production line (multiline for two or more products).

One important aspect of this design approach, is to establish so called design rules. These are actually design quidelines, which although varying from company to company in terms of specific applications like material handling, production line and production equipment, etc., in a broad sense apply to all types of industry. The purpose of these quidelines is, of course, overall reduction of the production cost. One approach to realize the above objectives, would be to standardize parts and subassembly so that the same equipment could be used for different products. With other words a chain reaction of time and cost reduction is set off. Departments like purchasing, accounting, stocking and material handling, field and service, which are not directly connected to the DFM (DFA), all of a sudden save time and money as a direct result of a successful DFM (DFA) activity. DFM (DFA) is able to decrease the cost on parts by up to 30% and up to 40% on labor. A design could be simplified to the extend that no automation equipment on the System Assembly line is needed. In order to prove this, justification for automation must be made.

IC. ASSEMBLY ORIENTED DESIGN PROCESS

During the development of a product, the assembly process must be analyzed and planned. The existing production line, the production equipment, etc., must be known and analyzed, in order to justify their use for a new product. If existing line and equipment cannot be used, a partly new or the use of a fully automated assembly line must be considered and evaluated in respect to the new product, which in turn has been developed into a 'perfect' design by means of standardization, etc.. This interrelationship of product and assembly is vital and both must match oneanother harmoniously before optimal overall assembling qualities are reached. As a result of this harmony between the standardization on products and subassembly, the same production line (possibly flexible line) can then be used not only for that particular product, but for other products as well. The end result being, production efficiency, the ultimate result being competitiveness with overseas manufacturers.

II. DESIGN STUDY

A. COMPUTER MOUSE
(APPLE COMPUTER MOUSE 1.5 VS 2.0)
PATENT NO.: 4,464,652

1. DESIGN APPROACH

The design approach in this case, was to create the perfect design in terms of its overall design, manufacturability, production yield, serviceability and cost efficiency. At the same time Mouse 2.0 had to match all previous Mouse Systems, all existing Apple Computer Systems as well as all new Apple Computer Systems. Furthermore, slight changes were incorporated into the overall industrial design. The completion time was set for 12 months.

2. ASSEMBLY PROCESS

The assembly process was simplified to a minimum. Although, the justification for automation was positive, Mouse 2.0 was designed for both automated and manual assembly as well. The Mouse 2.0 and its automated production line were being developed concurrently. The automation equipment was ready in 12 months, the Mouse 2.0 was ready for production in only 9 months.

During these three remaining months, the Mouse 2.0 was being assembled manually, whereby this assembly was monitored as to its efficiency. The results were positive and it was concluded that the design was 'perfect' in terms of both manual as well as automated assembly. Thus the Mouse 2.0 design demonstrated DFM's (DFA) success in producing the 'perfect' design in its first attempt. Production yield on Mouse 2.0 was 99.9% to 100% in the beginning production stages. The savings on direct labor were 42.85% and additional labor savings were achieved with the elimination of the rework line. The Assembly Process proposal was accurate in all aspects, no corrections were ever necessary.

3. STANDARDIZATION

Standardization was implemented on several parts as well as on subassemblies. This enabled the use of several identical parts on both the Mouse 1.5 and on the Mouse 2.0. A number of new parts and subassemblies, especially developed and designed for the Mouse 2.0, were designed to match existing dimensions on major functional parts on the Mouse 1.5 in order to eliminate any design changes. It was because of these design improvements, that production yield on the Mouse 1.5 was enormously improved from 40.1% by up to 99.9%. The production efficiency on the Mouse 2.0 was so great that the entire production was reduced by appr. 30% in labor as to the production line of the Mouse 1.5.

4. COST SAVINGS OVERVIEW

The following cost overview reveals major cost reductions. Due to the confidential nature of this issue, I am unable to disclose details on this matter. However, I believe that the data, I am able to reveal, will prove sufficient in reflecting the enormous savings resulting directly from a successful DFM (DFA).

Cost savings - overview:

Material cost:

Part description: (Mouse)	1.5 (old design) P.C.	Y.%	T.M.	2.0 (new design) P.C.	Y.%	T.M.
Ribcage	0.45	85	I Mo	0.15	99.5	6 Mo
Ballbearing	1.00	85	-	0.10	100	-
Ball	0.84	65	-	0.35	100	-
Cable Assy	4.70	95	-	3.87	99.5	-
Top Housing	0.80	99.5	-	0.67	99.5	-
PCB	1.35	-	-	0.27	-	-
Encoder Disc Assy	0.48	-	-	0.30	-	-
Screw	0.04	-	-	0.02	-	-
Total. (in $):	9.66			5.73		
Assembly Yield:	38 to 43%			99.9 to 100%		
Rework Cost. (in $):	0.95			0.05		

Manufacturing Cost: Number of Operators:

	1.5 (old design)	2.0 (new design)
Quick Test	1	1
B/H + PCB	2	1
P/B + T/H	2	1
T/H + B/H	2	1
Total:	7	4

Cost savings (in US $):

Material Cost:	3.93
Rework	0.90
Material scrap	0.15
Manufacturing cost	0.40
Total cost savings:	**5.38**

P.C. - part cost
T.M.- reg. sch. maint.
B/H - Bottom Housing
P/B - Push Button
T/H - Top Housing

INDEX

A

Adhesives, 242
American automobile industry, 132
Analysis of variance, 94
Artificial intelligence, 29, 145, 153-154
Assemblability, 205-214
Assembly
  agitator, 12, 185, 222, 225, 162, 265
  automatic, 237-243
  base drive, 230
  brake, 11
  capacitor, 231
  clamping, 243
  costs, 217, 218, 219, 229, 233
  door, 231
  ease of, 206
  evaluation, 238
  fan, 231
  final, 225
  housing, 219
  ideal-driven, 168
  manual, 91, 226, 229
  mechanical, 234
  minimum, 241
  "pancake," 242
  principles of, 240-241
  spindle/housing, 217
  time, 241
Assembly agitator, 265
  and fastening, 185
  final, 225
  principles, 185
  systems, 12
  time, 222
  workers, 262
Authors
  Allen, C. Wesley, 63-68
  Boothroyd, G., 217-221, 225-229
  Budill, Edward J., 73-81
  Bussey, John, 56-60
  Coffman, Cathy, 82-84
  Coleman, John, 85-91, 222-224
  Dewhurst, P., 217-221, 225-229
  Evans, Bill, 3-4
  Fabrycky, Wolter Jr., 5-7
  Forman, John, 92-108
  Frances, Philip H., 8-13
  Funk, Jeffrey L., 230-236
  Gager, Russ, 237-243
  Goldstein, Gina, 244-246
  Gordon, Fred, 14-16
  Griffin, J. David, 106-108
  Hinckley, John P., Jr., 109-126
  Holbrook, A.E.K., 152-155
  Inglesby, Tom, 247-253
  Isenhour, Robert, 14-16
  Kirkland, Carl, 127-131
  Kunak, D.V., 36-49
  Madsen, Claus, 254-259
  Martin, John M., 17-22, 260-263
  Miller, Bernie, 264-267
  Myers, Charles F., 106-108
  Norman, Rick, 132-140
  Rooks, Brian, 141-144
  Rouse, Nancy E., 145-150
  Sackett, P.J., 152-155
  Schuch, Linda K., 156-159
  Sease, Douglas R., 56-60
  Smith, Michael, 151
  Stauffer, Robert N., 268-271
  St. Charles, David P., 160-164
  Stoll, Henry W., 23-29, 165-171
  Szakonyl, Robert, 172-175
  Tanner, John P., 30-32
  Thurmond, R.C. 36-49
  Turtle, Quentin C., 176-183
  Vasilash, Gary S., 50-55
  Waterbury, Robert, 184-186
  Yamada Takuro, 187-204
  Zajicek, Lada, 272-278
  Zorowski, Carl F., 205-214
Automotive industry, 82, 109-126, 132, 187

B

Base drive assembly, 230
Bidding process, 59
Brake assembly, 11
Brazing, 242
Break-even times, 80
Business discipline, 57
Business strategies, 100

C

CAD/CAM, See: Computer-aided design/
  computer-aided manufacturing
Capacitor chassis assembly, 231
Capital savings, 270
Changeover, 199
Check fixtures, 163
Chief tool engineers, 255
CNC, See: Computer numerical control
Commercialization, 176
Communications, 63, 74, 89, 103, 171, 178
Company alignment, 38
Company commitment, 155
Company communications, 89
Competition, 73, 112, 115, 125, 181
Competitive advantage, 67, 92-105,
Competitive benefits, 68
Competitive edge, 69
Competitive pressures, 71

Competitors, 159, 175, 182
Computer-aided design/
  computer-aided manufacturing, 7, 9, 11, 83, 132-140, 151, 218
Computer input screens, 219
Computer integrated manufacturing, 9, 86, 97, 160
Computerized simulation, 256
Computer numerical control, 54, 160-164, 187-204, 224
Computer technology, 145
Concurrent design methodology, 133
Consolidated manufacturing, 244
Continuous improvement programs, 101
Control systems, 203
Corporate manager of manufacturing engineering, 255
Costs
  assembly, 217, 218, 219, 229, 233
  budgetary, 256
  design, 220
  development, 60
  effectiveness, 50
  estimating, 256
  handling, 267
  labor, 117, 236
  machining, 221
  manufacturing, 217, 221, 230, 254, 278
  material, 221, 235, 278
  models, 230
  mold, 235
  operational, 235
  product, 131, 179
  production control, 217
  reduction, 114, 129
  rework, 278
  savings, 100, 277-278
  tool, 71
  for two-spindle housing assembly designs, 219
Creative capabilities, 120
Creativity, 26
Cross-disciplinary teamwork, 109-126
Cultural changes, 86, 139
Customer requirements, 96
Customer satisfaction, 139
Cycle times, 16, 190
Cylindrical turning, 221

D

Database elements, 76
Database management, 148
Databases, 230
Database sources, 77
Delivery, 165
Design
  alternatives, 209
  axions, 25
  cell, 11
  concurrent, 133
  deficiencies, 152
  electro-mechanical, 272-278
  engineering, 173
  fit, 3
  form, 3
  function, 3
  geometry, 161
  information, 137
  modular, 123, 157
  practices, 171
  process, 3
  product, 3
  for quality, 30
  for reliabilitty, 30
  rules, 88
  schedules, 87
  standardization, 94, 199
  strategy, 170
  systems, 11
Designers, 69
Development costs, 60
Development cycles, 56, 59
Development/production transition, 34
Direct labor, 238, 265
Door assembly, 231
Durability and reliability, 112

E

Ease of assembly, 206
Economics, 7, 10
Effective organizations, 65
Efficiency, 9, 22, 147, 196
Electrical products, 230-236, 231
Electro-mechanical design, 272-278
Electronic data interchange, 97
Energy, 10, 191
Engineering databases, 161
Engineering methods, 129
Environment, 205, 209
Experimentation, 151
Expert systems, 29, 145, 153

F

Factory floor process simulation, 124
Failure mode and effects analysis, 28
Failures, 52, 182
Fan assembly, 231
Fasteners, 265
Feature-based modeling, 151
Feedback, 178
Feeder bowls, 239
Final assembly, 225
First-time-through capability, 124
Fixed budget approach, 254
Fixtures, 163
Flexible assembly systems, 24, 141
Flexible manufacturing, 175, 187-204
Flexible production lines, 275
Fly winding, 239
Forecasts, 176
Free flow manufacturing systems, 201
Functional requirements, 25

G

Geometry, 149
Global improvement plans, 244
Governmental requirements, 73
Government support, 9
Graphics workstations, 150
Grippers, 228
Group technology, 11, 28, 98

H

Hand-off cycle, 118
Hands-on experience, 173
Hard tools, 163
Housing assembly designs, 219
Human relationship, 202
Human resources, 63-68
Human systems, 101

I

Ideal driven design, 168
IGES, See: Initial Graphics Exchange Specification
Implementation, 19, 22, 26, 93
Increased communication, 92
Industrial engineers, 257
Information transfer, 11
Initial Graphics Exchange Specification, 91, 97
Injection molding, 220, 234
Inspection, 12, 160-164, 203, 244, 247
Integrated circuits, 248
Integrated engineering, 51
Integrated product/process design methods, 166
Integrated teams, 69
Integration, 4, 10, 11, 131
Interchangeability, 157
Interfaces, 92
Internal design capabilities, 71
Investment, 114, 142, 191
Irreversible corrective action, 115
Iterative process, 74

J

Japanese automobile manufacturers, 187
Japanese competitors, 56
Japanese industries, 195
Jigs, 247
Just-in-time, 11, 12, 117, 199, 200, 222, 223, 261, 266

K

Kitting, 233
Knowledge-based systems, 154

L

Labor, 8, 10, 80, 102
Lead-time, 57
Liberty approach, 109
Liberty feasibility rating, 125
Liberty objectives, 114
Liberty Team, 109-126
Life cycles, 5, 36
Limit stacking, 83
Linear process, 160
Line efficiency, 195

M

Machinability, 193, 203
Machine cycle time, 190
Machined backing plate, 13
Machine tools, 187-204, 269
Machine vision, 247
Machining, 187-204, 221, 269
Maintainability, 156
Management
  changing, 10
  commitment, 136, 257
  methodologies, 26
  philosophies, 269
  and quality, 109-126
  strategy, 50
  support, 19, 179
  techniques, 100
  tools, 8
  top, 137
  understanding, 67
  upper, 258
Manual assembly, 91, 226, 229
Manufacturing Automation Protocol, 11
MAP, See: Manufacturing Automation Protocol
Marketing, 20, 22, 96
Market models, 135
Market needs, 38
Material flow, 79
Materials database, 131
Materials handling, 12
Materials processing, 12
Mechanical assembly, 234
Merit analyses, 212
Metal cutting facility, 188
Methodologies, 16, 26, 29, 70, 95, 132-140, 166
Methodology comparison, 27
Modeling techniques, 145
Modular design, 123, 157
Multi-functional involvement, 183
Multi-station robots, 229

N

National Science Foundation, 142
NC, See: Numerical control
Nesting features, 158
Non-linear processes, 160
Non-synchronous systems, 192
Numerical control, 12, 13, 18

O

Objectives, 23
Operational methods, 67
Operation theories, 152
Organizational flexibility, 102

P

Packaging, 179, 262
Parallel engineering, 58
Parameter assumptions, 226
Parameter design, 95
Part design, 149
Part geometry, 147
Part handling, 158
Partnership approaches, 270
Part proliferation, 28
Part redundancy, 206
Parts reduction, 166
Paybacks, 251
Performance data, 72
Philosophy, 58, 92, 105
Pick and place machines, 239
Pioneers, 72
Pitfalls, 87
Planning, 26, 73, 78, 81, 92-105, 178, 196
Plant engineers, 257
Plant layouts, 78, 116
Plant manufacturing engineers, 255
Plant strategy, 80
Politics, 22
Price spreads, 91
Printed circuit boards, 232
Problem curves, 16
Procedural methods, 152
Process control, 11, 96, 172
Process driven design, 109-126
Process logistics planning, 73
Process planning, 73, 92-105
Process simulation, 269
Producibility, 39, 156-166
Product alignments, 36
Product concepts, 131
Product confidence index, 124
Product costs, 85
Product design, 14, 63-68, 86, 92-105, 205
Product design alternatives, 86
Product design merit, 205
Product driven design, 167
Production efficiency, 188
Production engineers, 204
Production lines, 275
Production methods, 167
Production scheduling, 117, 200
Productivity Estimation Method, 153
Product performance, 177
Product transfer, 20
Profitability, 106, 110
Prototype evaluation, 96
Prototypes, 96, 118
Purchase orders, 256
Purchasing agents, 255

Q

Quality
  of life, 247
  matrix, 46
  objectives, 135
  performance, 112
  programs, 261
  transition matrix, 46
  worker participation in, 262
Quality assurance, 160
Quality control, 160-164
Quality function deployment, 96, 133, 136

R

Research and development, 172, 174, 176-183
Resin prices, 71
Risk analysis, 269
Risk taking, 251
Robotic assembly, 90
Robots, 90, 159, 227, 229, 249, 262, 266
Rule-based processors, 146
Rule-based systems, 146, 148

S

Semi-automatic inserts, 233
Sensors, 13
Sequential product development process, 118
Serial segmentation, 63-68
Simulation and analysis, 97
Simulation studies, 79
Simulation tools, 138
Software, 150, 160-164, 200, 206, 237, 250, 269
Soldering, 242
Solids modeling, 99, 269
Specifications, 59, 118, 161, 217
Spindle/housing assembly, 217
Standardization, 21, 92-105, 196, 277
Standardized information, 20
Statistical analyses, 92-105
Statistical process control, 86, 116, 223, 250, 259
Statistical quality control, 65, 84, 132
Stock removal, 189, 200
Strategy, 9, 21, 50, 95, 132, 146, 168, 175
Structural analysis, 269
Subassembly, 52, 184
Suppliers, 69, 125, 222
Surface-mount technology, 248
Surrogate databases, 76
Synchronous engineering, 176-183
Synchronous manufacturing, 100
Synchronous transfer machines, 193
Synergistic team approach, 119
Systems engineering, 92-105, 187-204

T

Taguchi methods, 94
Teamwork, 53, 87, 106, 109-126, 245
Technology transfer, 183
Test strategies, 95
Theoretical analyses, 204
Time schedules, 27, 162
Tolerance design, 95
Tool costs, 71
Tool engineering supervisors, 257

Tool engineers, 257
Tooling, 264, 269, 271
Tool management, 194
Total quality control, 94, 108
Training, 26, 107, 174, 198, 273
Transition quality matrix, 46
Turning, 221

U

Ultrahigh-speed machining, 269

V

Vacuum cleaners, 240
Value engineering, 98
Vendors, 54, 150
Volume decomposition, 147

W

Wave solder, 233
Welding, 242
Wire preparation, 233